OPTICS FOR MATERIALS SCIENTISTS

OPTICS FOR
MATERIALS SCIENTISTS

Myeongkyu Lee, PhD

AAP | APPLE ACADEMIC PRESS

Apple Academic Press Inc.	Apple Academic Press Inc.
3333 Mistwell Crescent	1265 Goldenrod Circle NE
Oakville, ON L6L 0A2	Palm Bay, Florida 32905
Canada	USA

© 2020 by Apple Academic Press, Inc.

First issued in paperback 2021

Exclusive worldwide distribution by CRC Press, a member of Taylor & Francis Group

No claim to original U.S. Government works

ISBN 13: 978-1-77463-440-0 (pbk)
ISBN 13: 978-1-77188-757-1 (hbk)

Library and Archives Canada Cataloguing in Publication

Title: Optics for materials scientists / Myeongkyu Lee, PhD.

Names: Lee, Myeongkyu, author.

Description: Includes bibliographical references and index.

Identifiers: Canadiana (print) 20190088117 | Canadiana (ebook) 20190088125 | ISBN 9781771887571 (hardcover) | ISBN 9780429425356 (ebook)

Subjects: LCSH: Optics. | LCSH: Materials science.

Classification: LCC QC355.3 .L44 2019 | DDC 535.02/462011—dc23

Library of Congress Cataloging-in-Publication Data

Names: Lee, Myeongkyu, author.

Title: Optics for materials scientists / Myeongkyu Lee, PhD.

Description: Toronto ; New Jersey : Apple Academic Press, 2020. | Includes bibliographical references and index.

Identifiers: LCCN 2019014399 (print) | LCCN 2019017540 (ebook) | ISBN 9780429425356 (ebook) | ISBN 9781771887571 (hardcover : alk. paper)

Subjects: LCSH: Optics. | Optical materials.

Classification: LCC QC355.3 (ebook) | LCC QC355.3 .L44 2020 (print) | DDC 535--dc23

LC record available at https://lccn.loc.gov/2019014399

Apple Academic Press also publishes its books in a variety of electronic formats. Some content that appears in print may not be available in electronic format. For information about Apple Academic Press products, visit our website at **www.appleacademicpress.com** and the CRC Press website at **www.crcpress.com**

About the Author

Myeongkyu Lee, PhD

Myeongkyu Lee, PhD, is currently a professor in the Department of Materials Science and Engineering at Yonsei University in Seoul, Korea. Formerly, he was a researcher at the National Institute for Materials Science, Japan, and a Postdoctoral Fellow in the Department of Electrical Engineering, Stanford University, California, where he received his PhD. Dr. Lee's research interests include dye-sensitized solar cell, non-lithographic thin-film patterning for electronics, transparent electrode, diffractive optics and holography, and laser materials processing. He is a member of several academic societies, including the Materials Research Society (MRS), the European Materials Research Society (E-MRS), and the Optical Society of America (OSA), and he is the former editor of *Electronic Materials Letters* (2013–2014).

Contents

Abbreviations

1D	one-dimensional
AR	antireflection
BCC	body-centered cubic
DDA	discrete dipole approximation
DPI	dots per inch
EMF	electromotive force
EMT	effective medium theory
FCC	face-centered cubic
FDTD	finite-difference time-domain
FEM	finite element method
FOM	figure of merit
FTIR	frustrated total internal reflection
IR	infrared
ITO	indium tin oxide
LSPRs	localized surface plasmon resonances
LSPs	localized surface plasmons
OPL	optical path length
PMMA	polymethylmethacrylate
PS	polystyrene
PVA	polyvinyl alcohol
RCWA	rigorous coupled-wave analysis
SERS	surface-enhanced Raman spectroscopy
s-Si	single-crystalline Si
SPP	surface plasmon polariton
SPR	surface plasmon resonance
SPs	surface plasmons
TA	transmission axis
TE	transverse electric
TEM	transmission electron microscope
TIR	total internal reflection
TM	transverse magnetic
UV	ultraviolet

Preface

Materials science, commonly termed materials science and engineering, is an interdisciplinary field emerging at the intersection of various fields such as metallurgy, ceramics, solid state physics, chemistry, chemical engineering, and mechanical engineering. Thus, many physicists, chemists, and engineers also work in materials science. Besides their traditional role of discovering new materials and improving the performance of existing materials, materials scientists/engineers nowadays play crucial roles in the design and fabrication of a variety of photonic devices such as solar cells, displays, and light-emitting devices. Therefore, knowledge of optics became indispensable to them. Since I joined Department of Materials Science and Engineering, Yonsei University in 2001, I have been teaching optics, optical properties of materials, and X-ray diffraction to undergraduate and graduate students. While teaching these courses and supervising PhD students in our laboratory, I have strongly felt the necessity of an optics textbook for those who are not familiar with the field. This book is intended to help materials scientists/engineers well comprehend the principles of optics/optical phenomena and effectively utilize them for the design and fabrication of optical materials and devices.

The book consists of 10 chapters. Chapter 1 treats the general properties of electromagnetic waves, including how they propagate in dielectrics and metals. Maxwell's equations form the foundation of classical optics as well as electromagnetics. An important consequence of the equations is that light is a transverse wave propagating with oscillating electric and magnetic fields. Chapter 2 describes a variety of phenomena occurring when light passes from one medium to another. Since light is an electromagnetic wave, most optical phenomena including reflection and refraction can be explained using the classical electromagnetic description of light. Chapter 3 discusses the superposition of waves and explains the concept of coherence after a brief description of Fourier series and integral. Interference and diffraction also result from the superposition of waves. Chapter 4 describes multibeam interference and introduces various interferometers. Interference in thin film is also briefly discussed as an introduction to thin-film optics, which is treated in Chapter 8 in more detail. Chapter 5 presents a mathematical description of light diffraction and includes X-ray and electron diffractions, which are both powerful tools for materials analysis. Chapter 6 deals with the behavior of light in anisotropic media. In isotropic media such as air and glass, light behaves the same way no matter which direction it is propagating in the medium. However, many crystalline materials are naturally anisotropic, with their optical properties dependent on the propagation direction of light and its polarization state. This optical anisotropy forms a basis for many of the current optical devices. Chapter 7 treats the polarization of light. This chapter uses Jones matrices to represent the different polarization states of light and describes in detail the physical mechanisms responsible for the production of polarized light.

While Chapter 1 through 7 deals with the fundamentals of optics, the remaining three chapters are application oriented. Chapter 8 introduces thin-film optics and shows how the reflection and transmission of light can be controlled using multilayer thin film coatings. Micro/nanoscale structures formed into the surface of a material are also presented as a means of achieving broadband and omnidirectional antireflection. With the advent of short wavelength and high-power lasers, interference lithography became a very powerful tool for fabricating periodic structures. Chapter 9 describes the basic concepts of interference lithography and shows how 1D, 2D, and 3D periodic structures can be created using this technique. Metal plasmonics is a rapidly growing active field that finds a wide range of application areas including electronic/photonic devices, photovoltaics, chemical and biological sensors, surface decorations, and plasmonic color printings. Chapter 10 describes the principles of surface plasmon resonance and localized surface plasmon resonance and presents some promising applications associated with the plasmon resonance of metal nanostructures.

Each chapter of the book has a problem and reference section, and many of the recent articles are cited in the final three chapters. The book is aimed at assisting materials scientists/engineers that must be aware of optics and optical phenomena. This book will also be a good textbook for students in materials science, physics, chemistry, and engineering throughout their undergraduate and early graduate years. The author wishes to thank J. Kim who helped in preparing the calculated intensity profiles presented in Chapter 9.

—**Myeongkyu Lee**
Professor/PhD
Department of Materials Science and Engineering
Yonsei University, Korea

CHAPTER 1

Electromagnetic Waves

1.1 MATHEMATICS OF WAVE MOTION

1.1.1 ONE-DIMENSIONAL WAVE EQUATION

A wave can be described as a disturbance that travels through a medium from one position to another position. In physics, waves are time-varying oscillations of a physical quantity around fixed locations. Wave motion transfers energy from one point to another. A wave can be *transverse* or *longitudinal*. Transverse waves such as electromagnetic waves occur when a disturbance generates oscillations that are perpendicular to the propagation of energy transfer. Longitudinal waves (e.g., sound waves) occur when the oscillations are parallel to the propagation direction. Waves are described by a wave equation that sets out how the disturbance proceeds over time. Consider a transverse pulsed wave traveling in the positive x-direction with a constant speed v, as shown in Figure 1.1. Since the disturbance is a function of both position and time, it can be written as

$$\psi = f(x,t).$$ (1.1)

We here deal with a wave whose shape does not change as it propagates through space. The shape of the disturbance at any time, say $t = 0$, can be obtained by holding time constant at that value.

$$\psi(x,t)\big|_{t=0} = f(x,0) = f(x).$$ (1.2)

Figure 1.1 shows the wave profiles at $t = 0$ and $t = t$, that is, $f(x,0)$ and $f(x,t)$. They represent the shapes of the disturbance taken at the beginning and end of a time interval t. We now introduce a new coordinate x' that is defined as

$$x' = x - vt.$$ (1.3)

In this new coordinate system, the disturbance $\psi = f(x')$ is no longer a function of time and looks the same at any value of t as it did at $t = 0$ in the original stationary coordinate system. It follows from Figure 1.1 that the disturbance can be represented in terms of the variables associated with the stationary system as

$$\psi(x,t) = f(x - vt). \tag{1.4}$$

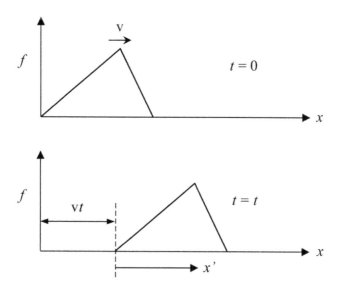

FIGURE 1.1   Profiles of a pulsed wave moving in the positive x-direction at $t = 0$ and $t = t$.

This is the most general form of the one-dimensional *wave function*. Equation 1.4 describes a wave propagating in the positive x-direction. The resulting expression for a wave propagating in the negative x-direction would be

$$\psi(x,t) = f(x + vt). \tag{1.5}$$

We can figure out that regardless of the shape of the disturbance, the variables x and t appear in the wave function as a single variable in the form of $(x - vt)$ or $(x + vt)$. If the pulsed wave depicted in Figure 1.1 propagated by $v\Delta t$ for an interval Δt, we find that

$$f[(x + v\Delta t) - v(t + \Delta t)] = f(x - vt). \tag{1.6}$$

That is, the disturbance at two different variable combinations is identical, as manifest from Figure 1.2. Note that the original shape of the pulse does not vary but is simply translated along the x-direction by $v\Delta t$.

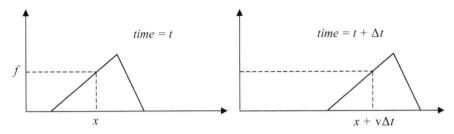

FIGURE 1.2 Equivalence of the disturbances at two variable combinations (x, t and $x + v\Delta t$, $t + \Delta t$).

Next we wish to develop the differential wave equation that is satisfied by all waves, regardless of the particular wave function. To this end, we make use of eqs 1.3 and 1.4. Taking the partial derivative of $\psi = f(x, t)$ with respect to x gives the following relation:

$$\frac{\partial \psi}{\partial x} = \frac{\partial f}{\partial x'}\frac{\partial x'}{\partial x} = \frac{\partial f}{\partial x'}$$

since

$$\frac{\partial x'}{\partial x} = 1 \cdot$$

Repeating the procedure to find the second derivative, we have

$$\frac{\partial}{\partial x}\left(\frac{\partial \psi}{\partial x}\right) = \frac{\partial}{\partial x'}\left(\frac{\partial f}{\partial x'}\right)\frac{\partial x'}{\partial x} = \frac{\partial^2 f}{\partial x'^2} \cdot$$

Similarly, the time derivatives can be found as follows:

$$\frac{\partial \psi}{\partial t} = \frac{\partial f}{\partial x'}\frac{\partial x'}{\partial t} = -v\frac{\partial f}{\partial x'}$$

$$\frac{\partial}{\partial t}\left(\frac{\partial \psi}{\partial t}\right) = \frac{\partial}{\partial x'}\left(-v\frac{\partial f}{\partial x'}\right)\frac{\partial x'}{\partial t} = v^2\frac{\partial^2 f}{\partial x'^2}$$

Combining these relations, we obtain

$$\frac{\partial^2 \psi}{\partial x^2} = \frac{1}{v^2}\frac{\partial^2 \psi}{\partial t^2} \cdot \tag{1.7}$$

This is the one-dimensional *differential wave equation*. Any wave of the form of eqs 1.4 or 1.5 satisfies eq 1.7, regardless of its physical nature. It is apparent from eq 1.7 that if two different wave functions are separate solutions, their sum or difference is also a solution. This constitutes the superposition principle of waves.

1.1.2   HARMONIC WAVES

Although a wave can take on any shape, of special importance are *harmonic waves* whose profile is a sine or cosine function. These periodic waves, often known as sinusoidal waves, represent smooth pulses that repeat themselves endlessly. Actual periodic waves may have many different shapes such as square, triangle, and saw-tooth. It follows from the theorem of Fourier series that any wave shape can be generated by a superposition of sinusoidal functions. Thus, combinations of harmonic waves can represent more complex wave forms including a series of rectangular pulses and square waves. Since the only difference between the sine and cosine functions is a relative translation of $\pi/2$ radians, it is sufficient to treat one of these functions at our convenience. We here consider a sine wave traveling at speed v in the positive *x*-direction, which is mathematically expressed as

$$y = \psi(x,t) = A \sin k(x - vt).\qquad(1.8)$$

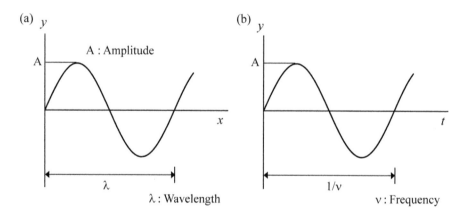

FIGURE 1.3   (a) Disturbance variation with *x* at a fixed time and (b) disturbance variation with *t* at a fixed position.

Equation 1.8 is surely a solution of the differential wave equation. The maximum disturbance (here, constant A) is known as the *amplitude* of the wave. Holding either *x* or *t* fixed result in a sinusoidal disturbance, so the wave is periodic in both position and time. Figure 1.3a shows a section of the wave at a fixed time. The disturbance variation with time at a fixed position is pictured in Figure 1.3b. The *spatial period* of the wave is known as the *wavelength* and is denoted by λ, as shown in Figure 1.3a. An increase in *x* by

λ reproduces the same wave. This is equivalent to changing the argument of the sine function by 2π. That is,

$$A \sin k(x - vt) = A \sin k[(x + \lambda) - vt] = A \sin[k(x - vt) + 2\pi].$$

Therefore, we have

$$k = 2\pi/\lambda, \tag{1.9}$$

where k is known as the *propagation constant*. When the wave is viewed from a fixed position as in Figure 1.3b, it is also periodic with a repetitive unit called the *temporal period*, τ. Increasing t by τ reproduces the wave, so that

$$A \sin k[x - v(t + \tau)] = A \sin[k(x - vt) + 2\pi].$$

It follows from this relation that $kv\tau = 2\pi$. Note that all these parameters are positive quantities. Rather than the temporal period, its reciprocal called the *frequency*, $v = 1/\tau$, is more commonly used. The relation between speed (v), frequency (v), and wavelength (λ) is then

$$v = v\lambda. \tag{1.10}$$

There are two other quantities that are often used in describing the wave motion. The combination $\omega = 2\pi v$ is called the *angular frequency*. The reciprocal of the wavelength, denoted by $\kappa = 1/\lambda$, is known as the *wave number*. The wavelength, propagation constant, temporal period, frequency, angular frequency, and wave number all describe the periodic nature of a wave in position and time. These concepts can be equally applied to non-harmonic waves, as long as each wave is in a regular periodic fashion. We can derive a variety of equivalent formulations of the harmonic wave, some of which are given below.

$$
\begin{aligned}
y &= \psi(x,t) = A \sin k(x - vt) \\
y &= \psi(x,t) = A \sin(2\pi x/\lambda - 2\pi t/\tau) \\
y &= \psi(x,t) = A \sin(2\pi x/\lambda - 2\pi vt) \\
y &= \psi(x,t) = A \sin(kx - \omega t)
\end{aligned}
\tag{1.11}
$$

Among them, the third and fourth of eq 1.11 will be encountered most frequently. It should be noted that these waves are of infinite extent. There is no limitation on x for a fixed value of t and vice versa. Each wave has a single frequency (and also a single wavelength) and is thus said to be *monochromatic*.

The argument of the sine or cosine function, which is an angle that depends on position and time, is called the phase (θ) of the wave, so that

$$\theta = (2\pi x/\lambda - 2\pi vt) = (kx - \omega t). \tag{1.12}$$

When x and t change together in such a way that the phase is constant, the disturbance $y = A \sin \theta$ is also constant. In a traveling harmonic wave, the origin can be arbitrarily chosen. Since $\sin \theta = \cos (\theta - \pi/2)$, the wave motion can be handled by either the sine or cosine function if an initial phase angle of $\pi/2$ is added to the phase. In many cases, the absolute phase value of the wave is not of interest. The disturbance of the wave is identical at the same phase. In Figure 1.4, two wave configurations have the same disturbance at x_1, t_1 and x_2, t_2, which means that the phases at x_1, t_1 and x_2, t_2 are identical, that is,

$$\frac{2\pi}{\lambda} x_1 - 2\pi vt_1 = \frac{2\pi}{\lambda} x_2 - 2\pi vt_2. \tag{1.13}$$

From eq 1.13, we obtain

$$\frac{x_2 - x_1}{t_2 - t_1} = \frac{dx}{dt} = v = v\lambda. \tag{1.14}$$

This is the speed at which the profile moves and is known as the wave velocity or *phase velocity*. The propagation of constant phase evidently describes the motion of a fixed point on the wave form, which moves at the phase velocity. Thus, if θ is constant,

$$d\theta = 0 = kdx - \omega dt$$

and

$$\frac{dx}{dt} = v = \frac{\omega}{k}. \tag{1.15}$$

It is to be noted that the above phase velocity has a negative sign when the wave moves in the negative x-direction.

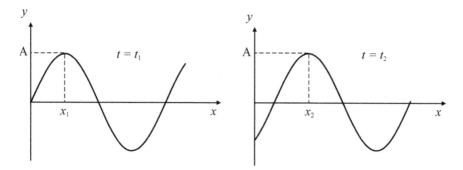

FIGURE 1.4   The phase and displacement of the wave at x_1, t_1 are the same as those at x_2, t_2.

1.1.3 COMPLEX REPRESENTATION

In many situations, it becomes more convenient to represent harmonic waves in complex-number notations. Especially when dealing with the superposition of harmonic waves, the complex representation is mathematically simpler than the trigonometric manipulation. In fact, the complex exponential form of the wave equation is extensively used in optics and quantum mechanics. A complex number z is expressed as the sum of its real and imaginary parts,

$$z = a + ib, \tag{1.16}$$

where $a = \mathrm{Re}\,(z)$ and $b = \mathrm{Im}\,(z)$ are real numbers. In Figure 1.5, the complex number z is represented in terms of its real and imaginary parts along the corresponding axes. The magnitude of this complex number, denoted by $|z|$, is given as

$$|z| = \sqrt{a^2 + b^2}\,. \tag{1.17}$$

Referring to Figure 1.5, $a = |z|\cos\theta$ and $b = |z|\sin\theta$. Thus, it is possible to express eq 1.16 as

$$z = |z|(\cos\theta + i\sin\theta), \tag{1.18}$$

where θ is the phase angle of z. By Euler's formula, the expression in parentheses is

$$e^{i\theta} = \cos\theta + i\sin\theta. \tag{1.19}$$

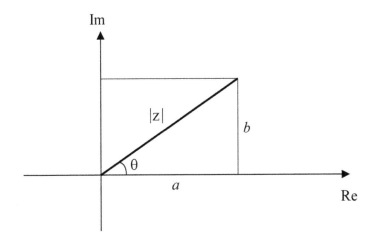

FIGURE 1.5   Representation of a complex number $z = a + ib$ along the corresponding axes.

We can then express the complex number in the following polar form:

$$z = |z|e^{i\theta} \tag{1.20}$$

The complex conjugate $z*$, indicated by an asterisk, is simply the complex number z with i replaced with $-i$, as shown below.

$$z^* = a - ib = |z|e^{-i\theta} \tag{1.21}$$

A useful relation is that the product of a complex number with its complex conjugate equals the square of its magnitude.

$$zz^* = |z|e^{i\theta}|z|e^{-i\theta} = |z|^2 = a^2 + b^2. \tag{1.22}$$

The complex representation is utilized to take advantage of the ease with which complex exponentials can be manipulated. The harmonic wave expressed in the form of eq 1.20 includes both cosine and sine functions. Calculations based on the complex form provide correct results for both cosine and sine waves. After such calculations (even during calculations), appropriate expressions for either wave can be extracted by taking the real or the imaginary parts of both sides of the equation. From the polar form where

$$\mathrm{Re}(z) = |z|\cos\theta \quad \text{and} \quad \mathrm{Im}(z) = |z|\sin\theta,$$

either part can be chosen to describe a harmonic wave. It is customary, however, to choose the real part so that a harmonic wave is written as

$$\psi(x,t) = \mathrm{Re}[Ae^{i(kx-\omega t)}]. \tag{1.23}$$

This is equivalent to

$$\psi(x,t) = A\cos(kx - \omega t). \tag{1.24}$$

The complex number $Ae^{i\theta}$ is represented in the complex plane by a vector of magnitude A inclined at an angle θ to the real axis. As θ increases, this vector is counterclockwise rotated and the real part of the complex number changes between A and $-A$. The wave given by eq 1.24 has a disturbance varying between these two values. Accordingly, it is quite common to write the harmonic wave function as

$$\psi(x,t) = Ae^{i(kx-\omega t)} = Ae^{i\theta}, \tag{1.25}$$

where the actual wave is the real part.

1.1.4 PLANE WAVES

Wavefront means an imaginary surface joining all points of equal phase in a wave. The wavefronts of a plane wave are planes. Any wave propagates in the direction normal to its wavefront. A plane wave thus has a straight propagation direction, as shown in Figure 1.6. A small stone vertically dropped into a tranquil lake will generate a two-dimensional circular wave on the water surface, in which the wavefronts are in the form of concentric circles. Similarly, the radiation emanating from a point source of light can be considered as a spherical wave spreading out in all radial directions. The resulting wavefronts, that is, the surfaces of constant phase, are spherical surfaces centered at the source. The surface of a sphere becomes more flattened with increasing radius. Thus, a spherical wave will behave like a plane wave when it is far away from the source. As illustrated in Figure 1.6a, the wavefronts are usually drawn along the crests of a wave, with the spacing between two adjacent wavefronts equal to one wavelength. When a plane-wave beam is focused by a lens, the beam size decreases and then increases after being minimized at focus. Since the wave propagates normal to its wavefront, planar wavefronts become spherical after going through the lens, as shown in Figure 1.7.

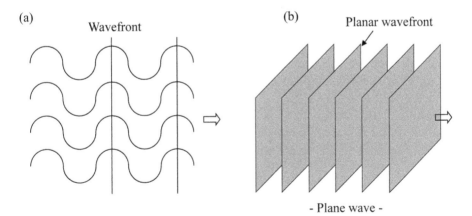

FIGURE 1.6 (a) Wavefront refers to the locus of points having the same phase. (b) Wavefronts of a plane wave.

It is not possible to have a plane wave of infinite extent. However, many waves are approximately plane waves in a localized region of space, as is the case for light in the field of optics. By using optical devices, we can produce light rays that behave like plane waves. We now wish to derive the general expression for a plane wave passing through some point (x, y, and z) in space. Equation 1.26 represents a harmonic wave propagating along the positive x-direction.

$$\psi = A\sin(kx - \omega t) \tag{1.26}$$

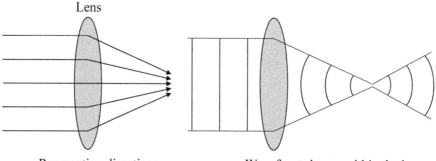

Lens

Propagation directions   Wavefront shapes within the beam

FIGURE 1.7   Wavefront shapes of a plane-wave beam focused by a lens.

At fixed time (for simplicity, $t = 0$), the spatial extent of the wave is

$$\psi = A\sin(kx). \tag{1.27}$$

When x = constant, the phase $\theta = kx$ = constant, since this wave was assumed to be a plane wave. Thus the surfaces of constant phase are the family of planes. A single planar wavefront is pictured in Figure 1.8a. Evidently, the wave disturbance given by ψ is the same for all points on this wavefront. The disturbance at an arbitrary point in space, defined by the position vector **r**, is thus the same as for the point x on the x-axis, where $x = r\cos\alpha$. Equation 1.27 is then rewritten as

$$\psi = A\sin(kr\cos\alpha). \tag{1.28}$$

Here we introduce a vector quantity **k** pointing in the direction of wave propagation. This vector **k** is known as the *wave vector*, whose magnitude is equal to the propagation constant, $k = 2\pi/\lambda$. At each point in space, the wave vector is normal to the wavefront that passes through the point. Since $\mathbf{k} \cdot \mathbf{r} = kr\cos\alpha$ in Figure 1.8a, the harmonic wave of eq 1.26 becomes

$$\psi(\mathbf{r}, t) = A\sin(\mathbf{k} \cdot \mathbf{r} - \omega t). \tag{1.29}$$

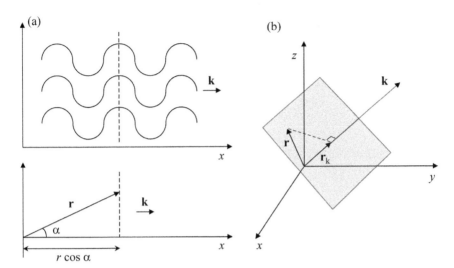

FIGURE 1.8 (a) Wave disturbance is the same for all points on the dotted wavefront. (b) A plane wave propagating in the *k*-direction.

This function can represent planes waves propagating in any arbitrary directions given by the wave vector **k**. The general case depicted in Figure 1.8b has the form

$$\mathbf{k} \cdot \mathbf{r} = k_x x + k_y y + k_z z, \tag{1.30}$$

where k_x, k_y, and k_z are the components of **k** and x, y, and z are the coordinates of the point where the disturbance is measured. The three-dimensional plane wave is also expressed in complex form either as

$$\psi(\mathbf{r}, t) = A e^{i(\mathbf{k} \cdot \mathbf{r} - \omega t)}$$

or

$$\psi(x, y, z, t) = A e^{i(k_x x + k_y y + k_z z - \omega t)}. \tag{1.31}$$

The *three-dimensional differential wave equation* is a generalization of eq 1.7 and is given by

$$\frac{\partial^2 \psi}{\partial x^2} + \frac{\partial^2 \psi}{\partial y^2} + \frac{\partial^2 \psi}{\partial z^2} = \frac{1}{v^2} \frac{\partial^2 \psi}{\partial t^2}. \tag{1.32}$$

Using the *Laplacian* operator:

$$\nabla^2 = \frac{\partial^2}{\partial x^2} + \frac{\partial^2}{\partial y^2} + \frac{\partial^2}{\partial z^2},$$

Equation 1.32 becomes simply

$$\nabla^2 \psi = \frac{1}{v^2} \frac{\partial^2 \psi}{\partial t^2}.$$ (1.33)

If we rotate the coordinate system in Figure 1.8b so that **k** is parallel to the x-axis or simply take the propagation direction as the x-direction, eq 1.31 is reduced to eq 1.25.

1.2 FUNDAMENTAL ELECTROMAGNETIC THEORY

In this section, we briefly review the fundamental electromagnetic theory required to develop the concept of electromagnetic waves. Light is also an electromagnetic wave. Many optical phenomena, particularly occurring at the interfaces, can be well described by applying electromagnetic boundary conditions. There are four vector quantities involving the basic laws of electromagnetic theory: electric field E, electric displacement D, magnetic field H, and magnetic induction B. The magnetic field can be defined in several different ways. The vector B is also sometimes called the *magnetic field*, since it is a more fundamental quantity than H. A particle of charge q in an electric field E experiences an electrostatic force $F = qE$. When a charged particle moves in the vicinity of a current-carrying wire, there is another force exerting on the particle so that the overall force satisfies the *Lorentz force law*, $F = q(E + v \times B)$. Here v is the particle's velocity and "×" denotes the vector product. The vector B is termed the magnetic field, which is defined as the vector field necessary to make the Lorentz force law correctly describe the motion of a charged particle. In a vacuum, B and H are proportional to each other. Many authors in physics textbooks use the term "magnetic field" to represent the H quantity, while calling B the magnetic induction or magnetic flux density. There are some alternative names for both. In this book, the "magnetic field" phrase will be used to represent B as well as H. When necessary to distinguish between them, however, either B-field or H-field will be specified instead of simply calling the magnetic field. Electric fields are generated by both electric charges and time-varying magnetic fields. By the same token, magnetic fields are generated by electric currents and time-varying electric fields.

1.2.1 *GAUSS'S LAW—ELECTRIC*

Gauss's law is a law relating the distribution of electric charge to the resulting electric field. This law is one of Maxwell's four equations, which form the basis of classical electrodynamics; the other three are Gauss's law for magnetism, Faraday's law of induction, and Ampere's law with Maxwell's correction. Gauss's law states that the net electric flux through a closed surface is equal to the net electric charge within the surface divided by the electric permittivity of free space. The electric flux Φ_E through a closed surface S is expressed as

$$\Phi_E = \oint_S \mathbf{E} \cdot d\mathbf{A}, \tag{1.34}$$

where E is the electric field and dA, a vector representing an infinitesimal element of surface area. The surface integral on the right means the difference between the flux flowing into and out of any closed surface. If the net flux is nonzero, it is due to the presence of sources or sinks of the electric field within S. Obviously, the integral should be proportional to the total enclosed charge. In integral from, Gauss's law is given by

$$\Phi_E = \oint_S \mathbf{E} \cdot d\mathbf{A} = Q/\varepsilon_o, \tag{1.35}$$

where Φ_E is the electric flux through a closed surface S enclosing any volume V. Q is the total charge enclosed within S, and ε_o ($= 8.854 \times 10^{-12}\ \text{C}^2\text{N}^{-1}\,\text{m}^{-2}$) is the electric permittivity of free space. The vector dA is in the direction of an outward normal, as shown in Figure 1.9. Since the electric flux through an area is defined as the electric field multiplied by the surface area projected onto a plane perpendicular to the field, Gauss's law is a general law applying to any arbitrarily-shaped closed surface. Another way of visualizing this is to consider the electric field component perpendicular to that area. Q represents the total net charge. Thus, if a positive charge $+q$ and a negative charge $-q$ coexist in space, the electric flux through a closed surface enclosing both charges becomes zero. When the charge Q is continuously distributed over the volume enclosed by S, we can express it in terms of the charge density ρ.

$$Q = \int_V \rho dV \tag{1.36}$$

By applying the divergence theorem:

$$\oint_S \mathbf{E} \cdot d\mathbf{A} = \int_V (\nabla \cdot \mathbf{E}) dV,$$

Equation 1.35 can be rewritten as

$$\nabla \cdot \mathbf{E} = \rho / \varepsilon_o. \tag{1.37}$$

This is *Gauss's law in differential form*. Both of the integral and differential forms carry the same message.

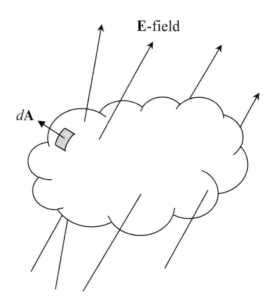

FIGURE 1.9   *E*-field through a closed surface.

1.2.2   GAUSS'S LAW—MAGNETIC

While the electric field E may diverge from a positive charge and converge toward a negative charge, the magnetic induction field B does not have any sources or sinks. No isolated magnetic poles have ever been found. Since there are no magnetic monopoles (i.e., magnetic charges), the lines of B are themselves continuous and closed. The magnetic flux Φ_B through a closed surface S is always zero. Then we have

$$\Phi_B = \oint_S \mathbf{B} \cdot d\mathbf{A} = 0, \tag{1.38}$$

where dA is a vector whose magnitude is the area of an infinitesimal element of the surface S and whose direction points to the surface normal. The differential form of Gauss's law for magnetism is

$$\nabla \cdot \mathbf{B} = 0. \tag{1.39}$$

Gauss's law for magnetism states that there is no net magnetic flux. It is equivalent to stating that for each volume element in space, the same number of B lines enters and exit it. This is due to the absence of any magnetic monopoles within the enclosed volume.

1.2.3   FARADAY'S LAW OF INDUCTION

Faraday's law of induction is a basic law that predicts how a magnetic field will interact with an electric circuit to produce an electromotive force (EMF). Faraday discovered that a time-varying magnetic flux through a conducting loop generates a current around that loop. Faraday's law states that the EMF induced in any closed circuit is equal to the negative of the time derivative of the magnetic flux enclosed by the circuit. Let us consider a conducting wire loop *C* that is represented as a bold line in Figure 1.10. The magnetic flux through an open surface *S* bounded by the conducting loop is given by

$$\Phi_B = \int_S \mathbf{B} \cdot d\mathbf{A}. \tag{1.40}$$

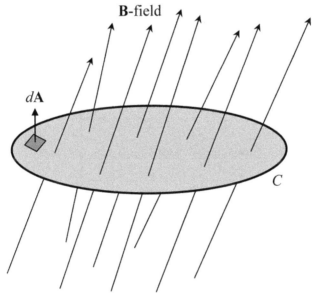

FIGURE 1.10   *B*-field through an open surface bounded by a conducting loop *C*.

The EMF generated in the loop is then

$$\text{EMF} = -d\Phi_B / dt. \tag{1.41}$$

It is the EMF that governs how much current will flow in a given loop. The EMF exists as a result of the presence of an entirely new kind of electric field *E*. Since the net effect is determined by the line integral of *E* around the circuit, we have

$$\text{EMF} = \oint_C \mathbf{E} \cdot d\mathbf{l} \ . \tag{1.42}$$

The integral of eq 1.42 is a line integral of E taken around the closed loop C. In a strict sense, EMF is not a force at all but an integral of a force per unit charge. It is to be noted that the electric field in eq 1.42 is different from the electrostatic field arising from the presence of electric charges. Evidently, it is due to the time-varying magnetic field. With no charge sources or sinks, the electric field lines close on themselves. A changing magnetic field also induces an electric field. Faraday's law in integral form is

$$\oint_C \mathbf{E} \cdot d\mathbf{l} = -d\Phi_B / dt \ . \tag{1.43}$$

The minus sign accounts for the fact that the induced EMF is always against the change of the magnetic flux. Equation 1.43 can be easily turned into a differential form by applying Stokes' theorem:

$$\oint_C \mathbf{E} \cdot d\mathbf{l} = \int (\nabla \times \mathbf{E}) \cdot d\mathbf{A} = -\frac{d}{dt} \int \mathbf{B} \cdot d\mathbf{A} = -\int \left(\frac{\partial \mathbf{B}}{\partial t} \right) \cdot d\mathbf{A}$$

so that

$$\nabla \times \mathbf{E} = -\frac{\partial \mathbf{B}}{\partial t} \ . \tag{1.44}$$

Since the line integral of an electrostatic field around a closed loop is always zero, it does not matter whether we include this field or not, in calculating the electromotive force. Thus, when B remains static, $\nabla \times E = 0$.

1.2.4   AMPERE'S CIRCUITAL LAW

Ampere's circuital law relates the integrated magnetic field around a closed loop to the electric current passing through the loop. This law determines the magnetic field associated with a given current and is mathematically expressed as

$$\oint_C \mathbf{B} \cdot d\mathbf{l} = \mu_o I_{\text{enc}}, \tag{1.45}$$

where μ_o is the magnetic permeability of free space ($\mu_o = 4\pi \times 10^{-7} \ \text{N s}^2 \text{C}^{-2}$). Equation 1.45 states that the line integral of B around a closed curve C is proportional to the total current I_{enc} passing through a surface enclosed by C. If the current flow is represented by a current density J, the enclosed current is given by

$$I_{\text{enc}} = \int_S \mathbf{J} \cdot d\mathbf{A},$$

where the integration is taken over the surface S bounded by the loop C. Applying Stokes' theorem to eq 1.45 gives

$$\nabla \times \mathbf{B} = \mu_o \mathbf{J}. \qquad (1.46)$$

This is the differential version of Ampere's law. There is a sign ambiguity in eq 1.45; which direction through the surface corresponds to a positive current? The right-hand rule is always applied. If the fingers of our right hand represent the direction of line integration, our thumb defines the direction of a positive current.

1.2.5 MAXWELL'S EQUATIONS

So far we have encountered four laws of eqs 1.37, 1.39, 1.44, and 1.46, specifying the divergence and curl of E and B fields. However, there is a fatal inconsistency in the original formulation of Ampere's circuital law: eq 1.46. The divergence of a curl is always zero for any vector. Hence,

$$\nabla \cdot (\nabla \times \mathbf{B}) = 0.$$

Therefore, the original Ampere's law predicts that

$$\nabla \cdot \mathbf{J} = 0.$$

The divergence of J is zero only for steady currents. In general, it is given by

$$\nabla \cdot \mathbf{J} = -\partial \rho / \partial t. \qquad (1.47)$$

This is the exact mathematical expression for local charge conservation, called the *continuity equation*. The shortcoming of the original Ampere's law can be seen in another way. Moving charges, that is, currents are not the only source of a magnetic field. Consider a capacitor consisting of two parallel plates in free space. When charging or discharging the capacitor, we can detect a B-field between the plates, even though no actual current traverses the capacitor. As the charge varies, the electric field applied across the plates also changes. Maxwell hypothesized that just as a changing magnetic field induces an electric field (Faraday's law), a time-varying electric field will induce a magnetic field. He corrected the defect in Ampere's law by purely theoretical arguments. Maxwell added an extra term called the *displacement current* to Ampere's law:

$$\mathbf{J}_d = \varepsilon_o \frac{\partial \mathbf{E}}{\partial t} \, . \tag{1.48}$$

When this extra term is included in the current density, eq 1.46 is rewritten as

$$\nabla \times \mathbf{B} = \mu_o \mathbf{J} + \mu_o \varepsilon_o \frac{\partial \mathbf{E}}{\partial t} \, . \tag{1.49}$$

If we take the divergence for eq 1.49 and apply Gauss's law, both sides become zero, satisfying the continuity equation. Ampere's law with Maxwell's correction is particularly important, since it shows that not only does a time-varying magnetic field induce an electric field, but also a changing electric field accompanies a magnetic field. As we will see in Section 1.3.1, it plays a crucial role in the propagation of electromagnetic waves. The speed calculated for electromagnetic waves exactly matches the speed of light, revealing that light is indeed one form of electromagnetic radiation. As a courtesy to his greatest contributions in electromagnetism and optics, the following four laws are altogether called *Maxwell's equations*.

$$\begin{aligned}
\nabla \cdot \mathbf{E} &= \rho / \varepsilon_o \\
\nabla \cdot \mathbf{B} &= 0 \\
\nabla \times \mathbf{E} &= -\frac{\partial \mathbf{B}}{\partial t} \\
\nabla \times \mathbf{B} &= \mu_o \mathbf{J} + \mu_o \varepsilon_o \frac{\partial \mathbf{E}}{\partial t}
\end{aligned} \tag{1.50}$$

Maxwell's equations in the form of eq 1.50 are themselves complete and correct. However, when we are dealing with a polarizable medium, they can be expressed in a more pertinent way. When an electric field is applied to a dielectric material, its molecules (or atoms) can form microscopic electric dipoles; while their atomic nuclei move a small distance in the direction of the applied field, their electrons move a distance in the opposite direction. This produces a macroscopic bound charge in the material. A convenient measure of the effect is the *polarization P* of the material, which is defined as its dipole moment per unit volume. For uniform P, a macroscopic separation of charge is produced only at the surfaces: a layer of positive bound charge on one end of the material and a layer of negative charge on the other end. The charge density in eq 1.50 includes both free and bound charges, as indicated by $\rho = \rho_f + \rho_b$. The free charge may represent any charge such as ions embedded in a dielectric or electrons on a conductor, as long as this does not result from polarization. It is easy to show that the bound charge density is related to the electric polarization by

$$\rho_b = -\nabla \cdot \mathbf{P}. \tag{1.51}$$

Just like the total charge density, the total current density J can be separated into free and bound current densities: $J = J_f + J_b$. The bound current density J_b also has two components. One is contributed from the electric polarization and the other, from magnetization, such that

$$\mathbf{J}_b = \partial \mathbf{P} / \partial t + \nabla \times \mathbf{M}. \tag{1.52}$$

Any change in the electric polarization P induces a flow of bound charge that should be included in the total current. The first term on the right-hand side of eq 1.52 accounts for this bound current (sometimes called polarization current). Similar to the electric dipole moments, the constituent atoms of all materials exhibit magnetic moments that involve the spin and orbital motion of their electrons. The state of magnetic polarization of a material is described by the magnetization M, which is defined as the magnetic dipole moment per unit volume. It plays a role analogous to the electric polarization P. The effect of magnetization is to establish bound currents within the material. The field due to magnetization of the material is just the field generated by these bound currents. The second term in eq 1.52 represents the magnetization-induced bound current.

Now, Gauss's law can be rewritten as

$$\nabla \cdot \mathbf{E} = \left(\rho_f - \nabla \cdot \mathbf{P}\right) / \varepsilon_o$$
$$\nabla \cdot \left(\varepsilon_o \mathbf{E} + \mathbf{P}\right) = \rho_f \qquad , \tag{1.53}$$
$$\nabla \cdot \mathbf{D} = \rho_f$$

where D is known as the *electric displacement* and is given by

$$\mathbf{D} = \varepsilon_o \mathbf{E} + \mathbf{P}. \tag{1.54}$$

Meanwhile, Ampere's law with Maxwell's correction becomes

$$\nabla \times \mathbf{B} = \mu_o \left(\mathbf{J}_f + \partial \mathbf{P} / \partial t + \nabla \times \mathbf{M}\right) + \mu_o \varepsilon_o \partial \mathbf{E} / \partial t$$
$$\nabla \times \mathbf{H} = \mathbf{J}_f + \partial \mathbf{D} / \partial t \qquad , \tag{1.55}$$

where the magnetic H-field is given by

$$\mathbf{H} = \mathbf{B} / \mu_o - \mathbf{M}. \tag{1.56}$$

Maxwell's equations can then be expressed in terms of *free* charge and current.

$$\nabla \cdot \mathbf{D} = \rho_f$$
$$\nabla \cdot \mathbf{B} = 0$$
$$\nabla \times \mathbf{E} = -\partial \mathbf{B} / \partial t \qquad (1.57)$$
$$\nabla \times \mathbf{H} = \mathbf{J}_f + \partial \mathbf{D} / \partial t$$

Gauss's law for magnetism and Faraday's law are not affected by the separation of charge and current into free and bound terms. Some people regard these as the true Maxwell's equations, since they involve only free charge and current that we can control. The relationship between D and E (also B and H) depends on the nature of the material. In general, D is not necessarily proportional to E. The electric polarization of a dielectric results from an electric field. In many materials, the polarization P is linearly proportional to the field E:

$$\mathbf{P} = \varepsilon_o \chi_e \mathbf{E} \qquad (1.58)$$

unless E is too strong. χ_e is a proportionality constant known as the *electric susceptibility* of the material. In linear media, we have

$$\mathbf{D} = \varepsilon_o \mathbf{E} + \mathbf{P} = \varepsilon_o \left(1 + \chi_e\right)\mathbf{E} = \varepsilon \mathbf{E}, \qquad (1.59)$$

where ε is called the *electric permittivity* of the medium. In a vacuum, there is no matter to polarize. Then, the susceptibility term vanishes, rendering the permittivity to ε_o. For magnetic materials, the linear proportionality of M to the magnetic field is customarily expressed with H rather than B:

$$\mathbf{M} = \chi_m \mathbf{H}, \qquad (1.60)$$

where χ_m is called the *magnetic susceptibility*. Equation 1.56 then becomes

$$\mathbf{B} = \mu_o \left(1 + \chi_m\right)\mathbf{H} = \mu \mathbf{H}. \qquad (1.61)$$

Here μ is the *magnetic permeability* of the material. Equation 1.57 describes the general Maxwell's equations in matter. For linear media, D and H in these equations can be replaced with the following relations:

$$\mathbf{D} = \varepsilon \mathbf{E} \quad \text{and} \quad \mathbf{H} = \mathbf{B}/\mu. \qquad (1.62)$$

Here, ε and μ are not always constant even for linear media. In homogeneous materials, they are constant throughout the material. For isotropic materials, ε and μ are scalars, while they become tensors for anisotropic materials. For dispersive media, ε and μ depend on the frequency (and wavelength) of incident electromagnetic wave.

1.2.6 BOUNDARY CONDITIONS

The differential forms of Maxwell equations are applicable when the fields *E, D, H,* and *B* are continuous and differentiable. These vector fields may be discontinuous at an interface between two media with different ε and/or μ. The behaviors of electromagnetic field vectors at a boundary separating two media can be deduced from the integral forms of Maxwell's equations. Consider a tiny, thin Gaussian pillbox extending into media 1 and 2 from the boundary (Fig. 1.11). Suppose that this box has an infinitely small thickness and a surface area *A* on either side of the boundary. Gauss's law in eq 1.57 can be alternatively expressed as

$$\int_V (\nabla \cdot \mathbf{D})dV = \oint_S \mathbf{D} \cdot d\mathbf{A} = Q_f. \tag{1.63}$$

Here, Q_f is the total free charge enclosed within the box. Applying eq 1.63 to this pill box, we obtain

$$D_{1n} - D_{2n} = Q_f / A = \sigma_f, \tag{1.64}$$

since the edge of the box contributes nothing to the close-surface integration in the limit as the thickness goes to zero. D_{1n} and D_{2n} represent the components of *D* normal to the boundary. σ_f is the free surface charge density. The positive directions for D_{1n} and D_{2n} are the same. Applying the same approach to the magnetic field *B* that does not have relevant charges, we have

$$B_{1n} - B_{2n} = 0. \tag{1.65}$$

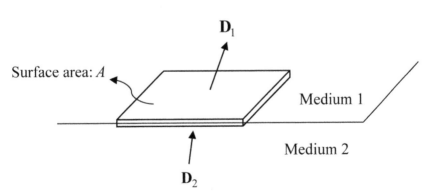

FIGURE 1.11 A thin pillbox extending into two media.

Equation 1.65 means that the normal component of the *B*-field should be conserved across any interface. We now construct a closed, rectangular loop that is perpendicular to the interface, as illustrated in Figure 1.12. This loop has a width l and its height is infinitely small approaching zero. Applying Faraday's law to the loop gives

$$\oint_C \mathbf{E} \cdot d\mathbf{l} = E_{1t}l - E_{2t}l = -\frac{d}{dt}\int \mathbf{B} \cdot d\mathbf{A}.$$

The third term of this relation contains an integral taken over the open surface area bounded by the loop. This integration area goes to zero in the limit as the height of the loop approaches zero. Therefore, the third term vanishes unless the magnetic flux is infinitely large. As a result, we have

$$E_{1t} - E_{2t} = 0. \tag{1.66}$$

The components of E parallel to the interface (i.e., tangential components) are continuous across the boundary. By the same token, we obtain

$$H_{1t} - H_{2t} = J_{fs}, \tag{1.67}$$

where J_{fs} represents the magnitude of free surface current density. Equations 1.64–1.67 constitute the general boundary conditions at an interface between two media. If there is no free charge or free current at the interface, they are reduced to

$$
\begin{aligned}
D_{1n} &= D_{2n} & B_{1n} &= B_{2n} \\
E_{1t} &= E_{2t} & H_{1t} &= H_{2t}
\end{aligned}. \tag{1.68}
$$

As we will see in Chapter 2, these boundary conditions determine the reflection and refraction of an electromagnetic wave at the interface.

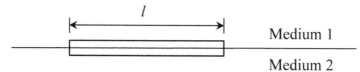

FIGURE 1.12 A closed, rectangular loop perpendicular to the interface.

1.3   ELECTROMAGNETIC WAVES

1.3.1   *ELECTROMAGNETIC WAVES IN FREE SPACE*

Electromagnetic radiation refers to the radiant energy released by certain electromagnetic processes. Classically, electromagnetic radiation consists of electromagnetic waves, which are produced whenever charged particles are accelerated. Suppose that a charge is somehow made to accelerate. When it is at rest, the associated E-field extends in all radial directions. When the charge begins to move, the E-field is modulated in the vicinity of the charge and this modulation propagates out into space at some speed. As discussed previously, a changing E-field induces a magnetic B field. Since the charge is accelerating, the time-variation of E is not constant, thus generating a time-dependent B field. The time-varying B-field also generates an E field. The process continues in this fashion, with synchronized oscillations of electric and magnetic fields that propagate through space. The fact that light is also an electromagnetic wave was revealed by Maxwell equations, which describe how electric and magnetic fields are generated and altered by each other. To derive the behaviors of electromagnetic waves in free space, we begin with the vector identity:

$$\nabla \times (\nabla \times \mathbf{A}) = \nabla (\nabla \cdot \mathbf{A}) - \nabla^2 \mathbf{A}, \tag{1.69}$$

which holds for any vector $\mathbf{A}$. For free space where there is no charge or current, Maxwell's equations are

$$\begin{aligned}
\nabla \cdot \mathbf{E} &= 0 \\
\nabla \cdot \mathbf{B} &= 0 \\
\nabla \times \mathbf{E} &= -\partial \mathbf{B} / \partial t \\
\nabla \times \mathbf{B} &= \mu_o \varepsilon_o \partial \mathbf{E} / \partial t
\end{aligned} \tag{1.70}$$

Then, the above vector identity for E can be described as

$$\nabla \times (\nabla \times \mathbf{E}) = \nabla (\nabla \cdot \mathbf{E}) - \nabla^2 \mathbf{E} = -\nabla^2 \mathbf{E}$$
$$\nabla \times (-\partial \mathbf{B} / \partial t) = -\partial (\nabla \times \mathbf{B}) / \partial t = -\nabla^2 \mathbf{E}$$

$$\mu_o \varepsilon_o \frac{\partial^2 \mathbf{E}}{\partial t^2} = \nabla^2 \mathbf{E}. \tag{1.71}$$

Likewise, substituting the B field vector into eq 1.69 and making use of Maxwell's equations leads to

$$\mu_o \varepsilon_o \frac{\partial^2 \mathbf{B}}{\partial t^2} = \nabla^2 \mathbf{B}. \tag{1.72}$$

Both eqs 1.71 and 1.72 satisfy the wave equation given by eq 1.33 and the resulting waves have a speed of

$$c = 1 / \sqrt{\mu_o \varepsilon_o} \approx 3 \times 10^8 \, \text{m/s}. \tag{1.73}$$

This theoretical value is exactly the speed of light, which was previously measured by Fizeau and thus known to Maxwell. It was a great intellectual triumph at that time to find out that light is an electromagnetic wave and that all electromagnetic waves propagate in free space at the same speed. It is customary to denote the speed of light in vacuum by the symbol c. The Laplacian, ∇^2, in eqs 1.71 and 1.72 operates on each component of E and B fields, so that the two wave equations lead to a total of six scalar equations. For an electric field given by $\mathbf{E} = \mathbf{i}E_x + \mathbf{j}E_y + \mathbf{k}E_z$, the following three scalar equations are obtained:

$$\partial^2 E_x / \partial x^2 + \partial^2 E_x / \partial y^2 + \partial^2 E_x / \partial z^2 = \mu_o \varepsilon_o \partial^2 E_x / \partial t^2$$
$$\partial^2 E_y / \partial x^2 + \partial^2 E_y / \partial y^2 + \partial^2 E_y / \partial z^2 = \mu_o \varepsilon_o \partial^2 E_y / \partial t^2 . \qquad (1.74)$$
$$\partial^2 E_z / \partial x^2 + \partial^2 E_z / \partial y^2 + \partial^2 E_z / \partial z^2 = \mu_o \varepsilon_o \partial^2 E_z / \partial t^2$$

The same type of scalar equations should also be satisfied for the magnetic field B. To prove that the electromagnetic wave is a transverse wave, consider the simple case of a plane E-field wave. If this plane wave is assumed to propagate in the positive x-direction, E is a function of only x and t and thus can be represented as $\mathbf{E}(x,t) = \mathbf{i}E_x(x,t) + \mathbf{j}E_y(x,t) + \mathbf{k}E_z(x,t)$. Here $\mathbf{i}$, $\mathbf{j}$, and $\mathbf{k}$ are the unit vectors along the x, y, and z directions, respectively. Applying Gauss's law to E leads to

$$\nabla \cdot \mathbf{E} = \partial E_x / \partial x + \partial E_y / \partial y + \partial E_z / \partial z = 0.$$

Since E and also its components are not a function of y or z, this equation is reduced to

$$\partial E_x / \partial x = 0. \qquad (1.75)$$

It means that E_x is independent of x. If E_x is not zero, it should be constant for all x values at any given time. This does not correspond to a wave traveling along the x-direction. Obviously $E_x = 0$, that is, the wave is transverse with no electric field component parallel to the direction of propagation. Now we like to know the direction of the associated magnetic field when the direction of the oscillating E field is fixed. We can adjust our coordinate axes so that the electric field is parallel to the y-axis. Then, both fields can be expressed as

$$\mathbf{E}(x,t) = \mathbf{j}E_y(x,t) = \mathbf{j}E_o e^{i(kx-\omega t)} \text{ and } \mathbf{B}(x,t) = \mathbf{B}_o e^{i(kx-\omega t)}, \qquad (1.76)$$

where E_o is the amplitude of the electric wave. $\mathbf{B}_o$ is the amplitude vector of the magnetic wave whose direction is unknown at present. Taking the curl of E gives

$$\nabla \times \mathbf{E} = \mathbf{k} \partial E_y / \partial x = \mathbf{k}(ik) E_o e^{i(kx - \omega t)}. \tag{1.77}$$

From Faraday's law, we have

$$\nabla \times \mathbf{E} = -\partial \mathbf{B} / \partial t = i\omega \mathbf{B}_o e^{i(kx - \omega t)}. \tag{1.78}$$

These two relations give

$$\mathbf{B}_o = \mathbf{k}(k / \omega) E_o. \tag{1.79}$$

Then, when the electric field is oriented along the y-axis, the magnetic field is directed along the z-axis. Its amplitude is proportional to the amplitude of the electric field. Evidently, E and B are *in phase* and *mutually perpendicular*, with their amplitudes related by $B_o = (k/\omega)E_o = E_o/c$. The spatial variations of these fields are graphically illustrated in Figure 1.13. A single-frequency electromagnetic wave exhibits a sinusoidal variation of electric and magnetic fields in space. As the electromagnetic wave propagates, it transports energy, stored in the propagating electric and magnetic fields. The energy flux density (energy per unit area, per unit time) is given by the Poynting vector $\mathbf{S}$:

$$\mathbf{S} = (\mathbf{E} \times \mathbf{B}) / \mu_o. \tag{1.80}$$

At optical frequencies, S is a very rapidly varying function of time, cycling from maxima to minima. This makes it practically impossible to measure its instantaneous value. So the time-averaged value of the magnitude of S, denoted by $\langle S \rangle$, is a measure of the energy flux density. The average energy flux density transported by an electromagnetic wave is called the *intensity* and is given by

$$I = \langle S \rangle = c^2 \varepsilon_o |\mathbf{E}_o \times \mathbf{B}_o| / 2 = c\varepsilon_o E_o^2 / 2. \tag{1.81}$$

The intensity, which has units of $j/m^2 \cdot s$, is therefore proportional to the square of the amplitude of the electric field. Although the two component fields of an electromagnetic wave always accompany each other, the force that the magnetic field exerts on charged particles (for instance, electrons) is negligibly small compared to that arising from the electric field. Therefore, the electric field is dominantly used in describing optical phenomena.

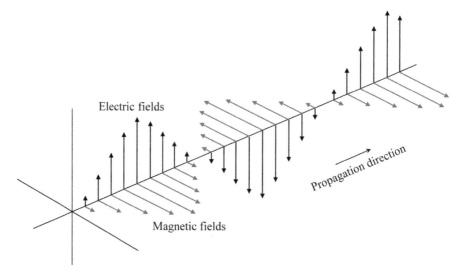

FIGURE 1.13 Electromagnetic wave.

1.3.2 *ELECTROMAGNETIC SPECTRUM*

Electromagnetic radiation is classified by its wavelength or frequency into radio wave, microwave, infrared (IR) light, visible light, ultraviolet (UV) light, X-rays, and γ-rays. However, there are no sharp boundaries between the regions. Figure 1.14 shows the whole spectrum of the electromagnetic radiation. Our naked eyes can sense a relatively small range of wavelengths (400–700 nm) called visible spectrum or simply light. Other wavelengths, especially nearby IR (longer than 700 nm) and UV (shorter than 400 nm), are also sometimes referred to as light. X-rays, located in between UV and γ-rays, have a wavelength range of 10^{-2} Å to 10 nm. X-rays used in diffraction experiments have wavelengths of 0.5–2.5 Å because the interatomic distances in materials are on the order of 1–10 Å. When Maxwell published his first extensive account of the electromagnetic theory in 1867, the frequency range was only known to extend from the IR, across the visible, to the UV. Although this range is of primary concern in optics, it is just a small segment of the whole electromagnetic spectrum. Ten years later, Hertz succeeded in producing and detecting electromagnetic waves. His transmitter was an oscillating discharge across a spark gap and he used an open loop of wire for a receiving antenna. A small spark induced between the two ends of the antenna indicated the detection of an incident electromagnetic wave. The waves used by Hertz are now classified in the radio frequency range, which extends from a few Hz to about 10^9 Hz (in wavelength, from several kilometers to 30 cm). These waves are usually generated by an assortment of electric circuits.

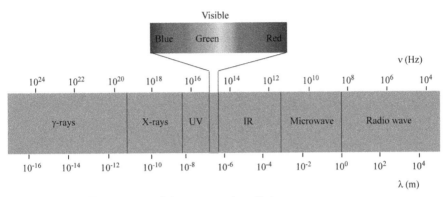

FIGURE 1.14 The spectrum of electromagnetic radiation.

The microwave region extends from about 10^9 Hz to 3×10^{11} Hz and the corresponding wavelengths lies in between 0.3 m and 1 mm. Radiation that can penetrate the Earth's atmosphere ranges from less than 1 cm to about 30 m. Microwaves are thus of interest in space-vehicle communications. Molecules can absorb and emit energy by altering the state of motion of their constituent atoms. They can be made to vibrate and rotate. The energy associated with each motion is quantized. Namely, molecules possess vibrational and rotational energy bands as well as their electronic energy bands. Only polar molecules experience forces via the electric field of an incident electromagnetic wave. The polar molecules can absorb a photon and make a rotational transition into an excited state. For instance, water molecules are polar. Thus, when exposed to an electromagnetic wave, they will swing around, trying to be lined up with the alternating electric field. This effect is particularly vigorous at any one of their rotational resonances (i.e., rotational energy bands). Consequently, water molecules efficiently absorb microwave at or near such a resonance frequency. The microwave oven (12.2 cm, 2.45 GHz) is a well-known application, which utilizes heating by water molecules contained in the food. On the other hand, nonpolar molecules such as carbon dioxide, hydrogen, oxygen, and methane cannot make rotational transitions via the absorption of electromagnetic wave. Microwaves are also widely used for wireless communications.

The IR region extends approximately from 3×10^{11} Hz to 4×10^{14} Hz. The IR is often divided into four subregions: the near IR, that is, near the visible (780–3000 nm), the intermediate IR (3000–6000 nm), the far IR (6000–15,000 nm), and the extreme IR (15,000 nm–1 mm). This is a rather loose division, and there is again no sharp boundary separating them. Any material radiates and absorbs IR via thermal agitation of its constituents. Although the molecules of any objects at a temperature above $T = 0$ K will radiate IR, it is abundantly emitted in a continuous spectrum from hot bodies such as electric heaters, burning coals, and house radiators. Approximately half of the electromagnetic energy from the Sun is IR. The human body also radiates IR, even though the radiant energy is quite weak. A molecule can not only rotate but also vibrate in several different ways, with its atoms moving in various directions. For the vibration mode, the molecule need not be polar. For example, CO_2 has three vibrational modes and many associated energy levels, each of which can be excited by photons. The corresponding absorption spectra lie in the IR region. A number of molecules have both vibrational and rotational resonances and are good IR absorbers. We can feel the resulting build-up of thermal energy when our face was put in the sunshine. IR energy is usually measured by a device that responds to the heat generated on absorption. A small difference in the temperatures of an object and its surroundings results in characteristic IR emission, which can be effectively utilized for medical diagnostics.

Visible light has a very narrow band of frequencies from about 3.8×10^{14} Hz to 7.5×10^{14} Hz. It is usually generated by a rearrangement of the outer electrons of atoms and molecules. The color of light is determined by its wavelength (and frequency). Newton was the first to recognize that white light is essentially a mixture of all the colors of the visible spectrum. Colors are actually the subjective human physiological responses to the various wavelengths ranging from about 650 nm for red, through orange, yellow, green, and blue, to violet at about 400 nm. A variety of different wavelength mixtures can evoke the same color response from the eye-brain sensor. For example, a beam of red light overlapping a beam of green light will result in the sensing of yellow light, even though the overlapped beam has no wavelengths belonging to the yellow band. That is why a display can be operated with only three light sources: red, green, and blue. Next to visible light in the electromagnetic spectrum is the UV region that ranges approximately from 8×10^{14} Hz to about 3.4×10^{16} Hz. A UV photon can be emitted by an atom when its electron makes a long jump down from a highly excited state. Photon energies in UV range from roughly 3 eV to 100 eV. UV rays from the Sun thus have more than enough energy to ionize atoms. Fortunately, ozone in the atmosphere substantially absorbs a lethal UV stream from the Sun.

X-rays was fortuitously discovered by W. Röntgen in 1895. They have extremely short wavelengths; most are smaller than the atom size. The most practical method for producing X-rays is the rapid deceleration of charged particles accelerated to a very high speed. A broad X-ray spectrum arises when an energetic electron beam collides with a target material, such as a Cu plate. The atoms of the target may also be ionized during the bombardment. If the ionization occurs by removal of an inner electron tightly bound to the nucleus, the atom will emit X-rays as the vacant level is occupied by one of higher-lying electrons. The resulting quantized emissions are characteristic of the target atom, and accordingly are called *characteristic* radiation. γ-rays are the highest-photon energy, lowest-wavelength electromagnetic radiations. These rays are emitted by particles undergoing transitions within the atomic nucleus. Since the wavelengths are so short, it is very difficult to observe any wave-like properties from the γ-rays.

1.3.3   WAVE-PARTICLE DUALITY

Electromagnetic radiation exhibits wave-like properties and particle-like properties at the same time (wave-particle duality). Therefore, the propagation of an electromagnetic wave can also be considered as a stream of light particles called *photons*. According to quantum theory, the energy of a photon (E_p) is related to the frequency and wavelength of the wave as follows:

$$E_p = h\nu = h\frac{c}{\lambda}, \tag{1.82}$$

where h is Plank constant (6.63×10^{-34} j·s). As a consequence of this duality, an electromagnetic wave with $\lambda = 100$ nm can be viewed as equivalent to a flow of photons with $E_p = 12.44$ eV. Various phenomena occurring with the electromagnetic radiations can be better explained sometimes with the wave concept, and sometimes, the particle concept. Wave characteristics are more apparent when the radiation is measured over relatively large time-scales and over large distances (e.g., interference and diffraction). On the contrary, particle characteristics will be more obvious in the case of absorption because it occurs quite fast in specific positions. When the wavelength and frequency are fixed, the intensity of an electromagnetic wave is proportional to the square of its amplitude. Thus if the amplitude is doubled, the intensity becomes four times. Viewed from the particle nature, this means that the number of photons crossing unit cross-section per unit time increases four-fold. Let's consider a case in which atoms or electrons are photoexcited to higher energy states by absorbing the incident electromagnetic wave. When the amplitude of the wave is doubled, four times more photons will be absorbed as a result of the four-fold intensity increase. It is to be noted, however, that the energy of individual photons is fixed and the excitation to energy levels exceeding the photon energy is thus impossible no matter how high the intensity may be.

1.4 LIGHT PROPAGATION IN DIELECTRICS

In Section 1.3.1, we have assumed that the electromagnetic waves are traveling through empty space. The effect of introducing a linear, homogeneous, and isotropic dielectric is simply to change ε_o to ε and μ_o to μ in Maxwell's equations.[1-7] Then, the speed in the medium becomes

$$v = 1 / \sqrt{\mu\varepsilon} \cdot \tag{1.83}$$

The ratio of the speed of an electromagnetic wave in vacuum to that in matter is defined as the *refractive index n* of the medium and is given by

$$n = \frac{c}{v} = \sqrt{\frac{\mu\varepsilon}{\mu_o\varepsilon_o}} \cdot \tag{1.84}$$

As the speed of the wave changes, its wavelength also changes within the medium. Since the frequency is invariant in linear optics, the speed and wavelength are reduced by the same proportion when the wave enters a higher-index medium. Except for such ferromagnetic materials as Fe, Ni, and Co, most substances are only weakly magnetic. For these media, μ is very close to μ_o but ε may be greatly different from ε_o. Therefore, the refractive index of a dielectric medium is simply expressed as

$$n = \sqrt{\varepsilon / \varepsilon_o} = \sqrt{K_e} \, , \tag{1.85}$$

where K_e is the dielectric constant of the medium. The intensity given by eq 1.81 is more generalized to

$$I = \varepsilon v E_o^2 / 2 = \varepsilon_o c n E_o^2 / 2 \cdot \tag{1.86}$$

In a linear, homogeneous, isotropic dielectric, the intensity of an electromagnetic wave is proportional to the square of its amplitude and also to the refractive index of the medium.

The dielectric constant and refractive index of a material are actually *frequency-dependent*. The dependence of *n* on the color (wavelength) of light is a well-known effect called *dispersion*. This effect accounts for the dispersion of white light into its constituent rainbow colors by a prism. To deal with the well-known frequency dependence of the refractive index, it is necessary to exploit the atomic nature of matter and its frequency-dependent aspect. The electrons in an atom are attached to the positive nucleus by some binding forces whose detailed structure may be very complicated. When the bound electron is displaced from equilibrium, a net force should exist that returns the system to equilibrium. For very small displacements, *y*, the force will be linear with *y*. Then, the restoring force is given by

$$F_{restoring} = -\kappa y = -m\omega_o^2 y , \qquad (1.87)$$

where κ is a force constant and *m*, the electron's mass. Once momentarily disturbed somehow, a bound electron will oscillate about its equilibrium opposition with a *natural* or *resonant frequency* given by $\omega_o = (\kappa/m)^{1/2}$, just like a simple harmonic oscillator. When the electron oscillates, there will be some damping force proportional to the speed of motion:

$$F_{damping} = -m\gamma \frac{dy}{dt} . \qquad (1.88)$$

Here γ is a damping or friction constant with dimensions of reciprocal time. The damping must be opposite in direction to the electron motion. In the presence of an electromagnetic wave of frequency ω, the electron is subject to a driving force

$$F_{driving} = -eE = -eE_o \cos \omega t . \qquad (1.89)$$

where E_o is the amplitude of the wave at the point *x* where the electron is situated. Since we are only interested in one fixed point, the electron position can be set to *x* = 0 here. Note that the force ($-ev \times B$) exerted on the electron by the magnetic field of the wave is omitted because it is negligible compared with the force ($-eE$) due to the electric field. Putting all these forces into Newton's second law leads to the following equation of motion:

$$m\frac{d^2 y}{dt^2} = -m\omega_o^2 y - m\gamma \frac{dy}{dt} - eE_o \cos \omega t \qquad (1.90)$$

This equation describes the electron as a damped harmonic oscillator driven at frequency ω (Fig. 1.15). Here we assumed that the much heavier nucleus remains at rest. Equation 1.90 is easier to handle if it is regarded as the real part of a complex equation:

$$m\frac{d^2\tilde{y}}{dt^2} + m\omega_o^2\tilde{y} + m\gamma\frac{d\tilde{y}}{dt} = -eE_o e^{-i\omega t}.$$ (1.91)

We can anticipate that the electron oscillates with the same frequency as the electric field. Then, the displacement can be expressed as

$$\tilde{y} = \tilde{y}_o e^{-i\omega t}.$$ (1.92)

It is to be noted that when the amplitude of the electric field, E_o, is a positive real value, $\tilde{y}_o$ can have either a positive or negative value. It may also be a complex number.

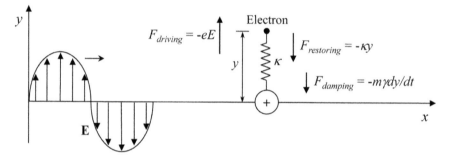

FIGURE 1.15 A damped harmonic oscillator model describing the motion of a bound electron under an oscillating electric field.

Inserting eqs 1.92 into 1.91 gives

$$\tilde{y}_o = \frac{(-e/m)}{(\omega_o^2 - \omega^2) - i\omega\gamma}E_o.$$ (1.93)

The dipole moment is equal to the product of the charge, $-e$, and its displacement. If there are N contributing electrons per unit volume, the electric polarization P becomes

$$P(t) = -e\tilde{y}N = \frac{Ne^2/m}{(\omega_o^2 - \omega^2) - i\omega\gamma}E_o e^{-i\omega t} = \frac{Ne^2/m}{(\omega_o^2 - \omega^2) - i\omega\gamma}E(t).$$ (1.94)

Since the proportionality constant between $P(t)$ and $E(t)$ is complex, the resulting polarization may be *out of phase* with the applied electric field. When the driving frequency ω is far from the resonance frequency ω_o, we may set $\gamma = 0$, corresponding to negligible damping. For $\omega \ll \omega_o$, P and E have the same sign and the dipoles are oscillating in phase with the electric field. Beyond resonance, that is, $\omega \gg \omega_o$, they have opposite signs, implying a phase difference of π. Near resonance ($\omega \cong \omega_o$), the damping term in the denominator is not negligible but makes a dominant contribution. This leads to a phase shift of 90° between E and P. The magnitude of P is given by

$$|P| = \frac{Ne^2 / m}{\sqrt{(\omega_o^2 - \omega^2)^2 + \omega^2 \gamma^2}} E_o. \qquad (1.95)$$

Cleary, it drastically increases as ω approaches ω_o. From eqs 1.58 and 1.59, we have

$$\varepsilon = \varepsilon_o + \frac{P(t)}{E(t)} = \varepsilon_o + \frac{Ne^2 / m}{(\omega_o^2 - \omega^2) - i\omega\gamma}. \qquad (1.96)$$

The dependence of n on ω is then

$$n^2 = 1 + \frac{Ne^2}{m\varepsilon_o} \frac{1}{(\omega_o^2 - \omega^2) - i\omega\gamma}. \qquad (1.97)$$

The denominator in eqs 1.96 and 1.97 contains both real and imaginary parts. If the real part has a very large absolute magnitude, the imaginary part, that is, the damping term can be ignored. For $\omega \ll \omega_o$, the electric polarization will be nearly in phase with the applied electric field. Then, the dielectric constant and the corresponding refractive index will both be greater than 1. For $\omega \gg \omega_o$, the resulting polarization becomes 180° out of phase with the electric field. Hence, the dielectric constant and the refractive index will both be smaller than 1. In these ordinary frequency ranges, the imaginary damping term is insignificant. However, when ω is very close to ω_o, it plays an important role. Since the refractive index n is now complex, the propagation constant k also has real and imaginary parts:

$$k^2 = \frac{\omega^2}{c^2} n^2 = \frac{\omega^2}{c^2} \left[1 + \frac{Ne^2}{m\varepsilon_o} \frac{1}{(\omega_o^2 - \omega^2) - i\omega\gamma} \right]. \qquad (1.98)$$

By defining the propagation constant as

$$k = k_R + ik_I \qquad (1.99)$$

and inserting it into the expression for a harmonic wave propagating in the x-direction, we have

$$E(x,t) = E_o e^{i(kx - \omega t)} = E_o e^{-k_I x} e^{i(k_R x - \omega t)}. \qquad (1.100)$$

Since the intensity is proportional to the square of the amplitude, eq 1.100 gives the following relation:

$$I(x) = I_o e^{-2k_1 x} = I_o e^{-\alpha x}, \tag{1.101}$$

where $\alpha = 2k_I$ is the *absorption coefficient* of the medium. Since $k = \omega n/c$, the real and imaginary parts of the complex refractive index are given as

$$n_R = \frac{c}{\omega} k_R, \ n_I = \frac{c}{\omega} k_I. \tag{1.102}$$

Here n_R is the usual refractive index and n_I is called the *extinction coefficient*. It is obvious from eq 1.100 that k_R and n_R determine the velocity of the wave. On the contrary, k_I and n_I are related to the amplitude attenuation of the wave. Equation 1.97 can be rearranged into

$$(n_R + in_I)^2 = 1 + \frac{Ne^2}{m\varepsilon_o} \left[\frac{\omega_o^2 - \omega^2}{(\omega_o^2 - \omega^2)^2 + \omega^2 \gamma^2} + i \frac{\omega \gamma}{(\omega_o^2 - \omega^2)^2 + \omega^2 \gamma^2} \right]. \tag{1.103}$$

From eq 1.103, we have

$$\begin{aligned} n_R^2 - n_I^2 &= 1 + \frac{Ne^2}{m\varepsilon_o} \left[\frac{\omega_o^2 - \omega^2}{(\omega_o^2 - \omega^2)^2 + \omega^2 \gamma^2} \right] \\ 2n_R n_I &= \frac{Ne^2}{m\varepsilon_o} \left[\frac{\omega \gamma}{(\omega_o^2 - \omega^2)^2 + \omega^2 \gamma^2} \right] \end{aligned}. \tag{1.104}$$

These equations can be simultaneously solved to give n_R and n_I. Typical variations of n_R and n_I near a resonant frequency ω_o are plotted in Figure 1.16. The absorption described by the extinction coefficient is represented by a peak at the resonance frequency. This absorption has nothing to do with conductivity. Ordinarily, the (usual) refractive index gradually increases with increasing frequency. However, the index of refraction sharply drops in the vicinity of a resonance. After passing through the resonance, it increases again, approaching the value $n_R = 1$ at higher frequencies. The dispersion curve has a narrow region where n_R decreases with frequency. Since this is contrary to the usual dispersion of transparent materials, it is called *anomalous dispersion*. The region of anomalous dispersion is coincident with the region of maximum absorption and the material may be opaque in this frequency region. As ω approaches a natural frequency, the electron oscillators will begin to resonate, with their amplitudes markedly increased. This will be accompanied by a strong absorption of energy from the incident wave and a large amount of energy dissipated by damping. As the amplitude of the oscillation gets larger, the energy lost by damping will correspondingly increase. The damping term obviously becomes dominant when $\omega = \omega_o$ in eq 1.97.

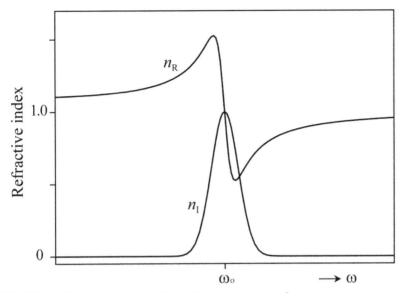

FIGURE 1.16 Typical variations of n_R and n_I near a resonant frequency ω_o.

When the radiation frequency is slightly greater than the resonance frequency, the index of refraction becomes less than unity (< 1), implying that the speed of light in the medium at that frequency exceeds the speed of light in vacuum. The refractive index lower than unity does not contradict the special theory of relativity; no information-carrying signal can travel faster than the speed of light in vacuum. The refractive index is defined with respect to the phase (or wave) velocity, which does not carry information. The phase velocity is not the same as the group velocity or the signal velocity. It means the speed at which the crests of the wave move and can be faster than the speed of light in vacuum, thereby giving a refractive index below 1. Equation 1.94 and Figure 1.16 have been derived considering electron oscillators only. Similar results may be obtained for ions bound to fixed atomic positions as well. The appearance of the mass m in the denominator of eq 1.104 shows that electron oscillations are more important than ionic oscillations in determining the refractive index. However, the contribution from ionic polarization may be significant in the region of resonance. The addition of other contributing terms usually increases the background level and in some cases, keeps $n > 1$ on both sides of the resonance. Although we have thus far considered only a single resonance frequency, there may be a series of resonant frequencies in a material. Then, eq 1.97 is more generalized to

$$n^2 = 1 + \frac{Ne^2}{m\varepsilon_o} \sum_j \frac{f_j}{(\omega_j^2 - \omega^2) - i\omega\gamma_j},$$ (1.105)

where f_j is a weighting factor known as *oscillator strength* and γ_j, a damping coefficient for the resonance ω_j. If any of the resonant frequencies lies in the visible range, the material absorbs a portion of the spectrum and looks colored. Transparent materials like glass have resonance frequencies in the UV and IR regions but not in the visible range. This is why they are color-less. If the frequency of the incident wave is appreciably different from a resonance or natural frequency, the resulting oscillations are small. Thus there is little dissipative absorption. At resonance, however, the oscillation amplitudes are markedly increased, with the wave field doing an increased work on the charges. The absorbed energy is mostly dissipated in the form of heat.

Now we wish to show how the index of refraction actually varies with frequency (or wavelength) in the region where absorption is negligible, that is, $\gamma = 0$. As stated earlier, n is an increasing function of ω under normal dispersion. Since significant resonances occur in the UV region for most transparent materials, we shall assume a single resonant frequency ω_o in the UV, such that $\omega \ll \omega_o$ holds for the visible frequencies ω. Then, eq 1.105 is simplified to take the form:

$$n^2 = 1 + \frac{Ne^2}{m\varepsilon_o} \frac{1}{(\omega_o^2 - \omega^2)}.$$ (1.106)

The refractive index, which is a real number in this case, slightly increases as ω moves toward ω_o from later. This behavior is characteristic of normal dispersion. The dispersion relation is more conventionally expressed in terms of the wavelength in vacuum ($\lambda = 2\pi c/\omega$). The frequency factor in eq 1.106 can be expanded in a binomial series:

$$\frac{1}{(\omega_o^2 - \omega^2)} = \frac{1}{\omega_o^2}\left(1 - \frac{\omega^2}{\omega_o^2}\right)^{-1} = \frac{1}{\omega_o^2}\left(1 + \frac{\omega^2}{\omega_o^2} + \frac{\omega^4}{\omega_o^4} + \cdots\right)$$

so that

$$n^2 = 1 + \frac{Ne^2}{m\varepsilon_o\omega_o^2}\left(1 + \frac{\omega^2}{\omega_o^2} + \frac{\omega^4}{\omega_o^4} + \cdots\right).$$ (1.107)

Equation 1.107 can be alternatively expressed as

$$n^2 = a + \frac{b}{\lambda^2} + \frac{c}{\lambda^4} + \cdots.$$

Taking the square root of both sides and then expanding the right-hand side in a binomial series again, we have

$$n = A + \frac{B}{\lambda^2} + \frac{C}{\lambda^4} + \cdots . \qquad (1.108)$$

This is known as *Cauchy's equation*, where A, B, and C are coefficients that can be determined by fitting the equation to refractive indices measured at different wavelengths. Cauchy's equation agrees well with the measured refractive indices over the visible region. An example is given in Figure 1.17.

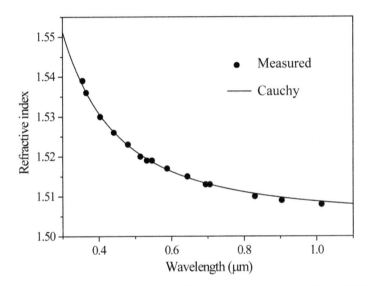

FIGURE 1.17 Cauchy's equation fitting to the measured refractive indices of BK7 glass (borosilicate glass). Measured data adapted from the (2003) optical parts catalogue of Sigma Koki Co., Ltd. The fitting coefficients in eq 1.108 are A = 1.5046 and B = 0.0042 (μm^2).

1.5 LIGHT PROPAGATION IN METALS

The free electrons in a conductor are not bound to any particular nuclei and thus have no restoring force. By setting the restoring force term in eq 1.91 to zero, we have the equation of motion for free electrons as follows:

$$m \frac{d^2 \tilde{y}}{dt^2} + m\gamma \frac{d\tilde{y}}{dt} = -eE_o e^{-i\omega t} \qquad (1.109)$$

Here γ is a damping factor that reflects the effect of many collisions between free electrons. Although there are also bound electrons, the free electrons dominate the electrical and optical properties of metals. The equation of motion can be conveniently expressed in term of the current density J, which is given by

$$J = -Nev = -Ne\frac{dy}{dt}. \tag{1.110}$$

Then, eq 1.109 is alternatively expressed as

$$\frac{dJ}{dt} + \gamma J = \frac{Ne^2}{m}E_o e^{-i\omega t}. \tag{1.111}$$

When the applied field is a harmonic wave $E = E_o e^{-i\omega t}$, we expect that the current density varies at the same frequency and can be written as $J = J_o e^{-i\omega t}$. Inserting this into eq 1.111 gives

$$(-i\omega + \gamma)J = \frac{Ne^2}{m}E. \tag{1.112}$$

For the static, DC field specified by $\omega = 0$,

$$J = \left(\frac{Ne^2}{m\gamma}\right)E = \sigma_o E. \tag{1.113}$$

Here σ_0 is the *static conductivity* defined by Ohm's law. From eqs 1.112 and 1.113, we have

$$J = \left(\frac{\sigma_o}{1 - i\omega/\gamma}\right)E. \tag{1.114}$$

The conductivity σ, which relates the applied electric field to the resulting current density, is generally frequency-dependent. The imaginary term in the denominator indicates that J may be out of phase with E. At low frequencies, the imaginary term is negligible and the conductivity is nearly independent of the frequency. For sufficiently high frequencies, however, this term cannot be ignored. As the frequency of the applied electric field increases, the inertia of electrons introduces a phase lag in the electron response to the field and the dynamic conductivity becomes complex.

An electromagnetic wave propagating in the conducting medium also satisfies Maxwell's equations given by eq 1.50. Any material remains electrically neutral unless we artificially implant charged particles into it. There is thus no net charge density. However, the current density in a conductor is proportional to the applied field and is certainly not zero. Thus, Maxwell's equations for linear, homogeneous, conducting media take the form:

$$\nabla \cdot \mathbf{E} = 0$$
$$\nabla \cdot \mathbf{B} = 0$$
$$\nabla \times \mathbf{E} = -\partial \mathbf{B} / \partial t \qquad (1.115)$$
$$\nabla \times \mathbf{B} = \mu_o \left(\frac{\sigma_o}{1 - i\omega/\gamma} \right) \mathbf{E} + \mu_o \varepsilon_o \partial \mathbf{E} / \partial t$$

Applying the vector identity of eq 1.69 as before, we have a modified wave equation for E.

$$\nabla^2 \mathbf{E} = \mu_o \varepsilon_o \frac{\partial^2 \mathbf{E}}{\partial t^2} + \mu_o \left(\frac{\sigma_o}{1 - i\omega/\gamma} \right) \frac{\partial \mathbf{E}}{\partial t} \qquad (1.116)$$

This equation can be solved using a harmonic wave function of $E = E_o$ $e^{i(kx - \omega t)}$ to obtain the propagation constant k given by

$$k^2 = \mu_o \varepsilon_o \omega^2 + i \left(\frac{\mu_o \sigma_o \omega}{1 - i\omega/\gamma} \right). \qquad (1.117)$$

For low frequencies ($\omega \ll \gamma$: this condition holds for visible and IR regions), eq 1.117 can be approximated to

$$k^2 \cong \mu_o \varepsilon_o \omega^2 + i\mu_o \sigma_o \omega. \cdot \qquad (1.118)$$

Then, the propagation constant is a complex number with its real and imaginary parts given by

$$k_R = \omega \sqrt{\frac{\mu_o \varepsilon_o}{2}} \left[\sqrt{1 + (\frac{\sigma_o}{\varepsilon_o \omega})^2} + 1 \right]^{1/2}$$
$$k_I = \omega \sqrt{\frac{\mu_o \varepsilon_o}{2}} \left[\sqrt{1 + (\frac{\sigma_o}{\varepsilon_o \omega})^2} - 1 \right]^{1/2} \qquad (1.119)$$

The imaginary part, k_I, of the propagation constant results in a high attenuation of the incident wave, as described in eqs 1.100 and 1.101. The amplitude of the electromagnetic wave exponentially decreases as it penetrates into the conductor. The distance to $1/e$ of its surface value ($1/e^2$ for the intensity) is called the *skin depth* defined as $\delta = 1/k_I$. For a good conductor, $\sigma_o \gg \omega \varepsilon_o$,

$$\delta = \sqrt{\frac{2}{\sigma_o \mu_o \omega}}. \qquad (1.120)$$

The skin depth is a measure of how deeply an electromagnetic wave penetrates into the conductor. Meanwhile, the wavelength and speed inside a conducting medium are determined by the real part of the propagation constant. It is to be noted that "wavelength" loses its geometrical significance in a good conductor, since the skin depth is often less than a single wavelength.

We next examine what happens in the case of high frequencies ($\omega \gg \gamma$). Under this high-frequency condition, the general relation of eq 1.117 is simplified to

$$k^2 \cong \mu_o \varepsilon_o \omega^2 - \mu_o \sigma_o \gamma = \mu_o \varepsilon_o (\omega^2 - \omega_p^2),$$

where

$$\omega_p = \left(\frac{Ne^2}{m\varepsilon_o} \right)^{1/2} \tag{1.121}$$

is known as the *plasma frequency*. The plasma frequency is a natural (or resonant) frequency for the oscillations of free electrons about their equilibrium positions. Plasma is an electrically neutral medium consisting of electrons and positive ions, where their number densities are the same. If the electrons are somehow displaced from a uniform background of ions, local electric fields will be built up in the plasma in order to restore the neutrality by pulling the electrons back to their original positions. Due to the inertia, however, the returning electrons will overshoot their original positions and oscillate with a characteristic frequency known as the plasma frequency. Free electrons and positive ions in a metal can be considered as a kind of plasma. When the frequency of the incident wave coincides with the plasma frequency, a strong absorption, known as *plasma resonance absorption*, is observed. The plasma frequency is a critical frequency determining whether the propagation constant k becomes real or imaginary. For $\omega < \omega_p$, the propagation constant is pure imaginary. Therefore, the incident wave cannot propagate in the medium and is mostly reflected from the surface. For $\omega > \omega_p$, the propagation constant is real and as a result, the wave propagates without attenuation. In reality, however, there are some absorptive losses due to inter-band transitions. The plasma frequencies for common metals lie in the UV region. This is why X-rays can deeply penetrate even into metals. Although metals are opaque to visible light, they are fairly transparent to X-rays. Some of the alkali metals are transparent even to UV. The attenuation of the electromagnetic wave in a conductor is caused by the imaginary part of the propagation constant. It is more basically due to the conductivity, as is manifest from eqs 1.119 and 1.121. When the driving force oscillates too rapidly, the free electrons cannot keep pace up with it since they also have a mass. This makes metals behave like nonconductors at high frequencies, for instance $\omega > \omega_p$. The plasma frequency of a metal can be regarded as the maximum frequency at which its free electrons can collectively oscillate.

The refractive index of metals is generally complex consisting of real and imaginary parts and is given by

$$n = n_R + in_I. \tag{1.122}$$

Both n_R and n_I are real numbers and are dependent on the frequency. As described before, the real index n_R determines the phase velocity of the wave, while n_I accounts for the attenuation of the wave. The refractive index is alternatively denoted as $n = n_R - in_I$, when the sinusoidal behavior of the wave is expressed as $\exp[i(\omega t - kx)]$ instead of $\exp[i(kx - \omega t)]$. Since $k = \omega n/c$, eq 1.121 can be expressed in term of the refractive index.

$$n^2 = 1 - \frac{\omega_p^2}{\omega^2}. \tag{1.123}$$

For $\omega < \omega_p$, the refractive index is complex and the wave is attenuated. On the contrary, the refractive index becomes real at $\omega > \omega_p$, making the medium transparent to the incident radiation. As ω further increases beyond ω_p, the refractive index approaches unity from below. This behavior is similar to that observed in dielectrics for $\omega > \omega_o$. In fact, all materials have a refractive index very close to 1 at X-rays wavelengths. For a material to be transparent, the skin depth should be larger than its thickness. Metals have a very small skin depth except for the case of $\omega > \omega_p$; it varies from less than 1 nm to tens of nm depending on the conductivity and frequency. This accounts for the general opacity of bulk metals. When an electromagnetic wave with $\omega < \omega_p$ is incident into the metal, most of the incoming energy is reflected from the surface. Of course, metals have a high absorption coefficient. However, the absorbed energy mostly reappears as the reflected wave, instead of being dissipated within the substance. This is responsible for the familiar metallic gloss. That some metals exhibit particular colors is because they reflect specific wavelengths (i.e., colors) more strongly than others. If a metal is made very thin, it can be partially transparent. It is to be noted that while the color of a bulk metal is determined by reflected light, it is determined by transmitted light for a partly transparent thin film. Bulk gold is reddish yellow in color because the imaginary refractive index increases with increasing visible wavelength and thus longer wavelengths are more strongly reflected. As a consequence, a thin gold film will appear greenish blue at our eyes since it will predominantly transmit shorter visible wavelengths corresponding to this color. Figure 1.18 shows the reflection and transmission spectra obtained from a 10-nm-thick gold film coated on a glass substrate. In usual, the addition of reflectance and transmittance does not become unity because some of the incident energy is inevitably dissipated even in a thin film.

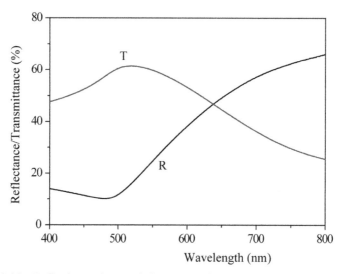

FIGURE 1.18   Reflection and transmission spectra for a 10-nm-thick gold film coated on a glass substrate.

PROBLEMS

1.1 Using the wave functions $f_1 = 4\sin2\pi(0.2x - 3t)$ and $f_2 = 0.2\sin(7x + 3.5t)$, determine in each case the values of (a) frequency, (b) wavelength, (c) period, and (d) direction of motion. t is in seconds and x is in meters.

1.2 Three different electromagnetic waves of the same frequency and polarization direction are spatially overlapped, interfering with one another. At a certain instant, each of the waves is represented as follows:

$$E_1 = E_o \cos\left(\frac{2\pi x}{\lambda}\right)$$

$$E_2 = 2E_o \cos\left(\frac{2\pi x}{\lambda} - \frac{\pi}{6}\right)$$

$$E_3 = E_o \cos\left(\frac{2\pi x}{\lambda} - \frac{\pi}{4}\right)$$

Then, express the combined wave as a single cosine function and find its intensity in terms of E_o.

1.3 The skin depth of visible light into metals ranges from less than 1 nm to tens of nm. If the skin depth of a certain metal is 6 nm at $\lambda = 632.8$ nm, how thin should it be to exhibit 50% transmittance at the same wavelength?

1.4 An X-ray beam of $\lambda = 1.54$ Å is propagating in free space. If this beam has an intensity of 100 W/cm^2, then how many photons are moving across unit area per unit time?

1.5 If the intensity of an X-ray-beam at $\lambda = 10$ nm is four times higher than that of a visible light beam at 600 nm, how much different is the number of photons moving across unit area per unit time in two radiations?

1.6 There are always equal numbers of electrons and positive ions in the plasma. Let N be the number density of the electrons, each having a charge of $-e$. Now we imagine that all of the electrons are displaced from their original positions by a small distance x, while the positive ions are held fixed. (a) Apply Gauss's law to find a local electric field induced in the plasma. (b) Using a frictionless harmonic oscillator model with the result of "(a)," prove that the plasma oscillation frequency is given by eq 1.121.

REFERENCES

1. Hecht, E. *Optics,* 5th ed; Pearson: San Francisco, 2016, San Francisco.
2. Pedrotti, F.; Pedrotti, L. M.; Pedrotti, L. S. *Introduction to Optics,* 3rd ed; Addison-Wesley: Boston, 2006
3. Griffiths, D. *Introduction to Electrodynamics,* 4th ed; Addison-Wesley: London, 2012.
4. Fowles, G. *Introduction to Modern Optics,* 2nd ed; Dover: New York, 1989.
5. Guenther, B. *Modern Optics,* 2nd ed; Oxford University Press: New York, 2015.
6. Born, M.; Wolf, E. *Principles of Optics,* 7th ed; Cambridge University Press: Cambridge, UK, 1999.
7. Saleh, B.; Teich, M. *Fundamentals of Photonics,* 2nd ed; Wiley: Hoboken, NJ, 2007.

CHAPTER 2

Reflection and Refraction

2.1 INTRODUCTION

In this chapter, we describe a variety of phenomena occurring when a light wave passes from one medium to another. Consider a wave impinging on the interface separating two different media. As we already know from daily life, a portion of the incident energy bounces off the interface in the form of a reflected wave, while the remainder is transmitted across the boundary. When the wave is obliquely incident onto the interface, it is partially reflected and partially refracted. Refraction refers to the change in propagation direction of a wave when it enters a medium of different refractive index. The phenomenon is mainly governed by the law of energy and momentum conservation. It is commonly observed when a wave passes from one medium to another at any angle other than 0° from the normal. Light refraction is the most commonly observed phenomenon, but any type of wave can be refracted when it interacts with a medium. Refraction is also responsible for rainbows and for the splitting of white light into a rainbow-spectrum as it passes through a glass prism. A long object such as a pencil or wood stick obliquely immersed in water looks bent due to refraction. This is because a light ray reflected from the tip of the object refracts as it leaves the surface of water. Thus, the ray conceived by our eyes looks as if it were reflected from a point other than the tip of the object. Understanding of this concept led to the invention of lenses and glasses.

The refraction of light is quantitatively described by *Snell's law*. This law represents the relationship between the angles of incidence and refraction, when a light wave passes through a boundary between two isotropic media with different refractive indices. Snell's law can be derived from *Fermat's principle* or the application of boundary conditions for electromagnetic waves. The detailed behaviors of reflection and refraction in a given situation are well explained by the *Fresnel equations*, which describe what fraction of the incident light is reflected and what fraction is refracted (i.e., transmitted) at a planar interface separating two optical media. They also describe the phase shift of the reflected light. It will be shown that all these quantities depend not only on the change in refractive index and the angle of incidence, but also the polarization state of the incident light. This chapter treats the general features associated with the propagation, reflection, and refraction of light in isotropic media. These media are assumed to be linear, homogeneous, and nonmagnetic. As discussed in Chapter 1, the refractive index of a substance varies with the wavelength of incident electromagnetic radiation. All materials have a refractive index very close to 1 at X-ray wavelengths. Therefore, no refraction occurs in this X-ray range. Here, we are concerned with the visible range, in which most transparent materials exhibit a nearly constant refractive index higher than unity.

2.2 LAWS OF REFLECTION AND REFRACTION

In geometric optics, the concept of a light ray is useful and also necessary. A ray is a line drawn along the direction of radiant energy flow. Thus, the ray can be regarded as the path along which light energy is transmitted from one point to another. For a plane wave traveling within homogeneous isotropic media, rays will be straight, parallel lines normal to its wavefronts. The fundamental laws of reflection and refraction are graphically illustrated in Figure 2.1 and these laws are described as follows:

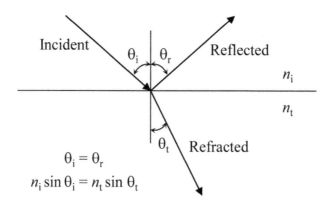

FIGURE 2.1 Laws of reflection and refraction.

Law of reflection: When a ray of light is reflected at an interface separating two media with refractive indices n_i and n_t, the angle of reflection is equal to the angle of incidence, that is, $\theta_i = \theta_r$. The law of reflection also states that the reflected ray remains within the *plane of incidence*, which refers to the plane containing the incident ray and the normal to the interface. Thus, the reflected ray shown in Figure 2.1 cannot have a directional component into or out of the page.

Law of refraction: When a ray of light is refracted at an interface separating two media with refractive indices n_1 and n_2, the refracted ray remains within the plane of incidence, and the angle of refraction is dependent on the angle of incidence by the relation; $n_i \sin \theta_i = n_t \sin \theta_t$. This relation is generally referred to as *Snell's law*.

The above laws also apply for curved interfaces. In this case, however, we must be careful to find the correct interface normal; it is the line perpendicular to the interface at the point of incidence. In Figure 2.1, n_t is assumed to be larger than n_i. When light is incident from a lower-index medium into a higher-index medium, the angle of refraction is smaller than the angle of incidence and the light is refracted toward the normal of the interface. When entering into a lower-index medium, the light is refracted away from the interface normal.

2.3 FERMAT'S PRINCIPLE

The laws of refraction and reflection can be derived in several different ways; for example, from Huygens's principle, Fermat's principle, and electromagnetic theory. Since Huygens's principle will be discussed in a later chapter on diffraction, we here describe the other two approaches and start with Fermat's principle; *the path taken between two points by a ray of light is the path that can be traversed in the least time*. This principle can be restated as "*light, in travelling between two points, takes the route having the smallest optical path length.*" A diagram is shown in Figure 2.2 to prove the law of refraction by applying Fermat's principle. Suppose that a ray of light travels from point P in a medium with refractive index n_i to point Q in another medium with refractive index n_t. There are numerous paths that the ray can take in going from P to Q. However, the ray path within each medium will be a straight line because it takes the least time. Therefore, the overall transit time, t, from P to Q can be represented with respect to the variable x. This time is mathematically expressed as follows:

$$t = \frac{\sqrt{a^2 + x^2}}{v_i} + \frac{\sqrt{b^2 + (c-x)^2}}{v_t}$$

Here v_i and v_t are the speeds of light in the incident and transmitting media, respectively. The minimum transit time will coincide with the actual path. Since the choice of path changes the position of intersecting point O and therefore the variable x, we can minimize the transit time by setting $dt/dx = 0$.

$$\frac{dt}{dx} = \frac{x}{v_i \sqrt{a^2 + x^2}} - \frac{(c - x)}{v_t \sqrt{b^2 + (c - x)^2}} = 0$$

This expression can be rewritten as

$$\frac{dt}{dx} = \frac{\sin \theta_i}{v_i} - \frac{\sin \theta_t}{v_t} = 0 .$$

From $n = c/v$, we arrive at Snell's law:

$$n_i \sin \theta_i = n_t \sin \theta_t . \tag{2.1}$$

As shown, a light beam traversing an interface does not take a straight line but travels along a path that takes the least time. This is because the refractive index changes in the middle of travel. The transit time given above can be rewritten as

$$t = \frac{s_i}{v_i} + \frac{s_t}{v_t} = \frac{1}{c}(n_i s_i + n_t s_t) ,$$

where s_i and s_t are the path lengths in the respective media. The summation within the parentheses is known as the *optical path length* (OPL) traversed by the ray, in contrast to the physical path length $s_i + s_t$. Thus, the actual path of a ray is a result of the minimum optical path length. In each medium of Figure 2.2, the ray travels along a line segment because not only the physical path length but also the optical path length is minimized along that segment. The two media in Figure 2.2 can be more generally viewed as a single medium whose refractive index is a function of position. Then, the OPL of a ray between arbitrary two points A and B is generally in the form of an integral:

$$OPL = \int_A^B n \, ds \tag{2.2}$$

Here ds is an infinitesimal displacement along the ray path. For an inhomogeneous medium with varying n, the ray path may be curved. When a light ray traverses several homogeneous media in sequence, the ray path is a sequence of straight-line segments. Only a case of two homogeneous media is depicted in Figure 2.2.

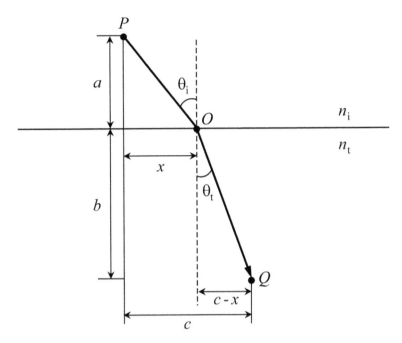

FIGURE 2.2 Fermat's principle applied to prove the law of refraction.

The law of reflection can also be derived by using the same principle of *least time*. Suppose that a light ray, starting from point A in Figure 2.3, arrives at point B after reflection from a planar surface. Three possible paths from A to B are shown in the diagram: ACB, ADB, and AEB. Of course, only one of them will be the correct path. If we view the ray as if it emanated from point A', which is the image of A, the distances to B are not altered. For example, the distance traversed by the ray from A to B via C is equal to the distance from A' to B via C. Obviously, the straight line $A'DB$ is the shortest distance from A to B. Therefore, the path ADB will be the actual one taken by the ray, which corresponds to $\theta_i = \theta_r$. To maintain $A'DB$ as a straight line, the reflected ray must remain within the plane of incidence, which is defined as the plane containing the incident ray and the surface normal, that is, points A, D, and B should lie in the plane of the page.

2.4 REFLECTION AND TRANSMISSION AT NORMAL INCIDENCE

It has been shown that the laws of refraction and reflection can be deduced using Fermat's principle. However, these laws say nothing but the directions of reflection and refraction. Even more information can be obtained by treating light as an electromagnetic wave. The electromagnetic approach provides not only the angles of reflection and refraction but also the fractions of reflected and transmitted energy and the phase change on reflection. We first consider the case of normal incidence to see how these quantities can be derived from the electromagnetic theory of light. Suppose that a plane wave of frequency ω is normally incident onto a planar interface formed by two linear, isotropic media with refractive indices n_1 and n_2, as shown in Figure 2.4. The wave is traveling in the x-direction and approaching the interface (the y–z plane, i.e., $x = 0$) from the left. We here assume that the incident wave is linearly polarized, with its electric field E along the y-direction. Once the E-field has a specific direction, the direction of the corresponding H-field is then determined to ensure that the Poynting vector $E \times H$ should aim in the direction of wave propagation. A fraction of the incident energy will be reflected from the interface, with the remainder being transmitted. Assuming the reference directions of the E-field for the incident, reflected, and transmitted waves as indicated in Figure 2.4, the directions of the H-field for the three waves will then be as shown in the figure. The E- and H-fields of the incident wave are expressed as

$$\mathbf{E}_i(x,t) = \mathbf{j}E_{0i}e^{i(k_1 x - \omega t)}$$
$$\mathbf{H}_i(x,t) = \mathbf{k}H_{0i}e^{i(k_1 x - \omega t)} ,$$

(2.3)

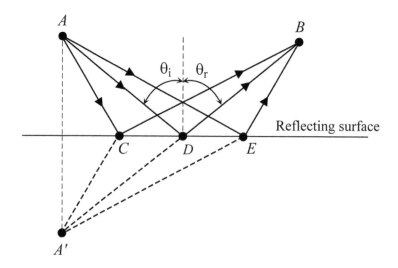

FIGURE 2.3   Law of reflection proved from the principle of *least time*.

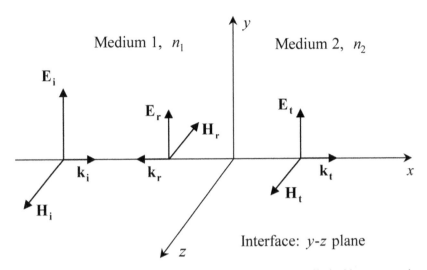

FIGURE 2.4 Reflection and transmission of a plane wave normally incident onto a planar interface formed by two media. The *E*-and *H*-field directions of incident, reflected, and transmitted waves are arbitrarily chosen.

where **j** and **k** are the unit vectors along the y and z directions, respectively. E_{oi} and H_{oi} are the amplitudes of the incident *E*- and *H*-field waves. The expressions of the reflected wave are:

$$\mathbf{E}_r(x,t) = \mathbf{j}E_{or}e^{i(k_1 x + \omega t)}$$
$$\mathbf{H}_r(x,t) = -\mathbf{k}H_{or}e^{i(k_1 x + \omega t)} \qquad (2.4)$$

Here E_{or} and H_{or} are the amplitudes of the reflected *E*- and *H*-field waves. Note the minus sign in $H_r(x, t)$. Since the incident and reflected waves are in the same medium (medium 1), their wave vectors, $\mathbf{k}_i$ and $\mathbf{k}_r$, have the same magnitude of $k_1 = 2\pi/\lambda_1 = n_1\omega/c$. The *E*- and *H*-fields of the transmitted wave are given by:

$$\mathbf{E}_t(x,t) = \mathbf{j}E_{ot}e^{i(k_2 x - \omega t)}$$
$$\mathbf{H}_t(x,t) = \mathbf{k}H_{ot}e^{i(k_2 x - \omega t)} \qquad (2.5)$$

where $k_2 = 2\pi/\lambda_2 = n_2\omega/c$ is the propagation constant in medium 2.

According to the boundary conditions of eq 1.68, the *E*-field component tangential to the interface should be continuous across it. That is, the net tangential component of *E* on one side of the interface should equal that on the other side; the same rule applies for *H*. These requirements must be satisfied at any instant at any point on the interface (i.e., for all t at $x = 0$). In the case of normal incidence, there are no E or H components perpendicular to the interface. Therefore, the application of the boundary conditions leads to the following relationships:

$$E_{oi} + E_{or} = E_{ot}$$
$$H_{oi} - H_{or} = H_{ot}$$
(2.6)

From eq 1.62 and eq 1.79, these relations can be rewritten as:

$$E_{oi} + E_{or} = E_{ot}$$
$$n_1 E_{oi} - n_1 E_{or} = n_2 E_{ot}$$
(2.7)

Equation 2.7 can be solved to give the amplitudes of the reflected and transmitted waves in terms of the incident amplitude:

$$\frac{E_{or}}{E_{oi}} = r_{12} = \frac{n_1 - n_2}{n_1 + n_2}$$
$$\frac{E_{ot}}{E_{oi}} = t_{12} = \frac{2n_1}{n_1 + n_2}$$
(2.8)

Here, r_{12} and t_{12} are defined as the *reflection amplitude coefficient* and *transmission amplitude coefficient*, respectively. It is important to note that the reflection amplitude coefficient r_{12} may be positive or negative, depending on whether n_2 is greater or less than n_1. The transmission coefficient t_{12}, however, is always positive. When the incident medium (medium 1 in Fig. 2.4) has a smaller refractive index than the transmitting medium, that is, $n_1 < n_2$, the reflection amplitude coefficient r_{12} has a negative value; E_{or} becomes negative when E_{oi} is positive. This means that at the interface, the electric field vector of the reflected wave is antiparallel to that of the incident wave. In other words, a phase shift of π is introduced on reflection. The phase change occurs only in the case of *external reflection* (i.e., $n_1 < n_2$). For the opposite situation of *internal reflection* ($n_1 > n_2$), r_{12} is positive and E_{or} has the same sign as E_{oi}. There is thus no phase change on reflection. Since t_{12} is always positive regardless of whether n_2 is greater or less than n_1, the transmitted wave is *in phase with* the incident wave in both cases.

The sign of r_{12} is related to the relative directions of E_i and E_r. In the configuration of Figure 2.4, the E-field of the reflected wave was arbitrarily chosen in the positive y-direction. If we had chosen it to be in the negative y-direction, the minus sign in eq 2.4 should have been introduced to E_r, instead of H_r. As a result, the reflection amplitude coefficient in eq 2.8 would have changed the sign, being $r_{12} = E_{or}/E_{oi} = (n_2 - n_1)/(n_2 + n_1)$. Now, both E_{or} and E_{oi} have positive values in the case of external reflection ($n_1 < n_2$). This means that when the E-field of the incident wave is along the positive y-direction, the E-field of the reflected wave is directed in the negative y-direction. That is, the incident and reflected waves are 180° *out of phase with* each other. Regardless of how we choose the relative directions of the E-fields, the results of the phase change are the same. There is always a phase shift of π on reflection when the wave is normally incident from a lower-index medium into a higher-index medium. On the contrary, no phase shift occurs in the wave reflected from a lower-index medium. The directional relationships between the electric fields at the interface are illustrated in Figure 2.5. When the incident medium has a lower index than the transmitting medium ($n_1 < n_2$), the electric field of the reflected wave is antiparallel to the incident field owing to a phase shift of π. The electric field in the transmitting medium is oriented in the same direction as the incident field. The wavelength as well as the amplitude of the field is reduced in this medium due to the increased index of refraction. If the incident medium has a higher index ($n_1 > n_2$), there is no phase difference between the incident and reflected electric fields, making them parallel to each other. In accordance with the boundary conditions, the amplitude of the transmitted field is larger than that of the incident field. At first glance, this may look contradictory to the conservation of energy. We need to remember that the (average) energy flux density of light is also proportional to the refractive index of the medium. The increased electric field is compensated for by the lowered refractive index.

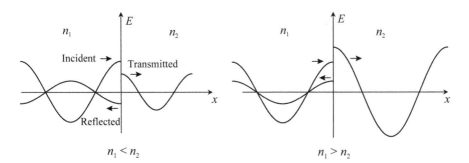

FIGURE 2.5 Directional relationships between the electric fields at the interface.

We next consider what fraction of the incident energy is reflected and what fraction is transmitted. From the conservation of energy, the incident power must equal the sum of the reflected and transmitted powers. We define the *reflectance R* as the ratio of the reflected power (P_r) to the incident power

(P_i) and the *transmittance* T as the ratio of the transmitted power (P_t) to the incident power:

$$R = \frac{P_r}{P_i}, \quad T = \frac{P_t}{P_i}. \tag{2.9}$$

Note that the power is the energy flow per unit time, while the intensity is the energy flow per unit time per unit area. For normal incidence, the cross-sectional areas of the incident, reflected, and transmitted beams are all the same. Therefore, R and T can be given in terms of the incident, reflected, and transmitted intensities. From eq 1.86, we have

$$
\begin{aligned}
R &= \frac{I_r}{I_i} = \left(\frac{E_{or}}{E_{oi}}\right)^2 = \left(\frac{n_1 - n_2}{n_1 + n_2}\right)^2 \\
T &= \frac{I_t}{I_i} = \frac{n_2}{n_1}\left(\frac{E_{ot}}{E_{oi}}\right)^2 = \frac{n_2}{n_1}\left(\frac{2n_1}{n_1 + n_2}\right)^2
\end{aligned}
\tag{2.10}
$$

It is obvious that $R + T = 1$. Note that while R is simply r^2, T is not equal to t^2. This is because the intensity of light (i.e., the average energy flux density) is proportional not only to the squared amplitude of the electric field but also to the refractive index of the medium through which it propagates (see eq 1.86). When light passes through a glass plate ($n = 1.5$), reflections occur twice, one at the front surface and another at the rear surface. The reflection from the front surface (i.e., air–glass interface) is an external reflection and thus the incident light undergoes a phase shift of π on reflection. On the contrary, there is no phase change in the light internally reflected from the glass–air interface. The reflectance is identical for both reflections, with $R = (0.5/2.5)^2 = 0.04$ at each interface. Thus, approximately 92% of the incident energy is transmitted through a lossless glass plate. Let's consider the internal reflection at the glass–air interface. This corresponds to the right-side configuration of Figure 2.5, where $n_1 = 1.5$ and $n_2 = 1.0$. When the incident electric field has amplitude E_o, the amplitudes of the reflected and transmitted fields are $0.2E_o$ and $1.2E_o$, respectively. Neglecting the common proportionality factor, the intensity of the incident wave is $1.5E_o^2$ and those of the reflected and transmitted waves are $1.5(0.2E_o)^2$ and $(1.2E_o)^2$, respectively. While the transmitted wave has larger amplitude than the incident wave, its intensity is 4% lower than that of the incident wave. Of course, this difference is carried by the reflected wave. The amplitude of the magnetic field is proportional to both the amplitude of the electric field and the refractive index of the medium. Thus, the amplitudes of the incident, reflected, and transmitted magnetic fields are

$1.5E_o$, $1.5 \times 0.2E_o$, and $1.2E_o$, respectively. However, when the incident and reflected electric fields are parallel to each other, the corresponding magnetic fields are antiparallel, as shown in Figure 2.4. Therefore, the net magnetic component on the left side of the interface is $1.5E_o - 1.5 \times 0.2E_o = 1.2E_o$. This value is equal to the magnetic component of the transmitted wave, satisfying the boundary condition for the H-field. For internal reflection, the H-field is stronger in the incident wave than in the transmitted wave, while the E-field is stronger in the latter. Nonetheless, the time-averaged magnitude of their cross product $E \times H$, which represents the intensity, is always smaller in the transmitted wave because a portion of the incident energy is reflected back at the interface.

2.5 REFLECTION AND REFRACTION AT OBLIQUE INCIDENCE

2.5.1 THE FRESNEL EQUATIONS

In the previous section, we treated reflection and transmission at normal incidence. We now turn to the more general case of oblique incidence, in which the incident wave intersects the interface at an arbitrary angle. Suppose that a plane wave of frequency ω_i is incident on a planar interface between two different media at an angle θ_i. In Figure 2.6, the y–z plane forms the interface and the wave vector $\mathbf{k}_i$ of the incident wave is on the x–y plane. Therefore, the x–y plane is the plane of incidence. At the interface, the wave will be partially reflected at an angle θ_r and partially refracted at θ_t. Here, all the angles θ_i, θ_r, and θ_t are measured from the normal to the interface. The incident, reflected, and refracted (i.e., transmitted) waves in Figure 2.6 can be expressed as:

$$\mathbf{E}_i = \mathbf{E}_{oi} e^{i(\mathbf{k}_i \cdot \mathbf{r} - \omega_i t)}$$
$$\mathbf{E}_r = \mathbf{E}_{or} e^{i(\mathbf{k}_r \cdot \mathbf{r} - \omega_r t)} . \qquad (2.11)$$
$$\mathbf{E}_t = \mathbf{E}_{ot} e^{i(\mathbf{k}_t \cdot \mathbf{r} - \omega_t t)}$$

Here $\mathbf{E}_{oi}$, $\mathbf{E}_{or}$, and $\mathbf{E}_{ot}$ are the amplitude vectors of the incident, reflected, and refracted waves, respectively. The directions of these vectors depend on the polarizations of the waves. In accordance with the boundary conditions of eq 1.68, the net tangential component of $\mathbf{E}$ on one side of the interface should equal that on the other side. The boundary condition has the general form:

$$(\quad) |\mathbf{E}_{oi}| e^{i(\mathbf{k}_i \cdot \mathbf{r} - \omega_i t)} + (\quad) |\mathbf{E}_{or}| e^{i(\mathbf{k}_r \cdot \mathbf{r} - \omega_r t)} = (\quad) |\mathbf{E}_{ot}| e^{i(\mathbf{k}_t \cdot \mathbf{r} - \omega_t t)}. (2.12)$$

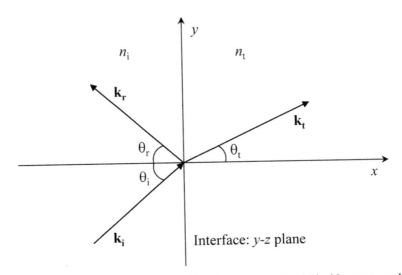

FIGURE 2.6   Reflection and refraction of a plane wave obliquely incident onto a planar interface.

The values filled in the parentheses are dependent on the directions of the amplitude vectors. If these vectors are directed parallel to the interface, either 1 or -1 will be inserted in the parentheses. When they are on the plane of incidence, some cosine terms should be filled within the parentheses. No matter what directions these amplitude vectors may have, the exponential factors of the three waves must be identical in order for the equality in eq 2.12 to be satisfied, that is,

$$(\mathbf{k}_i \cdot \mathbf{r} - \omega_i t) = (\mathbf{k}_r \cdot \mathbf{r} - \omega_r t) = (\mathbf{k}_t \cdot \mathbf{r} - \omega_t t). \tag{2.13}$$

The boundary condition must hold at all points on the interface for all times. At the interface point $\mathbf{r} = 0$, $\omega_i = \omega_r = \omega_t$. Thus, the frequency remains unaltered. Then we have

$$\mathbf{k}_i \cdot \mathbf{r} = \mathbf{k}_r \cdot \mathbf{r} = \mathbf{k}_t \cdot \mathbf{r}. \tag{2.14}$$

Important conclusions can be deduced from the relations of eq 2.14. The above relations can be expressed in terms of the components:

$$k_{iy} y + k_{iz} z = k_{ry} y + k_{rz} z = k_{ty} y + k_{tz} z. \tag{2.15}$$

Note that the boundary condition is satisfied on the interface, that is, the y–z plane in Figure 2.6 and thus the position vector $\mathbf{r}$ does not have the x-component. Equation 2.15 should hold for all y and all z. For $y = 0$, we have

$$k_{iz} = k_{rz} = k_{tz}. \tag{2.16}$$

For $z = 0$, we get

$$k_{iy} = k_{ry} = k_{ty}. \tag{2.17}$$

In Figure 2.6, the wave vector $\mathbf{k}_i$ of the incident wave is assumed to lie in the x–y plane and does not have the z-component. From eq 2.16, the wave vectors of the reflected and refracted waves do not have any z-components either, that is, $k_{rz} = k_{tz} = 0$. Therefore, all three wave vectors $\mathbf{k}_i$, $\mathbf{k}_r$, and $\mathbf{k}_t$ are coplanar in the x–y plane. In other words, the reflected and refracted waves also lie in the plane of incidence. The first equality in eq 2.17 gives

$$k_i \sin \theta_i = k_r \sin \theta_r$$
$$\theta_i = \theta_r \tag{2.18}$$

Since both of the incident and reflected waves propagate in the same medium, their propagation constants are identical, $k_i = k_r$. Therefore, we have the law of reflection: $\theta_i = \theta_r$. From the first and third members of eq 2.17, we have

$$k_i \sin \theta_i = k_t \sin \theta_t. \tag{2.19}$$

Since $k_i = n_i \omega/c$ and $k_t = n_t \omega/c$, we again arrive at Snell's law of refraction:

$$n_i \sin \theta_i = n_t \sin \theta_t. \tag{2.20}$$

As shown above, the laws of reflection and refraction can be derived by the electromagnetic approach. A significant consequence of this approach is that *the wave vector components tangential to the interface should be conserved*. Equation 2.17 requires that the reflected and refracted waves have the same tangential (wave vector) component as the incident wave. This statement may be better described with Figure 2.7a, in which a plane wave of wave vector $\mathbf{k}_i$ is incident from air into a medium of refractive index n at an angle θ_i. The tangential component of $\mathbf{k}_i$ has a magnitude of $k_i \sin \theta_i$. The absolute magnitude of a wave vector is fixed when the refractive index is fixed. However, the magnitude of its tangential component is dependent on the propagation direction. Since the tangential components of the wave vectors need to be conserved, the incident wave should be refracted in such a way that the wave vector $\mathbf{k}_t$ of the refracted wave has a tangential component whose magnitude is equal to $k_i \sin \theta_i$. When the wave is refracted at an angle θ_t, its wave vector in the medium has a tangential component of magnitude $k_t \sin \theta_t$. Since $k_i \sin \theta_i = k_t \sin \theta_t$, we arrive at Snell's law of $\sin \theta_i = n \sin \theta_t$.

Once the incident angle and the refractive index of the medium are fixed, the angle of refraction is invariantly determined. If the refractive index of the medium changes, the magnitude and direction of the wave vector within the medium also change, thereby altering the refraction angle. Nevertheless, the tangential component of the wave vector remains unaltered as long as the incident angle is fixed (Fig. 2.7b). This is to say that the wave adjusts its refraction direction so as to maintain the initial wave vector component carried by the incident wave.

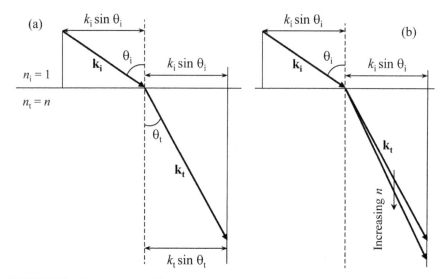

FIGURE 2.7 Conservation of the tangential component of the wave vector. (a) The refracted wave has the same tangential (wave vector) component as the incident wave. (b) Although the magnitude and direction of $\mathbf{k}_t$ change with varying n, its tangential component remains unaltered as long as the incident angle is fixed.

We have just found that the incident, reflected, and refracted waves are all in the plane of incidence. In Figure 2.6, the incident wave may be arbitrarily polarized. No matter what the polarization of the wave, we can resolve its E- and H-fields (or B-field) into components parallel and perpendicular to the plane of incidence. When the wave has its E-field perpendicular to the plane of incidence and H-field parallel to it, this mode of polarization is called the *transverse electric* (*TE*) mode. The opposite is called the *transverse magnetic* (*TM*) mode, in which the H-field is perpendicular to the plane of incidence and the E-field lies in the plane of incidence. TE and TM polarizations are sometimes called s- and p-polarizations, respectively. Figure 2.8a,b show the directions of E- and H-fields in these two polarization modes. An arbitrary polarization may be regarded as a linear combination of the two modes. The reflectance and transmittance of an obliquely incident wave strongly depend on whether it is TE- or TM-polarized. We begin with the behaviors of TE polarization mode. With respect to the directions of E chosen in Figure 2.8a, eq 2.11 can be rewritten as

$$
\begin{aligned}
\mathbf{E}_i &= \mathbf{k} E_{oi} e^{i(\mathbf{k}_i \cdot \mathbf{r} - \omega_i t)} \\
\mathbf{E}_r &= \mathbf{k} E_{or} e^{i(\mathbf{k}_r \cdot \mathbf{r} - \omega_r t)} , \\
\mathbf{E}_t &= \mathbf{k} E_{ot} e^{i(\mathbf{k}_t \cdot \mathbf{r} - \omega_t t)}
\end{aligned}
\tag{2.21}
$$

where $\mathbf{k}$ is the unit vector in the positive z-direction. Note that all three E-fields are parallel to the interface plane and in the $+ z$-direction. The boundary condition for the electric field requires

$$
E_{oi} + E_{or} = E_{ot}.
\tag{2.22}
$$

The requirement for the corresponding magnetic fields yields

$$
-H_{oi} \cos \theta_i + H_{or} \cos \theta_r = -H_{ot} \cos \theta_t .
\tag{2.23}
$$

The left and right sides of eq 2.23 are the total components of H parallel to the interface in the incident and transmitting media, respectively. The minus signs are inserted because the H_i and H_t components are along the negative y-direction.

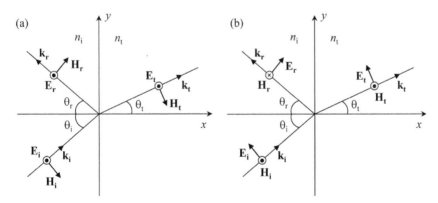

FIGURE 2.8 E- and H-field directions in (a) TE and (b) TM modes.

Equation 2.23 can be rewritten as

$$
-n_i E_{oi} \cos \theta_i + n_i E_{or} \cos \theta_r = -n_t E_{ot} \cos \theta_t .
\tag{2.24}
$$

From eqs 2.22 and 2.24, we have the reflection and transmission amplitude coefficients for TE polarization mode as follows:

$$\frac{E_{or}}{E_{oi}} = r^{TE} = \frac{n_i \cos\theta_i - n_t \cos\theta_t}{n_i \cos\theta_i + n_t \cos\theta_t}$$

$$\frac{E_{ot}}{E_{oi}} = t^{TE} = \frac{2n_i \cos\theta_i}{n_i \cos\theta_i + n_t \cos\theta_t} \tag{2.25}$$

A similar pair of equations can be derived for the TM mode. Using the diagram of Figure 2.8b, we have

$$E_{oi} \cos\theta_i + E_{or} \cos\theta_r = E_{ot} \cos\theta_t$$

$$H_{oi} - H_{or} = H_{ot} \tag{2.26}$$

These relations are correct for the E directions chosen in Figure 2.8b. If we make a different choice, for example, by reversing the direction of the $\mathbf{E}_r$ vector, the signs in front of E_{or} and H_{or} in eq 2.26 will be reversed. However, the physical meaning of these relations is the same regardless of whatever directions may be chosen. Equation 2.26 can be arranged to give

$$\frac{E_{or}}{E_{oi}} = r^{TM} = \frac{n_i \cos\theta_t - n_t \cos\theta_i}{n_i \cos\theta_t + n_t \cos\theta_i}$$

$$\frac{E_{ot}}{E_{oi}} = t^{TM} = \frac{2n_i \cos\theta_i}{n_i \cos\theta_t + n_t \cos\theta_i} \tag{2.27}$$

Equations 2.25 and 2.27 are known as the *Fresnel equations*, which describe the ratios of reflected and transmitted E-field amplitudes to the incident E-field amplitude. We take a brief look at the conservation of energy in the case of oblique incidence. Suppose that the incident, reflected, and refracted beams overlap on an area A at the interface, as depicted in Figure 2.9. The conservation of energy requires that the energy per unit time (i.e., power) reaching this area is equal to the energy per unit time leaving it:

$$P_i = P_r + P_t.$$

This can be represented in terms of the intensity:

$$I_i A \cos\theta_i = I_r A \cos\theta_r + I_t A \cos\theta_t. \tag{2.28}$$

Note that the cross-sectional areas of the three beams are all related to the area A intercepted by the beams through the cosines of the incidence, reflection, and refraction angles. The relation of eq 2.28 is alternatively expressed by

$$n_i E_{oi}^2 \cos \theta_i = n_i E_{or}^2 \cos \theta_i + n_t E_{ot}^2 \cos \theta_t$$

or

$$1 = \left(\frac{E_{or}}{E_{oi}} \right)^2 + \left(\frac{n_t \cos \theta_t}{n_i \cos \theta_i} \right) \left(\frac{E_{ot}}{E_{oi}} \right)^2. \tag{2.29}$$

By introducing the reflection and transmission amplitude coefficients r and t, R and T become

$$R = \left(\frac{E_{or}}{E_{oi}} \right)^2 = r^2$$

$$T = \left(\frac{n_t \cos \theta_t}{n_i \cos \theta_i} \right) \left(\frac{E_{ot}}{E_{oi}} \right)^2 = \left(\frac{n_t \cos \theta_t}{n_i \cos \theta_i} \right) t^2 \tag{2.30}$$

Since the magnitudes of r and t are polarization-dependent, R and T also depend on the polarization state of the incident wave. For normal incidence, $\theta_i = \theta_t = 0°$ and the distinction between TE and TM vanishes. This reduces eq 2.30 to eq 2.10.

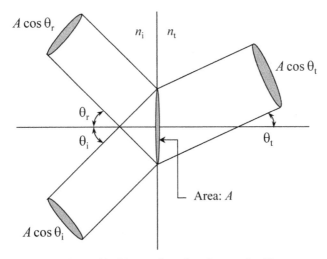

FIGURE 2.9 Cross-sections of incident, reflected, and transmitted beams.

2.5.2 EXTERNAL AND INTERNAL REFLECTIONS

The Fresnel equations describe the reflection and transmission of electromagnetic waves at a planar interface. They give the reflection and transmission amplitude coefficients for waves polarized parallel and perpendicular to the plane of incidence. As shown in eqs 2.25 and 2.27, the two coefficients r and t are influenced by many factors such as the polarization state of the wave, the incident and refracting angles, and the refractive indices of two media separated by the interface. As already discussed, the angle of refraction (i.e., the angle of transmission) is related to the angle of incidence by Snell's law. Therefore, it can be stated that the Fresnel equations describe how r and t vary with the incident angle, when a specifically polarized wave passes from one medium (incident medium with n_i) to another (transmitting medium with n_t). While the transmission amplitude coefficient t is always positive, the value of r can be either positive or negative depending on the incident angle. The variation of r with the incident angle is also strongly dependent on the reflection type, that is, whether it is an external ($n_i < n_t$) or internal reflection ($n_i > n_t$). It is thus necessary to distinguish between these two different situations for the interpretation of the Fresnel equations.

i) External Reflection ($n_i < n_t$)

Figure 2.10a plots r^{TE} and t^{TE} in the case of external reflection with $n_t/n_i =$ 1.5. At normal incidence ($\theta_i = 0°$), both $\cos \theta_i$ and $\cos \theta_t$ in eq 2.25 becomes one because θ_t is also zero from Snell's law. Thus we have $r^{TE} = (n_i - n_t)/$ $(n_i + n_t)$ and $t^{TE} = 2n_i/(n_i + n_t)$ at $\theta_i = 0°$. They have negative and positive values, respectively, indicating a phase change of the E-field on reflection. The phase change on reflection will be discussed presently. In this case of $n_i < n_t$, $\theta_i > \theta_t$ and $\cos \theta_i < \cos \theta_t$. Thus, as θ_i increases, r^{TE} gets more and more negative and becomes -1 at 90°. As a whole, r^{TE} changes from $(n_i - n_t)/(n_i + n_t)$ to -1 and t^{TE}, from $2n_i/(n_i + n_t)$ to 0, when θ_i varies from 0° to 90°. For example, at an air ($n_i = 1.0$)-glass ($n_t = 1.5$) interface, r and t are -0.2 and 0.8, respectively, at normal incidence. The reflection and transmission amplitude coefficients for TM polarization mode are also plotted in Figure 2.10b.

At normal incidence, r^{TM} is the same as r^{TE}. However, its variation with the incident angle is much different. While r^{TE} is negative at all values of θ_i, r^{TM} starts from a negative value of $(n_i - n_t)/(n_i + n_t)$ at $\theta_i = 0°$ and increases gradually until it becomes zero at an angle $\theta_i = \theta_B$. This particular incident angle is known as the *Brewster's angle* (θ_B) or *polarization angle* (θ_P), at which there is no reflection, that is, $R = 0$. As the incident angle increases beyond θ_B, r^{TM} becomes more and more positive, reaching $+1$ at $\theta_i = 90°$. It follows from Snell's law that when $n_i < n_t$, $\cos \theta_i < \cos \theta_t$. The numerator of r^{TM} in eq (2.27), ($n_i \cos \theta_t - n_t \cos \theta_i$), can be either negative or positive. For small θ_i values at which $n_i \cos \theta_t$ is less than $n_t \cos \theta_i$, r^{TM} is negative.

At relatively large values of θ_i, $n_i \cos \theta_t$ is greater than $n_t \cos \theta_i$ and thus r^{TM} becomes positive. These two quantities are equal to each other at $\theta_i = \theta_B$, resulting in $r^{TM} = 0$. t^{TM} exhibits a very similar behavior to t^{TE} and also changes from $2n_i/(n_i + n_t)$ to 0. $r^{TM} = 0$ occurs when the sum of the incident and refracted angles is equal to 90°, that is, $(\theta_B + \theta_t) = \pi/2$. This relation can be easily derived using Snell's law, which can be rewritten as

$$n_i \cos(\pi/2 - \theta_i) = n_t \cos(\pi/2 - \theta_t).\qquad(2.31)$$

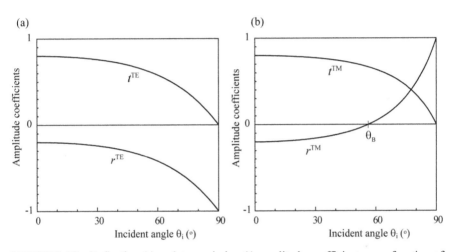

FIGURE 2.10 Reflection (r) and transmission (t) amplitude coefficients as a function of incident angle in (a) TE and (b) TM polarization modes for the case of external reflection with $n_t/n_i = 1.5$.

If $(\theta_i + \theta_t) = \pi/2$, we have $n_i \cos \theta_t = n_t \cos \theta_i$. This leads to $r^{TM} = 0$. Keeping in mind that $n_i \sin \theta_B = n_t \cos \theta_B$, the Brewster's angle can be expressed in terms of the refractive indices:

$$\theta_B = \tan^{-1}\left(\frac{n_t}{n_i}\right).\qquad(2.32)$$

The Fresnel equations predict that TM-polarized light (with its electric field parallel to the plane of incidence) will not be reflected if the angle of incidence satisfies eq 2.32. The physical mechanism for this can be qualitatively understood from the electron-oscillator model. When a light wave is incident on the interface, it is refracted and enters the transmitting medium at some angle θ_t. The electric field of the incoming wave oscillates the bound electrons of that medium, which in turn re-radiate. The oscillating dipoles generate a transmitted (refracted) wave. In general, a portion of the radiated energy appears in the form of a reflected wave. The polarization of light is always perpendicular to the direction in which the light is traveling. The dipoles that produce the transmitted (refracted) wave oscillate in the polarization direction of that wave. The reflected wave is also generated by the same dipoles. However, dipoles do not radiate any energy in the direction of the dipole moment. If the refracted wave is TM-polarized and propagates in the direction of $\theta_t = 90° - \theta_i$, the dipoles producing it oscillate exactly in the same direction that the reflected wave is expected to propagate (Fig. 2.11). Since they do not radiate any energy in this direction, there is no reflected light. In the case of TE-polarization, the dipole motion is perpendicular to the plane of incidence, while the propagation direction is parallel to it. Therefore, such a situation of $r = 0$ does not occur in this case. The Brewster's angle is sometimes called the polarization angle, because unpolarized light that reflects from an interface at this incident angle can be polarized completely perpendicular to the plane of incidence (TE-polarized or s-polarized). A glass plate or a stack of plates placed at the Brewster's angle can be used as a polarizer. When the incident medium is air ($n_i = 1.0$) and the transmitting medium is glass ($n_t = 1.5$), the polarization angle is about 56°. Therefore, if an unpolarized beam enters a glass at an angle of 56°, the reflected beam will be completely polarized with its E-field perpendicular to the plane of incidence, that is, parallel to the glass surface. However, a single plate is not effective as a polarizer. It is to be noted that although fully TE-polarized, the reflected beam is weak because r^{TE} at the Brewster's angle is not unity. The transmitted beam is stronger and it still contains both TM- and TE-components. A more useful polarizer can be obtained by placing a stack of glass plates at the Brewster's angle, as shown in Figure 2.12. Then, a fraction of the TE-polarized light is reflected from each surface of each plate (air–glass and glass–air interfaces). For a stack of plates, each reflection depletes the TE-component present in the beam, leaving a greater fraction of TM-polarized light in the transmitted beam. Although completely TE-polarized, the reflected beam is spread out and may not be so useful. Polarized sunglasses utilize the principle of Brewster's angle to suppress glare from the sun light reflecting off horizontal surfaces such as water, ground, and road. In some range of angles near the Brewster's angle, the reflection of TM-polarized light is lower than that of TE-polarized light. Thus, if the sun radiating unpolarized light is low in the sky, the sun light reflected from these horizontal surfaces is mostly TE-polarized. Polarized glasses use a polarizing material to transmit vertically polarized light, thereby blocking reflections from horizontal surfaces.

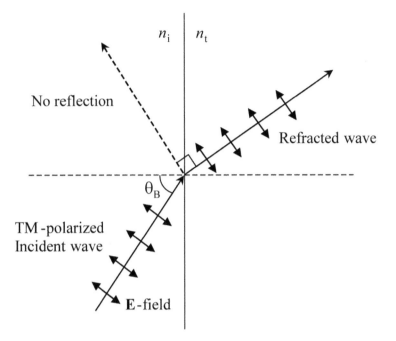

FIGURE 2.11 *E*-field directions at the Brewster's angle $\theta_B = \pi/2 - \theta_t$.

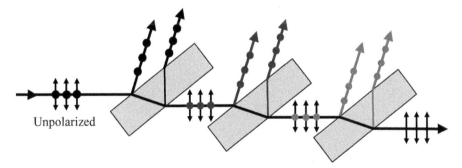

FIGURE 2.12 Brewster polarizer consisting of a stack of glass plates.

The Fresnel equations of eq 2.25 and eq 2.27 were derived based on the reference directions shown in Figure 2.8a,b. The negative values of *r* in Figure 2.10a,b imply that the electric field vector should reverse its direction on reflection at certain incident angles. Evidently, there is a phase shift of π on reflection in such situations. This is mathematically expressed in the following equation:

$$E_r = -|r| E_i = e^{i\pi} |r| E_{oi} e^{i(\mathbf{k} \cdot \mathbf{r} - \omega t)} = |r| E_{oi} e^{i(\mathbf{k} \cdot \mathbf{r} - \omega t + \pi)} \qquad (2.33)$$

A phase shift of π occurs at any incident angle for the TE mode and for $\theta_i < \theta_B$ for the TM mode. When θ_i exceeds θ_B, the TM-polarized wave is reflected without a phase change. Phase shifts for both polarization modes are plotted in Figure 2.13. At the interface, the E-fields of the incident, reflected, and refracted waves have such directional relationships as shown in Figure 2.14. The directions of the corresponding H-fields satisfy the condition that the vector $E \times H$ in each wave points to the propagation direction. R and T are plotted in Figure 2.15 for the case of $n_t/n_i = 1.5$. At any incident angle, $R + T = 1$. An arbitrary polarization may be resolved into the TE and TM modes and regarded as a linear combination of the two modes. If the incident light is unpolarized (i.e., randomly polarized), it will be equally mixed with TE and TM components. Therefore, the reflectance is $R = (R^{TE} + R^{TM})/2$. For normal incidence, there is no distinction between the two components. Evidently, we have $R^{TE} = R^{TM}$ and $T^{TE} = T^{TM}$.

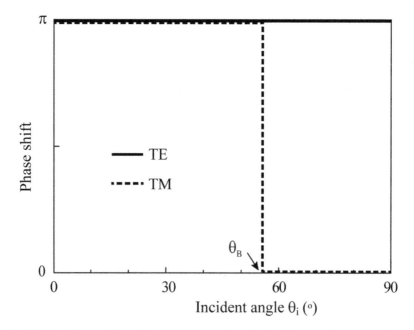

FIGURE 2.13 Phase shifts of the E-fields on reflection for the case of external reflection ($n_t/n_i = 1.5$).

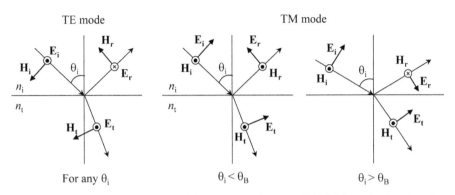

FIGURE 2.14 Directional relationships between the *E*- and *H*-fields at the interface for external reflection (i.e., $n_i < n_t$).

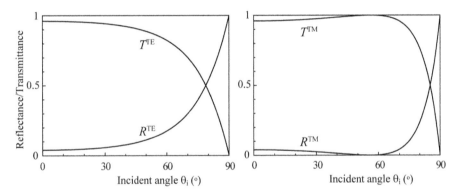

FIGURE 2.15 Reflectance (*R*) and transmittance (*T*) as a function of incident angle for the case of external reflection with $n_t/n_i = 1.5$.

ii)   *Internal Reflection ($n_i > n_t$)*

Before discussing internal reflection in more detail, we go back to Snell's law for a moment. The law predicts that when the incident medium has a higher refractive index ($n_i > n_t$), the angle of refraction is larger than the angle of incidence. Thus, there is a certain angle of incidence that makes the angle of refraction 90°. This particular angle of incidence is called the *critical angle* (θ_c) and is given by

$$\sin \theta_c = n_t / n_i . \tag{2.34}$$

Snell's law is then rewritten as

$$\sin \theta_i = \sin \theta_c \sin \theta_t . \tag{2.35}$$

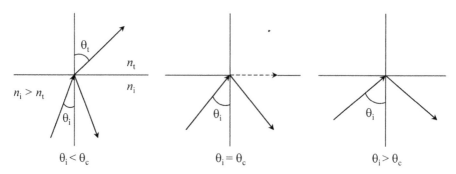

FIGURE 2.16 Internal reflection and the critical angle (θ_c).

For $\theta_i > \theta_c$, we have $\sin \theta_t > 1$. Thus, the angle of incidence, θ_t, does not exist when the incident angle exceeds the critical angle. If the incident wave is precisely at the critical angle, the refracted wave is tangential to the interface. For θ_i greater than θ_c, the refracted wave disappears and the incident wave is completely reflected back into the incident medium (Fig. 2.16). This is known as *total internal reflection* (TIR). It does not occur unless $n_i > n_t$. TIR is the operating principle of optical fibers, which are used in telecommunications and endoscopes. Light is transmitted along glass fibers by a series of TIR.

Let us examine the variations of r with θ_i in the case of internal reflection. The reflection amplitude coefficients of eq 2.25 and eq 2.27 can be rewritten as

$$r^{\mathrm{TE}} = \frac{n_i \cos \theta_i - n_t \cos \theta_t}{n_i \cos \theta_i + n_t \cos \theta_t} = \frac{n_i \cos \theta_i - \sqrt{n_t^2 - n_i^2 \sin^2 \theta_i}}{n_i \cos \theta_i + \sqrt{n_t^2 - n_i^2 \sin^2 \theta_i}} \qquad (2.36)$$

and

$$r^{\mathrm{TM}} = \frac{n_i \cos \theta_t - n_t \cos \theta_i}{n_i \cos \theta_t + n_t \cos \theta_i} = \frac{n_i \sqrt{n_t^2 - n_i^2 \sin^2 \theta_i} - n_t^2 \cos \theta_i}{n_i \sqrt{n_t^2 - n_i^2 \sin^2 \theta_i} + n_t^2 \cos \theta_i}. \qquad (2.37)$$

When $\theta_i = 0°$, r^{TE} and r^{TM} have an equal positive value of $(n_i - n_t)/(n_i + n_t)$. As θ_i increases, r^{TE} increases but r^{TM} decreases and becomes negative beyond a certain incident angle. This angle of incidence is the Brewster's angle (θ_B') for internal reflection. It can be easily shown that $\theta_B' + \theta_B = \pi/2$, where θ_B is the Brewster's angle for external reflection at the interface between the same media. Figure 2.17a shows the variations of r^{TE} and r^{TM} with the incident angle θ_i. Note that at $\theta_i = \theta_c$, r^{TE} and r^{TM} reach +1 and −1, respectively, because the terms within the square roots in eq 2.36 and eq 2.37 become zero at this critical angle. For $\theta_i > \theta_c$, this term has a negative value and both r^{TE} and r^{TM} are complex. However, their magnitudes are unity, making $R = 1$ in this range. We recall that the product of a complex number with its complex conjugate equals the square of its magnitude: $|r|^2 = rr^*$. Reflectance for internal reflection is plotted in Figure 2.17b.

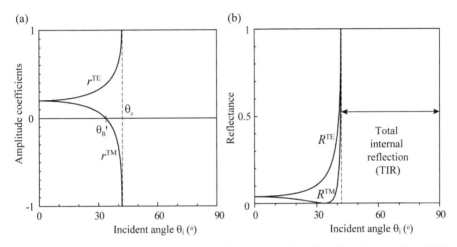

FIGURE 2.17 (a) Reflection amplitude coefficients as a function of incident angle and (b) the corresponding reflectance for the case of internal reflection with $n_t/n_i = 1/1.5$.

When r has a positive value, there is no phase difference between the incident and reflected E-fields. If it has a negative value, there is definitely a phase shift of π, as shown in eq 2.33. Then, what happens if r is a complex number? In the range of $\theta_i > \theta_c$, eq 2.36 and eq 2.37 can be expressed as

$$r^{TE} = \frac{n_i \cos\theta_i - i\sqrt{n_i^2 \sin^2\theta_i - n_t^2}}{n_i \cos\theta_i + i\sqrt{n_i^2 \sin^2\theta_i - n_t^2}} \qquad (2.38)$$

and

$$r^{TM} = \frac{-n_t^2 \cos\theta_i + in_i\sqrt{n_i^2 \sin^2\theta_i - n_t^2}}{n_t^2 \cos\theta_i + in_i\sqrt{n_i^2 \sin^2\theta_i - n_t^2}}. \qquad (2.39)$$

Equation 2.38 has the form of $r = (a - ib)/(a + ib)$. Since the numerator and denominator have the same magnitude, r^{TE} can be written in the following polar form:

$$r^{TE} = \frac{e^{-i\alpha}}{e^{i\alpha}} = e^{-i2\alpha} = e^{i\phi}, \qquad (2.40)$$

where

$$\tan\alpha = \frac{\sqrt{n_i^2 \sin^2\theta_i - n_t^2}}{n_i \cos\theta_i}$$

and $\phi = -2\alpha$ is the phase of r^{TE}, that is, the phase difference between the TE-polarized incident and reflected E-fields. A similar manipulation can be carried out for r^{TM}:

$$r^{TM} = \frac{e^{i(\pi-\beta)}}{e^{i\beta}} = e^{-i(2\beta-\pi)} = e^{i\phi}, \qquad (2.41)$$

where

$$\tan\beta = \frac{n_i\sqrt{n_i^2 \sin^2\theta_i - n_t^2}}{n_t^2 \cos\theta_i}.$$

The phase shifts in TE and TM modes, ϕ^{TE} and ϕ^{TM}, can be calculated from the following relations:

$$\tan\left(-\frac{\phi^{TE}}{2}\right) = \frac{\sqrt{n_i^2 \sin^2\theta_i - n_t^2}}{n_i \cos\theta_i} \qquad (2.42)$$

$$\tan\left(\frac{\pi}{2} - \frac{\phi^{TM}}{2}\right) = \frac{n_i\sqrt{n_i^2 \sin^2\theta_i - n_t^2}}{n_t^2 \cos\theta_i} \qquad (2.43)$$

As θ_i changes from θ_c to $90°$, ϕ^{TM} varies from π to 0 while ϕ^{TE} changes from 0 to $-\pi$. Phase shifts for internal reflection are plotted in Figure 2.18. In the case of external reflection, the phase shift is either 0 or π. For $n_i > n_t$, however, it may take on values other than 0 and π at incident angles between θ_c and $90°$. When $n_t/n_i = 1/1.5$, the relative phase shift, $(\phi^{TM} - \phi^{TE})$, is $3\pi/4$ near $\theta_i = 50°$. Thus, two consecutive TIRs will produce a phase difference of $\pi/2$ between the **E**-field components parallel and perpendicular to the plane of incidence. It can be utilized to convert a linearly polarized light wave to a circularly polarized one. The matter of polarization conversion will be described in a later chapter on polarization.

2.5.3 EVANESCENT WAVE

When TIR occurs at $\theta_i > \theta_c$, the incident and reflected waves at the interface have the following amplitude relationship:

$$E_{or} = rE_{oi} = |r|e^{i\phi}E_{oi} = e^{i\phi}E_{oi} \qquad (2.44)$$

Here, ϕ is the phase of the amplitude coefficient r. Note that the magnitude of r is unity, that is, $|r| = 1$. As shown in Figure 2.18, ϕ has an intermediate value between 0 and π for $\theta_i > \theta_c$. Thus, if there is no transmitted wave, it is impossible to satisfy the boundary conditions. For simplicity, consider a TE-polarized wave totally reflected from the interface. When the amplitude of the incident wave is E_{oi}, the amplitude of the reflected wave is $e^{i\phi}E_{oi}$. The boundary conditions of eq 1.68 require that the E-field component tangential to the interface be continuous across it. That is, the net tangential component of E on one side of the interface should equal that on the other side. In the above case, the tangential component of E on the incident medium side is $E_{oi} + e^{i\phi}E_{oi}$, which is nonzero except at $\theta_i = 90°$. Therefore, there must be an E-field component on the transmitting medium side to meet the boundary conditions. Under TIR, however, all the incoming energy is reflected back into the incident medium (i.e., $R = 1$). To solve this seemingly contradictory puzzle requires further investigation into the TIR. Let's consider an internal reflection illustrated in Figure 2.19.

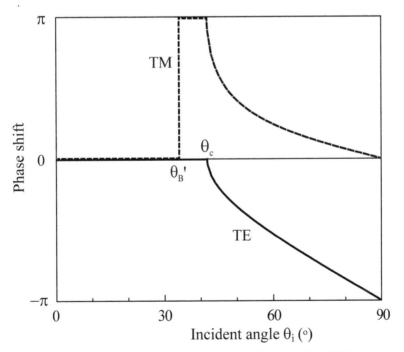

FIGURE 2.18   Phase shifts of the E-fields for internal reflection. $n_t/n_i = 1/1.5$.

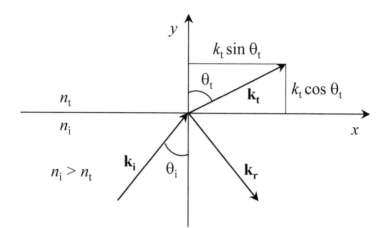

FIGURE 2.19   Wave vectors for internal reflection.

The electric field of the transmitted wave has the form

$$\mathbf{E}_t = \mathbf{E}_{ot} e^{i(\mathbf{k}_t \cdot \mathbf{r} - \omega_t t)}, \tag{2.45}$$

where

$$\mathbf{k}_t \cdot \mathbf{r} = k_{tx} x + k_{ty} y.$$

There is no z-component of $\mathbf{k}_t$ and we have $k_{tx} = k_t \sin \theta_t$ and $k_{ty} = k_t \cos \theta_t$. Using Snell's law, we find that

$$k_{tx} = k_t \sin \theta_t = k_i \sin \theta_i \tag{2.46}$$

and

$$k_{ty} = k_t \cos \theta_t = \pm k_t \sqrt{1 - \sin^2 \theta_t} = \pm k_t \sqrt{1 - (n_i^2 / n_t^2) \sin^2 \theta_i}. \tag{2.47}$$

Under TIR, $\sin^2 \theta_i$ is larger than (n_t^2/n_i^2). Thus, eq 2.47 can be expressed as

$$k_{ty} = \pm i k_t \sqrt{(n_i^2 / n_t^2) \sin^2 \theta_i - 1} = \pm i \beta. \tag{2.48}$$

The transmitted wave is then given by

$$\mathbf{E}_t = \mathbf{E}_{ot} e^{-\beta y} e^{i(k_i \sin \theta_i x - \omega_t t)}. \tag{2.49}$$

Equation 2.49 represents a wave traveling in the x-direction. The term $e^{-\beta y}$ describes an exponential decay in the amplitude of the wave. The positive exponential term of $e^{+\beta y}$ was neglected because it is physically unattainable.

Thus we have a surface wave whose amplitude drops off exponentially as it penetrates the less dense, transmitting medium. This is the so-called *evanescent wave*, in which "evanescent" means "tending to vanish." As apparent in eq 2.49, the propagation constant of the evanescent wave varies with the angle of incidence. Its amplitude rapidly decays in the y-direction, becoming negligible at a distance of a few wavelengths from the interface (Fig. 2.20). The quantity β in eq 2.49 determines how far the evanescent wave extends into the less dense medium. The penetration depth of the evanescent field is usually defined as $\delta = 1/\beta$. Even though there is a non-zero field in the less dense medium, the incident energy circulates back and forth across the interface, resulting in a zero net flow into this medium. Under TIR, the reflected wave has an angle-dependent phase shift with respect to the incident wave. It is mathematically stated that the Fresnel reflection coefficient r becomes a complex number. This phase shift is also polarization-dependent. Since the incident and reflected waves do not cancel each other completely, there must be an oscillatory **E**-field in the less dense medium with a component tangential to the interface.

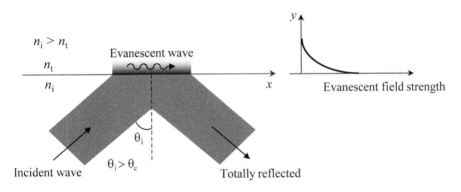

FIGURE 2.20 Evanescent wave and its rapid decay with distance from the interface.

TIR occurs whenever a wave propagating in an optically denser medium is incident into a less dense medium at an angle greater than the so-called critical angle. Under ordinary conditions, the evanescent wave transmits no net energy across the interface. However, if a third medium with a higher refractive index than the low-index second medium is placed in close proximity to the interface between the first and second media, the evanescent wave can pass some energy across the second into the third medium (Fig. 2. 21). This is known as *frustrated total internal reflection*

(FTIR). The occurrence of FTIR is due to the fact that the field of the evanescent wave has a finite penetration depth into the low-index second medium. When the gap between the first and third media (i.e., the width of the second medium) is quite small, the evanescent field can be coupled into the third medium. This frustrates the totality of the reflection, diverting some energy into the third medium. FTIR is remarkably analogous to the quantum mechanical *tunneling* effect. In quantum mechanics, it is possible for a particle to tunnel through a potential barrier because its wave function has a small but finite value in this classically forbidden region. We can regard FTIR as an optical analog of this quantum mechanical phenomenon; the low-index medium can be thought of as a potential barrier through which the incident wave tunnels. For this reason, FTIR is often called *optical tunneling*. The degree of FTIR is highly sensitive to the spacing between the first and third media, so this phenomenon can be effectively used to modulate optical transmission and reflection with a large dynamic range. As a typical example, Figure 2.22 shows a beam splitter cube consisting of two glass prisms. A 45°–90° prism will deflect a beam of light by TIR for $\theta_i > \theta_c$ at the glass–air interface, as shown. When two such prisms are sandwiched with a thin air gap between them, some of the light energy will tunnel through this gap, splitting the incident beam into reflected and transmitted beams. The extent of energy division between the two beams depends on the thickness of the air gap. No reflection can be achieved by squeezing the prisms together until they make intimate contact.

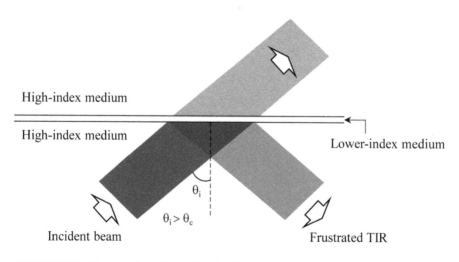

High-index medium

High-index medium

Lower-index medium

θ_i

$\theta_i > \theta_c$

Incident beam

Frustrated TIR

FIGURE 2.21 Frustrated total internal reflection.

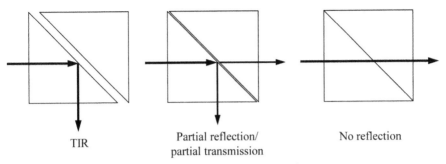

| TIR | Partial reflection/
partial transmission | No reflection |

FIGURE 2.22 Beam splitter cube consisting of two glass prisms.

2.5.4 STOKES RELATIONS

Fermat's principle of least time predicts the principle of reversibility. The general statement of this principle is that any actual ray of light in an optical system will trace the same path backward, if reversed in direction. Suppose that a plane wave of amplitude E_{oi} is impinging on a planar interface separating two lossless media, as shown in Figure 2.23a. The incident wave is partially reflected and partially transmitted at the interface. In the diagram shown, the incident and transmitting media have refractive indices of n_1 and n_2, respectively. The angle of refraction (θ_2) is related to the angle of incidence (θ_1) by Snell's law. As we have seen earlier in this chapter, the amplitudes of the reflected and transmitted waves will be rE_{oi} and tE_{oi}, respectively. Let's consider the situation depicted in Figure 2.23b, where all the ray directions are reversed. According to the principle of reversibility, this situation must also be physically possible. While a single wave is incident to generate reflected and transmitted waves in Figure 2.23a, two incident waves of amplitudes rE_{oi} and tE_{oi} produce a single wave of amplitude E_{oi} in Figure 2.23b. Since the latter is the reverse process of the former, it should also be valid. There are now two incident waves in Figure 2.23b. Each of the waves will give a reflected wave and a transmitted wave, as illustrated in Figure 2.23c. The incident wave of amplitude rE_{oi} will produce a reflected wave of amplitude r^2E_{oi} and a transmitted wave of amplitude trE_{oi}. Another wave of amplitude tE_{oi} has a different incident medium (n_2) and angle (θ_2). So its reflection and transmission amplitude coefficients are different from those of the incident wave rE_{oi}. They are here represented with r' and t'. The incident wave of amplitude tE_{oi} will then give a reflected wave of amplitude $r'tE_{oi}$ and a transmitted wave of amplitude $t'tE_{oi}$.

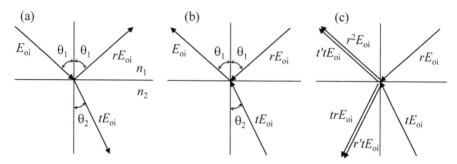

FIGURE 2.23   Diagrams used to derive the Stokes relations.

The situations depicted in Figure 2.23b,c are physically equivalent. This leads to the following relations:

$$E_{oi} = r^2 E_{oi} + t't E_{oi}$$
$$0 = tr E_{oi} + r't E_{oi} \tag{2.50}$$

Hence, we have

$$tt' = 1 - r^2$$
$$r' = -r \tag{2.51}$$

These two equations are the *Stokes relations*. The amplitude coefficients are functions of the incident angles. Therefore, the Stokes relations might be better expressed by

$$t(\theta_1)t'(\theta_2) = 1 - r^2(\theta_1)$$
$$r'(\theta_2) = -r(\theta_1) \tag{2.52}$$

Here, θ_1 and θ_2 are pairs of angles that are related by Snell's law. It is important to note that eq 2.52 is valid regardless of whether n_1 is greater or less than n_2. Consequently, θ_1 may be larger or smaller than θ_2, depending on the relative magnitudes of n_1 and n_2. The second equation of eq 2.52 states that there is a phase difference of π between the waves internally and externally reflected. This is manifest from $r' = -r = e^{i\pi}r$. To better understand the physical meaning of the Stokes relations, let's recall one of the Fresnel coefficients:

$$r^{TE} = \frac{n_i \cos\theta_i - n_t \cos\theta_t}{n_i \cos\theta_i + n_t \cos\theta_t}, \tag{2.53}$$

where n_i is the refractive index of the incident medium and n_t, the refractive index of the transmitting medium. θ_i and θ_t are the angles of incidence and refraction, respectively. When a TE-polarized wave is incident from the upper medium in Figure 2.23 at an incident angle θ_1, we have $n_i = n_1$ and $\theta_i = \theta_1$. Then, the reflection amplitude coefficient of this wave becomes

$$r(\theta_1) = \frac{n_1 \cos\theta_1 - n_2 \cos\theta_2}{n_1 \cos\theta_1 + n_2 \cos\theta_2}. \tag{2.54}$$

On the other hand, if the wave is incident from the lower medium of index n_2 at an incident angle θ_2, its reflection amplitude coefficient is expressed as

$$r'(\theta_2) = \frac{n_2 \cos\theta_2 - n_1 \cos\theta_1}{n_2 \cos\theta_2 + n_1 \cos\theta_1}, \tag{2.55}$$

because $n_i = n_2$ and $\theta_i = \theta_2$ in this case. Comparison of eqs 2.54 and 2.55 gives the relation of $r'(\theta_2) = -r(\theta_1)$. The same relation is obtained for TM polarization by interchanging θ_i and θ_t in r^{TM}. The Stokes relations given in eq 2.52 are applicable as long as the incident angle remains smaller than the critical angle, that is, $\theta_i < \theta_c$. The values of r for internal and external reflections at the same incident angle are obviously different. It should be noted again that θ_1 and θ_2 in eqs 2.52–2.55 are pairs of angles that are related by Snell's law. When θ_1 is the angle of incidence, θ_2 is the angle of refraction and vice versa. The Stokes relations will be needed later, when we deal with multi-beam interference.

2.6 REFLECTION AT THE BOUNDARY OF AN ABSORBING MEDIUM

In Figure 2.6, we assumed that both media are lossless and have real refractive indices. An absorbing medium has a complex refractive index. Suppose that a plane wave is incident on the boundary of a medium having a complex index of refraction: $n_t = n_R + in_I$. The imaginary part of the refractive index, n_I, is often called the *extinction coefficient*. The wave vector in the absorbing medium is also complex, denoted by

$$\mathbf{k}_t = \mathbf{k}_R + i\mathbf{k}_I. \tag{2.56}$$

For simplicity, we here consider only the case in which the incident medium is lossless with $n_i = n_o$. Then, the incident, reflected, and refracted waves (without the amplitude factors) can be written as

$$E_i = e^{i(\mathbf{k}_i \cdot \mathbf{r} - \omega t)}$$
$$E_r = e^{i(\mathbf{k}_r \cdot \mathbf{r} - \omega t)} \tag{2.57}$$
$$E_t = e^{i(\mathbf{k}_t \cdot \mathbf{r} - \omega t)} = e^{-\mathbf{k}_I \cdot \mathbf{r}} e^{i(\mathbf{k}_R \cdot \mathbf{r} - \omega t)}$$

The phase matching condition of eq 2.14, which is also valid in this case, leads to the following relations at the boundary:

$$\mathbf{k}_i \cdot \mathbf{r} = \mathbf{k}_r \cdot \mathbf{r}$$
$$\mathbf{k}_i \cdot \mathbf{r} = \mathbf{k}_t \cdot \mathbf{r} = \mathbf{k}_R \cdot \mathbf{r} + i\mathbf{k}_I \cdot \mathbf{r}$$

(2.58)

The first relation describes the usual law of reflection. The second relation yields $\mathbf{k}_i \cdot \mathbf{r} = \mathbf{k}_R \cdot \mathbf{r}$ and $\mathbf{k}_I \cdot \mathbf{r} = 0$. This result indicates that $\mathbf{k}_R$ and $\mathbf{k}_I$ generally have different directions. The normal to the planes of constant amplitude is defined by the vector $\mathbf{k}_I$. $\mathbf{k}_I \cdot \mathbf{r} = 0$ means that $\mathbf{k}_I$ is always perpendicular to the boundary. Namely, the planes of constant amplitude are parallel to the boundary. On the other hand, the vector $\mathbf{k}_R$, which defines the planes of constant phase (i.e., wavefronts), may have any direction. When the wave moves in the direction of $\mathbf{k}_R$, its amplitude diminishes exponentially with the normal distance from the boundary.

Let's denote the angle of incidence by θ and the angle of refraction by φ. By analogy with eq 2.20, we can express the law of refraction in terms of the complex refractive index: $n_t = n_R + in_I$.

$$n_o \sin \theta = n_t \sin \varphi$$

(2.59)

Here the angle φ is a complex number. Although it has no simple physical interpretation, the complex angle φ is very useful in simplifying the equations related to reflection and refraction by an absorbing medium. From eq 2.59, we have

$$\cos \varphi = \sqrt{1 - \frac{n_o^2}{n_t^2} \sin^2 \theta}.$$

(2.60)

By applying the boundary conditions stating the continuity of the tangential components of the electric and magnetic fields, we can obtain the reflection amplitude coefficients for TE and TM polarizations as follows[1]:

$$r^{TE} = \frac{n_o \cos \theta - n_t \cos \varphi}{n_o \cos \theta + n_t \cos \varphi}.$$
$$r^{TM} = \frac{n_o \cos \varphi - n_t \cos \theta}{n_o \cos \varphi + n_t \cos \theta}$$

(2.61)

The equations are of the same form as those for the dielectric case of eqs 2.25 and 2.27. The only difference is that n_t and ϕ are now complex. Since r^{TE} and r^{TM} are complex, phase changes occur on reflection. Equation 2.61 can be expressed as

$$r^{\mathrm{TE}} = \left| r^{\mathrm{TE}} \right| e^{i\phi^{\mathrm{TE}}}$$
$$r^{\mathrm{TM}} = \left| r^{\mathrm{TM}} \right| e^{i\phi^{\mathrm{TM}}} .$$

(2.62)

Here, $|r^{\mathrm{TE}}|$ and $|r^{\mathrm{TM}}|$ are the absolute magnitudes of r^{TE} and r^{TM}, and ϕ^{TE} and ϕ^{TM} are the corresponding phases. For a typical metal, the reflectance ($R = |r^{\mathrm{TE}}|^2$) for TE polarization increases monotonically from its value at normal incidence ($\theta = 0$) to unity at grazing incidence ($\theta = 90°$). On the other hand, the reflectance for TM polarization shows a shallow minimum at some incidence angle, which depends on the optical constants of the metal. This angle is called the *principal angle of incidence*, which corresponds to the Brewster angle for lossless media. TE- and TM-polarized waves also exhibit different phase shifts on reflection. Thus if linearly polarized light, whose polarization is neither TE nor TM, is incident onto a metal, the reflected light will generally be elliptically polarized. Therefore, it is possible to determine the complex refractive index of an absorbing medium by measuring the intensity and polarization of the reflected light. This method is known as *ellipsometry*. The technique can be applied to find the optical constants of thin films as well as bulk materials. More information can be found in References [2–5]. For the case of normal incidence, there is no distinction between TE and TM polarizations. Then we have

$$r^{\mathrm{TE}} = r^{\mathrm{TM}} = \frac{n_o - n_t}{n_o + n_t} = \frac{(n_o - n_{\mathrm{R}}) - in_{\mathrm{I}}}{(n_o + n_{\mathrm{R}}) + in_{\mathrm{I}}}.$$

(2.63)

The reflectance is

$$R = \frac{(n_o - n_{\mathrm{R}})^2 + n_{\mathrm{I}}^2}{(n_o + n_{\mathrm{R}})^2 + n_{\mathrm{I}}^2}.$$

(2.64)

For metals, the extinction coefficient n_I is large. This results in a large R value. For lossless media ($n_I = 0$), eq 2.64 reduces to the reflectance equation for dielectrics.

PROBLEMS

2.1 Suppose that a light beam is incident onto a thin film of thickness d and refractive index n_f, which is deposited on a substrate of refractive index n_s (Figure 2.24).

(a) Prove that the optical path difference, $\Delta = n_f(AB + BC) - n_0(AD)$, between reflected rays 1 and 2 is equal to $\Delta = 2n_f d \cos \theta_t$. (In this problem, the incident medium is air. Thus its refractive index n_0 is 1).

(b) A thin film with $n_f = 1.38$ is deposited on a glass substrate ($n_s = 1.52$) so that a normally incident beam is antireflecting at 580 nm. What wavelength will be minimally reflected when the beam is incident at 45°? Assume that the refractive indices are constant in the visible range.

(c) In Figure 2.24, the reflected intensity will depend on the phase difference between the reflected rays 1 and 2. The total phase difference has two components: one originating from the optical path difference as described above and the other, from the phase change on reflection. The latter component is also dependent on the polarization state of the incident beam. For the case of "(b)," however, the result will be unchanged regardless of whether the incident beam is TE or TM polarized. Explain why?

2.2 A light beam is reflected from a plane surface of fused silica glass with $n = 1.5$.

(a) Determine the reflectance and transmittance of a TE-polarized beam incident at $\theta_i = 30°$ and $\theta_i = 60°$, respectively.

(b) Repeat (a) for a TM-polarized beam.

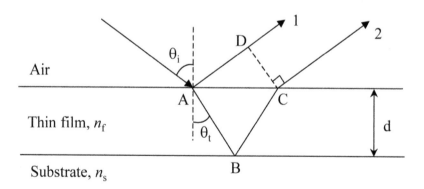

FIGURE 2.24   Reflection from a thin film deposited on a substrate.

2.3 The Fresnel equations given in eq 2.27 were derived based on the reference directions of E and H shown in Figure 2.8b. Prove that if we had reversed the direction of E_r (and also H_r), the corresponding expressions would be

$$\frac{E_{or}}{E_{oi}} = r^{TM} = \frac{n_t \cos\theta_i - n_i \cos\theta_t}{n_t \cos\theta_i + n_i \cos\theta_t}$$

$$\frac{E_{ot}}{E_{oi}} = t^{TM} = \frac{2n_i \cos\theta_i}{n_i \cos\theta_t + n_t \cos\theta_i}.$$

2.4 Calculate the reflectance of water ($n = 1.33$) for both TE and TM polarizations when the angles of incidence are $0°$, $30°$, and $60°$.

2.5 Light is incident into the interface between air and an isotropic material ($n = 2.2$).

(a) Determine the critical angle for TIR.
(b) Determine the Brewster's angle for internal and external reflection, respectively.

2.6 A TM-polarized beam is incident at the Brewster angle of a non-absorbing material with $n = 1.70$. What is the ratio between the incident intensity and the transmitted intensity?

2.7 Let's consider that a TE-polarized light beam of unit cross-sectional area is externally incident into the air-diamond ($n = 2.4$) interface at $45°$.

(a) Calculate the reflection amplitude coefficient and transmission amplitude coefficient for the above configuration.
(b) The intensity "I" refers to the energy flowing per unit area per unit time and is proportional to the refractive index and the square of the electric field amplitude, that is, $I = nE^2$. Let the amplitude of the incident light be $E_{inc} = E_o$. Then, express I_{inc}, I_{ref}, and I_{tran} in terms of E_o.
(c) Do you have $I_{inc} = I_{ref} + I_{tran}$? If not, what is the correct relation between the intensities that satisfies the energy conservation law?

2.8 The permittivity and permeability of medium 1 are ε_o and μ_o and those of medium 2 are ε and μ_o, where $\varepsilon > \varepsilon_o$. Consider a light beam normally incident from the medium 1 into the medium 2. Prove that if this beam is entirely transmitted with no reflection, it violates the electromagnetic boundary conditions.

REFERENCES

1. Fowles, G. *Introduction to Modern Optics,* 2nd ed.; Dover: New York, 1989.
2. Born, M.; Wolf, E. *Principles of Optics,* 7th ed.; Cambridge University Press: Cambridge, UK, 1999.

3. Tompkins, H.; Irene, E. *Handbook of Ellipsometry*; William Andrews: New York, 2005.
4. Gorlyak, A.; Khramtsovky, I.; Solonukha, V. Ellipsometry Method Application in Optics of Inhomogeneous Media. *Sci. Tech. J. Inform. Tech. Mech. Optics* **2015**, *15*, 378–386.
5. Ohlidal, I.; Franta, D. *Ellipsometry of Thin Film Systems*. In *Progress in Optics*; Wolf, E., Ed.; Elsevier: Amsterdam, 2000; Vol. 41, pp 181–282.

CHAPTER 3

Superposition of Waves

In the previous chapter, we described various phenomena occurring when a light wave of a given amplitude, wavelength, and frequency passes from one medium to another. We are here interested in what happens when two or more waves are superposed at some point in space. Interference and diffraction, which will be treated in subsequent chapters, also result from the superposition of waves. Of course, the combined effects of two or more waves are influenced by the specifications (frequency, amplitude, phase, etc.) of each constituent wave. When harmonic waves of different amplitudes and phases but with the same frequency are combined, the composite wave is another harmonic wave having the same frequency. While the principle of superposition is equally applicable to waves differing in frequency, the resultant may not be expressed by a single harmonic wave. The superposition of waves with some range of frequencies leads to the concepts of coherence and bandwidth. The term *coherence* is used to describe the phase correlation of monochromatic waves. If the phase of any portion of a wave is predictable from any other portion of the wave, it is said to be *perfectly coherent*. Since there are no perfectly monochromatic sources, no light waves are perfectly coherent. A perfectly coherent wave expressible by a single harmonic wave (sine or cosine wave) of infinite extent is just an ideal case far from reality. A real wave exists only over a finite duration of time and can be represented as a sequence of harmonic wave trains of finite length. The average time duration and length of the wave trains are referred to as *coherent time* and *length*, respectively. For an electromagnetic wave, the coherent length is the distance over which a propagating wave may be considered coherent. A more monochromatic wave has a longer coherent length. Interference is most prominent when the path lengths taken by the interfering waves are different by less than the coherence length. Coherent length is therefore an important parameter in interference and optical metrology.

3.1 SUPERPOSITION OF WAVES OF THE SAME FREQUENCY

Consider two electromagnetic waves E_1 and E_2 propagating along the positive x-direction, with their electric fields directed along the y-direction (Fig. 3.1a). When these waves of the same frequency have amplitudes E_{01} and E_{02} and a phase difference of δ, the resultant disturbance is the linear superposition of the two waves and is expressed as

$$E = E_1 + E_2 = E_{01}\cos(kx - \omega t) + E_{02}\cos(kx - \omega t - \delta). \tag{3.1}$$

In dealing with the superposition of harmonic waves, the complex-number representation is mathematically simpler than the trigonometric manipulation. Keeping in mind that $e^{i\theta} = \cos\theta + i\sin\theta$, the relation of eq 3.1 can be alternatively described by the following complex notations:

$$E_{\text{com}} = E_{01}e^{i(kx-\omega t)} + E_{02}e^{i(kx-\omega t-\delta)} = (E_{01} + E_{02}e^{-i\delta})e^{i(kx-\omega t)} \tag{3.2}$$

Since E is simply the real part of E_{com}, we can take it after manipulation. The term in parentheses of eq 3.2 is also complex, and thus it can be expressed as follows:

$$E_{\text{com}} = (E_{01} + E_{02}e^{-i\delta})e^{i(kx-\omega t)} = E_o e^{i\alpha}e^{i(kx-\omega t)} = E_o e^{i(kx-\omega t+\alpha)}, \tag{3.3}$$

where E_o is the magnitude of E_{com}. Now we know that $E = E_o \cos(kx - \omega t + \alpha)$. Thus, the superposition of two harmonic waves gives another harmonic wave of the same frequency. α is a value depending on E_{01}, E_{02}, and δ. However, the intensity of the composite wave can be obtained without calculating this value. Multiplying each side of $(E_{01} + E_{02}e^{-i\delta}) = E_o e^{i\alpha}$ by its complex conjugate leads to $E_o^2 = E_{01}^2 + E_{02}^2 + 2E_{01}E_{02}\cos\delta$. The intensity of an electromagnetic wave is proportional to its amplitude squared. Therefore, the superposition of two waves may yield a resultant intensity that is not simply the sum of the component intensities. If we are concerned only with relative intensities, we can neglect the common proportionality constant and represent the component intensities as $I_1 = E_{01}^2$ and $I_2 = E_{02}^2$. The total intensity is then

$$I = I_1 + I_2 + 2\sqrt{I_1 I_2}\cos\delta, \tag{3.4}$$

where $2\sqrt{I_1 I_2}\cos\delta$ is called *interference term*. The resultant intensity may be greater, equal to, or less than $I_1 + I_2$, depending on the value of the interference term. This is plotted in Figure 3.1b. The maximum intensity is obtained at $\cos\delta = 1$, when δ is an integer multiple of 2π. For $0 < \cos\delta < 1$, we have $I > I_1 + I_2$. The minimum intensity occurs when the waves are 180° out of phase, that is, when δ is an odd multiple of π. Interference refers to the interaction of waves that are coherent with each other. A crucial factor determining the interference result is the difference in phase between the involved waves. When the phase difference between two harmonic waves is $\delta = 0, 2\pi, 4\pi, \ldots$, the resulting amplitude is maximized, whereas $\delta = \pi, 3\pi, 5\pi, \ldots$ yield a minimum amplitude (Fig. 3.2). The former situation is referred to as *constructive interference*, in which the waves are in phase with each other. In the latter *destructive interference*, the waves are 180° out of phase. If the two waves are mutually incoherent, the phase difference δ varies with time in a random fashion. Then, the mean value of the interference term is zero, and there is no interference. The two waves must have the same polarization in order to maximize the interference effect. In particular, if the polarizations are mutually orthogonal, there is no interference again.

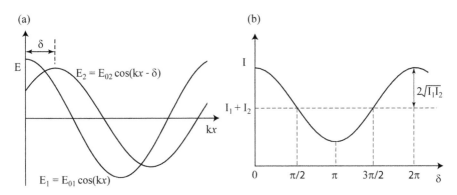

FIGURE 3.1 (a) Superposition of two harmonic waves. (b) Resultant intensity versus phase difference of δ.

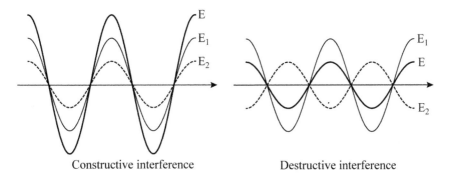

Constructive interference   Destructive interference

FIGURE 3.2 Constructive and destructive interference.

With the procedure used above, it can be easily shown that the superposition of any number of coherent harmonic waves of a given frequency leads to a harmonic wave with the same frequency. Consider N such waves that are expressed as

$$E = E_{01}\cos(kx - \omega t + \delta_1) + E_{02}\cos(kx - \omega t + \delta_2) + \cdots + E_{0N}\cos(kx - \omega t + \delta_N).\,(3.5)$$

Here $\delta_1, ..,$ and δ_N represent the relative phases of the constituent waves, that is, the initial phase angles of the N waves. The composite wave E may be written in its complex exponential form as follows:

$$\begin{aligned}E_{\text{com}} &= E_{01}e^{i(kx - \omega t + \delta_1)} + \cdots + E_{0N}e^{i(kx - \omega t + \delta_N)} \\ &= (E_{01}e^{i\delta_1} + \cdots + E_{0N}e^{i\delta_N})e^{i(kx - \omega t)} = E_o e^{i\alpha} e^{i(kx - \omega t)}\end{aligned} \qquad (3.6)$$

The expression in parentheses, which has N component terms, can be graphically pictured using the concept of *phasors*. Its magnitude (E_o) and phase angle (α) can be found by plotting the magnitude and phase angle of each component term and adding them, as shown in Figure 3.3. E is the real part of E_{com} and then we arrive at

$$E = E_o \cos(kx - \omega t + \alpha). \qquad (3.7)$$

The magnitude of E_{com} becomes the amplitude of E. From eq 3.6 and Figure 3.3, we have the relations:

$$\begin{aligned}E_o \cos\alpha &= E_{01}\cos\delta_1 + \cdots + E_{0N}\cos\delta_N \\ E_o \sin\alpha &= E_{01}\sin\delta_1 + \cdots + E_{0N}\sin\delta_N\end{aligned}. \qquad (3.8)$$

The amplitude E_o of the composite wave can be obtained by the Pythagorean theorem.

$$E_o^2 = (E_{01}\cos\delta_1 + \cdots + E_{0N}\cos\delta_N)^2 + (E_{01}\sin\delta_1 + \cdots + E_{0N}\sin\delta_N)^2 \qquad (3.9)$$

The phase angle (α) is clearly given by

$$\tan\alpha = \frac{E_{01}\sin\delta_1 + \cdots + E_{0N}\sin\delta_N}{E_{01}\cos\delta_1 + \cdots + E_{0N}\cos\delta_N}. \qquad (3.10)$$

Consider a harmonic wave of the following form:

$$\psi = A\cos\theta \qquad (3.11)$$

The complex notation for this wave is

$$\psi_{\text{com}} = Ae^{i\theta}. \qquad (3.12)$$

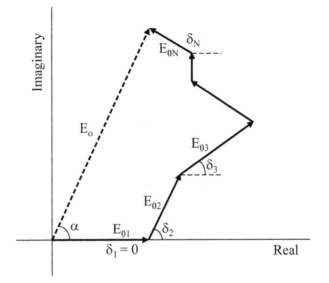

FIGURE 3.3   Addition of waves using the concept of *phasors*.

The complex number $Ae^{i\theta}$ is represented in the complex plane by a vector of magnitude "A" inclined at an angle θ to the real axis (Fig. 3.4). As θ increases, this vector is counterclockwise rotated and the real part of the complex number changes between A and $-A$. The wave given by eq 3.11 has a disturbance varying between these two values. Thus if we consider only the real part of eq 3.12, it is physically the same expression as eq 3.11. In many situations, it is more convenient to represent harmonic waves in complex-number notations. Especially when dealing with the superposition of harmonic waves, the complex notation is mathematically simpler than the trigonometric manipulation. As an example, suppose that the following three harmonic waves are superposed:

$$E_1 = \cos(kx - \omega t)$$
$$E_2 = 2\cos(kx - \omega t - 30°)$$
$$E_3 = 2\cos(kx - \omega t - 60°)$$

Figure 3.5a shows the addition of these three waves in the complex plane. When the magnitude and phase angle of the composite wave are determined, we can neglect the term $(kx - \omega t)$ simply by setting it to zero. This term can be inserted in the final expression of the composite wave later. It is easily found from Figure 3.5a that the magnitude and phase angle of the composite wave are 4.625 and $-36.2°$, respectively. Then, the superposition of the three constituent waves leads to a harmonic wave given by

$$E = 4.625\cos(kx - \omega t - 36.2°).$$

The real-space superposition of the waves is depicted in Figure 3.5b for comparison. In summary, the superposition of multiple harmonic waves of identical frequency produces a harmonic wave of the same frequency. The amplitude (and also intensity) of the resultant wave depends on the amplitudes and phase angles of the constituent waves.

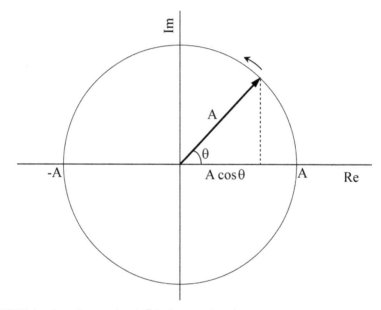

FIGURE 3.4   Complex number $Ae^{i\theta}$ in the complex plane.

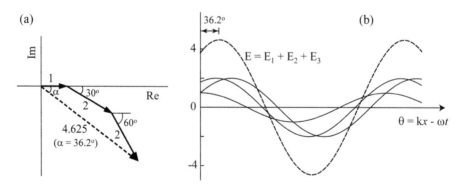

FIGURE 3.5   (a) Addition of three harmonic waves in the complex plane. (b) Real-space superposition of the three waves.

3.2 STANDING WAVES

A special case of superposition arises when two harmonic waves of the same frequency propagate in opposite directions. This type of superposition most frequently occurs when the wave reflected backward off a mirror overlaps the incident wave. Suppose that an incoming wave traveling in the negative x-direction strikes a mirror at $x = 0$ and is reflected into the positive x-direction (Fig. 3.6). We assume an ideal situation in which the incident energy is neither lost on reflection nor absorbed by the mirror. Then, the incident and reflected waves can be written as

$$E_I = E_{0I} \sin(kx + \omega t)$$
$$E_R = E_{0R} \sin(kx - \omega t + \delta)$$
$$(3.13)$$

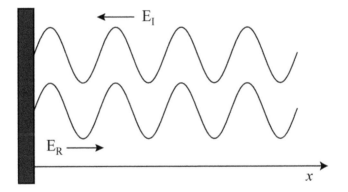

FIGURE 3.6   A typical standing wave situation.

Here, δ is introduced due to a potential phase shift on reflection. Since there is no energy loss, $E_{0I} = E_{0R}$. If the mirror is made of a perfect conductor, there should be no electric field inside it. The boundary conditions then require that $E = E_I + E_R = 0$ at $x = 0$. It follows from eq 3.13 that δ is zero. The composite wave is then

$$E = E_{0I} \{\sin(kx + \omega t) + \sin(kx - \omega t)\}.$$

Applying the trigonometric identity:

$$\sin\alpha + \sin\beta = 2\sin\frac{1}{2}(\alpha + \beta)\cos\frac{1}{2}(\alpha - \beta),$$

we obtain

$$E = 2E_{0I}\sin(kx)\cos(\omega t). \qquad (3.14)$$

This represents a *standing wave*, as opposed to a traveling wave. Unlike traveling waves, standing waves do not move through space. Consequently, there is no net transfer of energy. Equation 3.14 can be regarded as either a sine wave with time-varying amplitude or a cosine wave with space-dependent amplitude. Figure 3.7 shows the spatial variations of E at various times. There are x values for which the disturbance is zero for all t. These values ($x = 0$, $\lambda/2$, λ, $3\lambda/2$....etc.) are called the *nodes* of the standing wave. The amplitude is maximized at $t = 0$, $\tau/2$, τ, $3\tau/2$....etc., where τ is the temporal period of the constituent waves. The disturbance E will be zero at all x whenever $\cos \omega t = 0$, that is, when $t = \tau/4$, $3\tau/4$....etc.

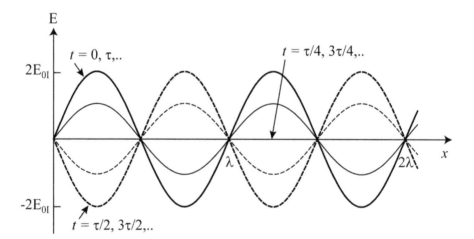

FIGURE 3.7   Spatial variations of a standing wave at various times.

In the above discussion, the incident and reflected waves were expressed by sine functions. They may also be written in terms of cosine functions, as shown below.

$$E_I = E_{0I}\cos(kx + \omega t)$$
$$E_R = E_{0R}\cos(kx - \omega t + \delta) \qquad (3.15)$$

Now, the boundary condition of $E = 0$ at $x = 0$ leads to $\delta = \pi$. Applying the trigonometric identity gives

$$E = -2E_{01}\sin(kx)\sin(\omega t). \tag{3.16}$$

Keeping in mind that sin $(\omega t - \pi/2) = -\cos(\omega t)$, eq 3.16 is physically the same expression as eq 3.14. There is only a time lag of $\tau/4$. The disturbance $E(x)$ at $t = \tau/2$ in eq 3.14 is equal to that at $t = \tau/4$ in eq 3.16. Therefore, the principal features of the standing wave remain unaltered. If the mirror is not a perfect reflector, as is often the case, it absorbs some of the incident energy. In this case, the two waves do not completely cancel out each other at the nodes. The resultant wave will then contain a traveling component along with the standing wave. Under such circumstances, there will be a net propagation of energy.

3.3   SUPERPOSITION OF WAVES OF DIFFERENT FREQUENCY

3.3.1   BEATS AND GROUP VELOCITY

Another case of superposition is that of waves of different frequency. Consider the combination of the following two waves that have equal amplitudes and zero initial phase angles:

$$\begin{aligned} E_1 &= E_o \cos(k_1 x - \omega_1 t) \\ E_2 &= E_o \cos(k_2 x - \omega_2 t) \end{aligned} \tag{3.17}$$

Since the waves differ in frequency, their propagation constants are also different. Applying the trigonometric identity, we have the resultant wave:

$$E = E_1 + E_2 = 2E_o \cos\frac{1}{2}[(k_1 + k_2)x - (\omega_1 + \omega_2)t]\cos\frac{1}{2}[(k_1 - k_2)x - (\omega_1 - \omega_2)t]. \tag{3.18}$$

Let

$$k_a = \frac{k_1 + k_2}{2}, \ k_s = \frac{k_1 - k_2}{2} \tag{3.19}$$

and

$$\omega_a = \frac{\omega_1 + \omega_2}{2}, \ \omega_s = \frac{\omega_1 - \omega_2}{2}. \tag{3.20}$$

Then we have

$$E = 2E_o \cos(k_a x - \omega_a t)\cos(k_s x - \omega_s t). \tag{3.21}$$

Equation 3.21 has the form of a product of two cosine waves. If the frequencies of the constituent waves are not much different from each other, that is, $\omega_1 \approx \omega_2$, then we have $\omega_a \gg \omega_s$ and $k_a \gg k_s$. The first cosine factor in eq 3.21 represents a wave with frequency ω_a and propagation constant k_a that are the averages of the frequencies and propagation constants of the constituent waves, respectively. The second cosine factor possesses a frequency and propagation constant that are much smaller than the first one. Therefore, the composite wave can be regarded as a wave of frequency ω_a with a slowly varying amplitude $E_o(x, t)$ such that

$$E = E_o(x,t)\cos(k_a x - \omega_a t), \tag{3.22}$$

where $E_o(x, t)$ is

$$E_o(x,t) = 2E_o \cos(k_s x - \omega_s t). \tag{3.23}$$

The low-frequency wave $E_o(x, t)$ in eq 3.22 serves as an envelope modulating the amplitude of the high-frequency (ω_a) wave, as shown in Figure 3.8. The dashed line in Figure 3.8c depicts the envelope of the resultant wave. The energy flux density carried by a wave is proportional to the square of its amplitude. Therefore, the combination of two waves of slightly different frequency exhibits *beats*. $E_o^2(x, t)$ oscillates with a frequency of $2\omega_s = \omega_1 - \omega_2$. This is known as the *beat frequency*. The beat frequency is twice the frequency of the modulating envelope. We find that the beat frequency is simply the difference between the frequencies of the two constituent waves.

The concept of group velocity plays an important role in optics and other physics fields, because any real harmonic wave, either electromagnetic or elastic, is actually a group of waves with close frequencies. Even so-called "monochromatic light" contains a very narrow range of wavelengths (frequencies). We showed in Chapter 1 that the dielectric constant and refractive index of a material are frequency-dependent. Thus, in such a dispersive medium, the phase velocity of a wave depends on its frequency. If the medium is nondispersive (e.g., vacuum), all harmonic waves present in the group propagate at the same phase velocity. The envelope of the group also propagates at the same velocity. That is, the group velocity is identical to the phase velocity. As a consequence, the envelope of the group maintains its original shape upon propagation. In dispersive media such as typical dielectrics, the harmonic waves forming the group propagate at different phase velocities and the relations among the phases of various harmonic waves continuously change upon propagation. This distorts the form of the group envelope. If the phase velocities of a wave group are very close to one another (i.e., if the signal has a very narrow spectrum), the shape of the propagating envelope is little distorted. Nevertheless, the effect of dispersion makes the group velocity of the waves differ from the phase velocity.

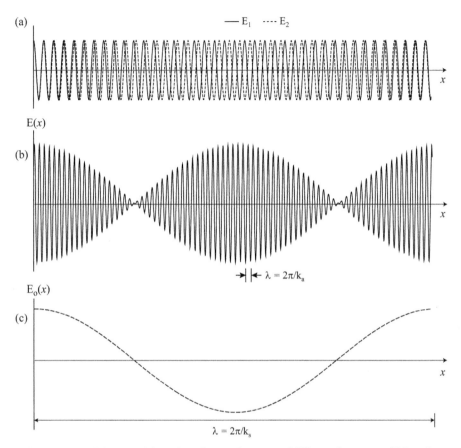

FIGURE 3.8 (a) Superposition of two harmonic waves of different frequency. (b) Resultant wave. (c) Envelope of the resultant wave.

The two harmonic functions in eq 3.17 may be regarded as two frequency components present in a group. The superposition of the two harmonic functions having different frequencies (and wavelengths) produces a resultant wave like the one shown in Figure 3.8b. It consists of a higher-frequency (ω_a) carrier wave, with its amplitude modulated by a lower-frequency (ω_s) cosine function.

$$E = 2E_o \cos(k_s x - \omega_s t)\cos(k_a x - \omega_a t). \tag{3.24}$$

The velocity of the higher-frequency carrier wave is the velocity at which each small peak in the carrier travels. This is the *phase velocity* and is given by

$$v_p = \frac{\omega_a}{k_a} = \frac{\omega_1 + \omega_2}{k_1 + k_2} \approx \frac{\omega}{k}. \tag{3.25}$$

The final approximation is valid in the case of $\omega_1 \approx \omega_2$ and $k_1 \approx k_2$. There is another motion to be considered. That is the propagation of the modulation envelope. The velocity at which the lower-frequency modulation envelope propagates is called the *group velocity* and is given by

$$v_g = \frac{\omega_s}{k_s} = \frac{\omega_1 - \omega_2}{k_1 - k_2} = \frac{\Delta\omega}{\Delta k} \approx \frac{d\omega}{dk}. \tag{3.26}$$

It was assumed again that the differences between angular frequencies and propagation constants are small. The group velocity $v_g = d\omega/dk$ is not necessarily equal to the phase velocity $v_p = \omega/k$. Suppose that the constituent waves $E_1(x, t)$ and $E_2(x, t)$ in eq 3.17 travel with the same velocity, that is, $v_1 = v_2$, which corresponds to $\omega_1/k_1 = \omega_2/k_2$. This condition is achieved when ω is linear with k. The phase velocity given by eq 3.25, the average of the velocities of the constituent waves, is thus equal to v_1 and v_2. Since ω is a linear function of k, $\omega/k = d\omega/dk$. Therefore we have $v_1 = v_2 = v_p = v_g$. This particular case applies only to *non-dispersive media*, in which the phase velocity is independent of wavelength (or frequency) so that the constituent waves can travel with the same speed. However, most dielectric materials are dispersive and exhibit a dispersion relation $\omega(k)$ that is nonlinear with k. Consequently, the group velocity at which the modulation envelope or signal propagates may be greater, equal to, or less than the phase velocity of the carrier wave. The difference between these two velocities can be found as follows. Substitution of eq 3.25 into eq 3.26 gives

$$v_g = \frac{d\omega}{dk} = \frac{d}{dk}(kv_p) = v_p + k\frac{dv_p}{dk}. \tag{3.27}$$

Equation 3.27 is valid for any group as long as the constituent waves have a narrow range of frequencies. If the phase velocity of a wave is independent of wavelength, $dv_p/dk = 0$ and $v_g = v_p$. This is the case of light propagation in a vacuum, where $\omega = kc$, $v_p = v_g = c$. In dispersive media, however, $v_p = \omega/k = c/n$, where the refractive index n is a function of λ or k.
From

$$\frac{dv_p}{dk} = \frac{d}{dk}\left(\frac{c}{n}\right) = \frac{-c}{n^2}\left(\frac{dn}{dk}\right),$$

Equation 3.27 can be written as

$$v_g = v_p \left[1 - \frac{k}{n} \left(\frac{dn}{dk} \right) \right]. \tag{3.28}$$

Using $k = 2\pi/\lambda$ and $dk = -(2\pi/\lambda^2)\, d\lambda$, eq 3.28 can be reformatted to

$$v_g = v_p \left[1 + \frac{\lambda}{n} \left(\frac{dn}{d\lambda} \right) \right]. \tag{3.29}$$

The refractive index of a material is usually measured in terms of the vacuum wavelength ($\lambda_o = 2\pi c/\omega$), not the wavelength in the medium. It would then be necessary to express the dispersion relation in eq 3.29 with respect to λ_o rather than λ. Since $\omega = 2\pi c/\lambda_o$ and $k = 2\pi n/\lambda_o$, we have

$$\frac{dk}{d\omega} = \frac{d(n/\lambda_o)}{d(c/\lambda_o)} = \frac{(-n/\lambda_o^2)d\lambda_o + (1/\lambda_o)dn}{(-c/\lambda_o^2)d\lambda_o} = \frac{n}{c} - \frac{\lambda_o}{c}\left(\frac{dn}{d\lambda_o}\right) = \frac{n}{c}\left[1 - \frac{\lambda_o}{n}\left(\frac{dn}{d\lambda_o}\right)\right]. \tag{3.30}$$

The group velocity is then

$$v_g = \frac{d\omega}{dk} = v_p \left[1 - \frac{\lambda_o}{n}\left(\frac{dn}{d\lambda_o}\right)\right]^{-1} \approx v_p \left[1 + \frac{\lambda_o}{n}\left(\frac{dn}{d\lambda_o}\right)\right]. \tag{3.31}$$

Since the second term in square brackets is much smaller than unity, the approximation is quite valid. Equations 3.31 and 3.29 are of the same form. As discussed in Section 1.4, the refractive index decreases with increasing wavelength in regions of normal dispersion, $dn/d\lambda_o < 0$, and thus $v_g < v_p$. Distinguished from the refractive index n, a *group index of refraction* is defined as

$$n_g = \frac{c}{v_g} = n - \lambda_o \frac{dn}{d\lambda_o}. \tag{3.32}$$

Equations 3.27 and 3.31 also hold for a large number of waves once their frequency range is narrow. The resultant wave is characterized both by the phase velocity, the average velocity of the constituent waves, and by the group velocity, which is the velocity of the modulation envelope. The group velocity is sometimes called the signal velocity, because a signal in the form of any modulated wave travels at the group velocity. The group velocity, which determines the speed with which energy or information is transmitted, is usually less than the phase velocity. When light pulses, consisting of a large number of harmonic waves, are transmitted in a dispersive medium, the velocity of the pulses is different from the velocity of the constituent waves. As shown in Section 1.4, the refractive index n of a medium can be less than unity (< 1) at certain frequencies, implying that the speed of light in the medium at that frequency exceeds the speed of light in vacuum. However, this does not contradict the special theory of relativity; no information-carrying signal can travel faster than the speed of light in vacuum. It is to be noted that the refractive index n is defined with respect to the phase (or wave) velocity, which does not carry information. The speed with which any form of signal travels is always less than c.

3.3.2 FOURIER ANALYSIS

The superposition of a number of harmonic waves of identical frequency, even though they differ in amplitude and phase, produces a harmonic wave of the same frequency, as already shown in Section 3.1. However, if the constituent waves have different frequencies (i.e., wavelengths), the composite wave is anharmonic and may take an arbitrary shape. As an example, Figure 3.9 shows a resultant that arises from the superposition of three harmonic waves having different wavelengths and amplitudes. Figure 3.9 implies that by combining a number of sinusoidal functions having different wavelengths, amplitudes, and relative phases, it would be possible to generate a great variety of wave profiles. *Fourier analysis* refers to the inverse process of decomposing a given profile into its harmonic components. Fourier series arises from the task of representing a given periodic function by a trigonometric series. When $f(x)$ is a periodic function of spatial period p, it can be represented as a Fourier series.

$$f(x) = \frac{a_o}{2} + \sum_{n=1}^{\infty} \left(a_n \cos\frac{2\pi n}{p}x + b_n \sin\frac{2\pi n}{p}x \right) \qquad (3.33)$$

Here, n takes on integers. The Fourier coefficients of $f(x)$ are given by

$$a_o = \frac{2}{p}\int_0^p f(x)dx$$

$$a_n = \frac{2}{p}\int_0^p f(x)\cos\frac{2\pi nx}{p}dx \ . \qquad (3.34)$$

$$b_n = \frac{2}{p}\int_0^p f(x)\sin\frac{2\pi nx}{p}dx$$

The integration in eq 3.34 can be done for any one period, from $x = 0$ to p or from $x = x_o$ to $x_o + p$. As an example, let's find the Fourier series of the periodic square wave shown in Figure 3.10. Over a period of $-L < x < L$, the wave is represented as follows:

$$f(x) = \begin{cases} 0 & -L < x < -L/2 \\ 1 & -L/2 < x < L/2 \\ 0 & L/2 < x < L \end{cases} \qquad (3.35)$$

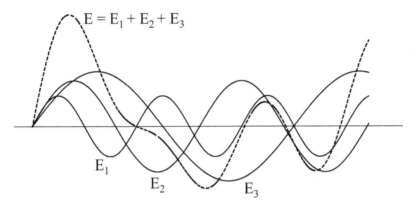

FIGURE 3.9 Superposition of three harmonic waves of different frequency.

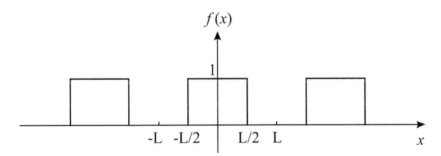

FIGURE 3.10 A periodic square wave.

By inserting eq 3.35 into eq 3.34, we obtain the Fourier coefficients. Since f (x) is even, $b_n = 0$. We find that

$$a_o = 1$$

$$a_n = \frac{2}{n\pi} \sin \frac{n\pi}{2}.$$

The corresponding Fourier series of eq 3.33 is then

$$f(x) = \frac{1}{2} + \frac{2}{\pi} \left(\cos \frac{\pi}{L} x - \frac{1}{3} \cos \frac{3\pi}{L} x + \frac{1}{5} \cos \frac{5\pi}{L} x - \cdots \right). \qquad (3.36)$$

Figure 3.11 depicts a few partial sums of the series. As more component terms are included in the summation, the series more closely reproduces the square wave of Figure 3.10. Only the first component term has the same spatial period as $f(x)$ and the succeeding terms have progressively decreased spatial periods. Each successive term make less contribution because its amplitude decreases. Although a finite number of terms may represent the function fairly well, an infinite number of components are required to completely reproduce it. It follows from eq 3.34 that a_o and a_n become zero when $f(x)$ is odd. Thus, the Fourier series of an odd periodic function will be represented as a summation of sine functions.

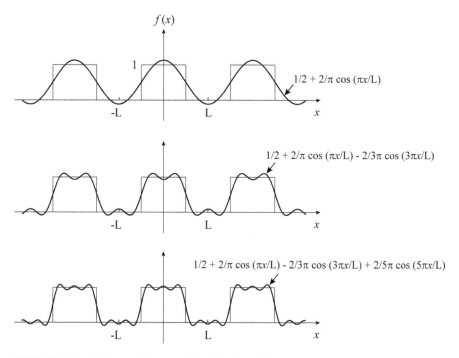

FIGURE 3.11   A few partial sums of the Fourier series.

The profile $f(x)$ of Figure 3.10 may correspond to a traveling wave f $(x - vt)$. We can find the time-domain profile $f(t)$ by simply changing x to t and replacing L with T, where $2T$ is the temporal period of the wave. So far, our analysis has been restricted to pure sinusoidal waves. Actual periodic waves may have arbitrary shapes including square, triangle, and saw-tooth. From the theorem of Fourier series, any anharmonic periodic wave can be envisioned as a superposition of harmonic components of different wave-length (and frequency) so that it can be written as

$$f(x-vt) = \frac{a_0}{2} + \sum_{n=1}^{\infty} \left\{ a_n \cos\frac{2\pi n}{p}(x-vt) + b_n \sin\frac{2\pi n}{p}(x-vt) \right\}. \quad (3.37)$$

Fourier series are powerful tools in treating various phenomena involving periodic functions. However, many practical problems involve *nonperiodic functions*. It is thus necessary to generalize the concept of Fourier series to include nonperiodic functions. This is of practical significance in a variety of physics fields, particularly in optics. Figure 3.12a shows a periodic square function whose period (*p*) and square width (*L*) are uncorrelated to each other. If we keep the width of the square peak constant while the period is made to approach infinity, the resulting function will be no longer periodic because one single square pulse remains, as shown in Figure 3.12b. However, the single pulse can be alternatively interpreted as a periodic function whose period is infinite. This provides a possible way to generalize the method of Fourier series to include nonperiodic functions. For a nonperiodic function f (*x*), the Fourier series is replaced with the so-called *Fourier integral*.

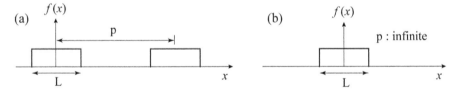

FIGURE 3.12 (a) A periodic function with a finite period. (b) A periodic function whose period is infinite.

$$f(x) = \frac{1}{\pi}\left[\int_0^{\infty} A(k)\cos kx\, dk + \int_0^{\infty} B(k)\sin kx\, dk\right], \quad (3.38)$$

where

$$A(k) = \int_{-\infty}^{\infty} f(x)\cos kx\, dx$$
$$B(k) = \int_{-\infty}^{\infty} f(x)\sin kx\, dx \quad (3.39)$$

Since an integral is the limit of a sum as the number of elements goes to infinity, an infinite number of harmonic components are required to reproduce the function. The coefficients $A(k)$ and $B(k)$ represent the amplitudes of these components. Let's find the Fourier integral of the square pulse shown in Figure 3.12b. The function is

$$f(x) = \begin{cases} 1 & |x| \le L/2 \\ 0 & |x| > L/2 \end{cases}.$$

Since $f(x)$ is an even function, $B(k)$ is zero. The remaining coefficient is

$$A(k) = 2\int_0^{L/2} \cos kx\, dx = L\frac{\sin(kL/2)}{(kL/2)} = L\operatorname{sinc}(kL/2). \tag{3.40}$$

The Fourier integral of this square pulse is given by

$$f(x) = \frac{1}{\pi}\int_0^\infty L\operatorname{sinc}(kL/2)\cos kx\, dk. \tag{3.41}$$

The Fourier transform of the square pulse, $A(k)$, is plotted in Figure 3.13. Note that as L increases, the spacing between successive zeroes of $A(k)$ decreases. k is a variable that determines both of the amplitude and spatial period of the constituent cosine functions. In the case of a Fourier series, the graphs of partial sums are approximate curves of the periodic function represented by the series (Fig. 3.11). Similarly, in the case of the Fourier integral of eq 3.41, approximations can be obtained by replacing ∞ with a certain number. Since $A(k)$ is a decreasing function of k, cosine functions of high k make less contribution. Thus, integration from zero to a finite value of k may represent the pulse fairly well.

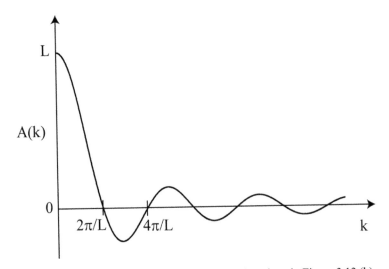

FIGURE 3.13   The Fourier transform A(k) of the function given in Figure 3.12 (b).

3.3.3 PULSES AND WAVE PACKETS

The Fourier series representation of a sinusoidal wave of infinite extent is fairly simple. It is represented by a single component term whose spatial period equals the actual period of the wave. However, this case is just a mathematical idealization. Practical sinusoidal waves have an end as well as a beginning, since they are turned on and off at finite times. The result will be a wave train of finite length, as shown in Figure 3.14. Obviously it cannot be represented by a single harmonic wave that has no beginning or end. Instead, numerous harmonic waves should be combined to produce the wave train exactly in the region where it exists and gives zero disturbance everywhere outside it. We here consider a harmonic wave pulse of finite length but the analysis to be done is also true of any isolated pulse regardless of its shape. Such pulses are called *wave packets* or *wave groups*. Let E_o and λ_p be the amplitude and wavelength of the harmonic wave train shown in Figure 3.14, respectively. The spatial profile of this wave pulse at $t = 0$ is

$$f(x) = \begin{cases} E_o \cos k_p x & |x| \le L \\ 0 & |x| > L \end{cases},$$

(3.42)

where k_p is $2\pi/\lambda_p$. Since $f(x)$ is even, $B(k) = 0$ and

$$A(k) = 2\int_0^L E_o \cos k_p x \cos kx\, dx = E_o L[\sin cL(k + k_p) + \sin cL(k - k_p)].$$

(3.43)

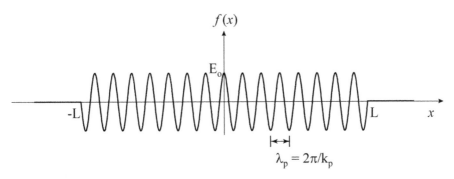

$$f(x)$$

$$\lambda_p = 2\pi/k_p$$

FIGURE 3.14 A wave train of finite length.

Here $A(k)$ consists of two sinc functions, one centered at $k = -k_p$ and the other, at $k = k_p$. It is plotted in Figure 3.15. If there are many wave oscillations in the train ($\lambda_p \ll L$), $2\pi/L \ll k_p$. Only positive values of k are considered in the Fourier integral, as shown in eq 3.38. Since the left-side sinc function is centered at $k = -k_p$, only its tail extending into the positive k region can contribute to $A(k)$. When $L \gg \lambda_p$, this sinc function rapidly falls off, hardly reaching the positive k region. As a consequence, we can neglect the first sinc function in eq 3.43 and simply write $A(k)$ as

$$A(k) = E_o L \ \mathrm{sinc} L(k - k_p). \tag{3.44}$$

The resulting Fourier integral is then

$$f(x) = \frac{1}{\pi} \int_0^\infty E_o L \ \mathrm{sinc} L(k - k_p) \cos kx dk. \tag{3.45}$$

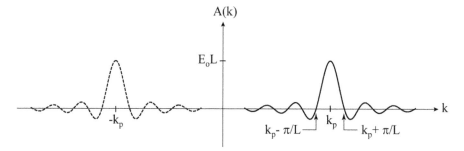

FIGURE 3.15   The Fourier transform of the wave train given in Figure 3.14.

Unless the wave train is infinitely long, an infinite number of harmonic waves are required to produce it. In other words, it must be synthesized from a continuous range of wavelengths. Clearly, the dominant contribution arises from $k = k_p = 2\pi/\lambda_p$. As the length L of the wave train increases, the central peak of $A(k)$ becomes sharper and narrower. When the train is infinitely extended, that is, $L = \infty$, the curve of $A(k)$ exhibits a single spike at k_p. This is the case of an ideal monochromatic wave, where the spectral width of k is zero. Actual monochromatic waves have a finite spectral width, however small. The time-domain expression of the wave train is as follows:

$$f(t) = \begin{cases} E_o \cos \omega_p t & |t| \le T \\ 0 & |t| > T \end{cases}$$

Here, $2T$ is the pulse duration length in t. Then we will have

$$A(\omega) = E_o T \ \mathrm{sinc} T(\omega - \omega_p). \tag{3.46}$$

As discussed, ω and k are related by the phase velocity. The spectrum $A(\omega)$ is the same as Figure 3.15, except for the notations. The first zeroes of the peak occur at $\omega = \omega_p \pm \pi/T$. Thus, the spectrum has a full width of $\Delta\omega = 2\pi/T$. However, the width at points where $A(\omega)$ drops to half its maximum value is customarily defined as the spectral width. Then we have $\Delta\omega \approx \pi/T$ and $\Delta k \approx \pi/L$, which are known as the *bandwidths* of the wave train. The spatial and temporal extents of the pulse are $\Delta x = 2L$ and $\Delta t = 2T$, respectively. The product of the bandwidth of the packet and its actual width is $\Delta k \, \Delta x \approx 2\pi$ and $\Delta\omega \, \Delta t \approx 2\pi$. From $\Delta\omega = 2\pi\Delta v$,

$$\Delta v \approx 1/\Delta t. \tag{3.47}$$

The frequency bandwidth is roughly the same as the reciprocal of the time extent of the pulse. If $\Delta t \to \infty$, which corresponds to a pulse of infinite length, a single frequency or wavelength suffices to produce the pulse. The shorter the pulse, the greater is the number of frequencies or wavelengths required to produce it. Equation 3.47 is analogous to the Heisenberg uncertainty principle in quantum mechanics, where particles are regarded as wave packets. When we analyze the light emitted by a monochromatic or quasi-monochromatic source, there is a spread in the frequencies because the emitted wave trains have a finite length. This is known as the *natural bandwidth* or *linewidth*, which is a measure of the width of a range of frequencies. Figure 3.16 shows a typical intensity versus frequency spectrum. The natural bandwidth, $\Delta\omega$, is usually defined as the spectral width at half the maximum intensity. Most of the energy is clearly carried by the frequency components present within this spectral width.

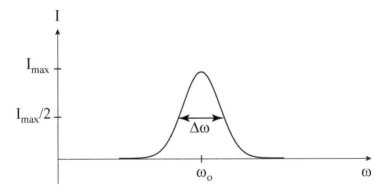

FIGURE 3.16 A typical intensity versus frequency spectrum of monochromatic light.

3.4   COHERENCE

Coherence describes the correlation between phases of a single wave, or between those of several waves. When a traveling electromagnetic wave is represented by a pure sinusoidal function with a well-defined propagation constant and an angular frequency, as shown in eq 1.11, we assume that the wave extends infinitely both in time and space. Such a perfectly monochromatic wave is perfectly coherent because all points on the wave are predictable. Therefore, perfect coherence means that the phase of any portion of the wave can be predicted from any other portion of the wave. As stated above, this case is just a mathematical idealization. Clearly there are no perfectly coherent sources. The light that we simply call "monochromatic light" consists of a sequence of harmonic wave trains of finite length, each separated from the others by a discontinuous change in phase, as illustrated in Figure 3.17. Excited atoms in a light source undergo electron transitions between energy levels, producing random radiation wave trains. The phase change reflects the electron transitions responsible for the generation of light. Since the emitted wave trains are finite, there will be a spread in the frequencies, known as the natural bandwidth. They can be characterized by an average wave train lifetime τ_o, called the *coherence time*. The *coherence length l* is defined as

$$l = c\tau_o, \qquad\qquad\qquad (3.48)$$

where c is the speed of light in a vacuum. The coherence time is the time interval over which the phase of a propagating light wave is predictable at a given point in space. The coherence length is the spatial extent over which we can predict the phase of the light wave at a given time. The coherence time is inversely proportional to the natural bandwidth (or linewidth) of the source. Equation 3.48 can then be rewritten as

$$l = \frac{c}{\Delta\nu} = \frac{\lambda^2}{\Delta\lambda}. \qquad\qquad (3.49)$$

If the light is perfectly monochromatic, $\Delta\nu$ will be zero, giving rise to infinite τ_o and l. But this is an idealization far from reality. An actual wave consists of a sequence of wave trains, characterized by its coherence time and length. However, for a time interval much smaller than τ_o, the wave behaves as if it were monochromatic. Temporal coherence measures the extent to which any two points, separated in time at a given location in space, can be correlated. P and Q in Figure 3.17 are within the same wave train and are always correlated because we can predict the phase of one point from the phase of the other. On the contrary, P and R, which belong to different trains, are too away in time and their phase relationship is totally unpredictable.

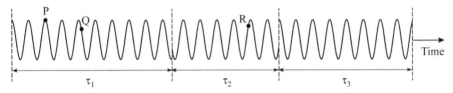

FIGURE 3.17 Sequence of harmonic wave trains of varying lifetime (τ) or length.

Spectral bandwidth and coherence are closely linked to one another. For example, the orange radiation at 589 nm emitted from a sodium lamp has a bandwidth of $\Delta v = 5 \times 10^{11}$ Hz. This leads to a coherence time of $\tau_o = 2 \times 10^{-12}$ s, which means that its coherent length is $l = 0.6$ mm. The green spectral line centered at 546 nm from a mercury lamp has a coherence length of 1.2 cm. Monochromatic sources such as lasers have substantial coherence lengths. The typical red emission from a He-Ne laser has a bandwidth of 1.5 $\times 10^9$ Hz, which corresponds to a coherence length of 20 cm. The coherence lengths of the lasers can be much increased under carefully controlled conditions. There are many commercially available lasers whose coherent lengths extend from meters to kilometers. White light has a wavelength range of 400–700 nm. In terms of the frequency, it ranges from approximately 4×10^{14} Hz to 7×10^{14} Hz, with a bandwidth of 3×10^{14} Hz. The coherence time is then 3.3×10^{-15} s, which corresponds to a coherence length of $l = 1$ μm. Thus, it is difficult to obtain interference fringes by white light. The concept of coherence is extremely important when we deal with the interference of waves. Interference is prominent only when the path lengths taken by the interfering waves are different by less than the coherence length. Consider a very thin transparent film deposited onto a glass substrate (Fig. 3.18a). For light incident on the film, constructive or destructive interference can occur between the waves reflected from the air-film and film-substrate interfaces, depending on their phase relationship. Constructive interference occurs when the two reflected waves are in phase with each other. In the case of destructive interference, the waves are 180° out of phase. The interference effect is observable under the ordinary Sun light, even though it is white light with a coherence length of about 1 μm. This is because the film is very thin and the optical path length difference between the two reflected waves is less than the coherent length of the incident light. In other words, the two waves originate from the same wave train. The optical path length difference depends on the refractive index and thickness of the film. Unless the film has an extremely high index of refraction, the optical path length difference will remain smaller than the coherence length of the incident light. For a bare substrate without the film, reflection occurs twice; one from the front surface and another, the rear surface. In this case, however, interference does not occur in the reflected waves. Typical glass substrates have a thickness much larger than the coherent length of white light. Therefore, the two reflected waves have different contributing wave trains, as shown in Figure 3.18b. Since the phase relationship between individual wave trains changes in a random fashion, the two waves are mutually incoherent, without a definite phase relation. Evidently, there is no interference. Of course, monochromatic light having a very long coherence length (such as a laser beam) will produce interference in the reflected beams.

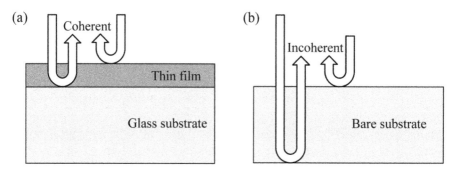

FIGURE 3.18 (a) Interference under white light is possible. (b) No interference occurs between the reflected waves because the substrate is much thicker than the coherent length of white light.

PROBLEMS

3.1 Three harmonic waves traveling together are given by

$$y_1 = \sin(kx - \omega t)$$
$$y_2 = 2\sin(kx - \omega t - \pi/3)$$
$$y_3 = 5\sin(kx - \omega t - \pi/4)$$

Write the resultant wave.

3.2 Suppose that in eq 3.5, the component waves have unit amplitude and their phases are $\delta_1 = 0$, $\delta_2 = -\delta,..,$ and $\delta_N = -(N-1)\delta$. Then prove that the resultant wave is given by

$$E = \frac{\sin(N\delta/2)}{\sin(\delta/2)}\cos[kx - \omega t - (N-1)\delta/2].$$

3.3 Show that the group velocity given in eq 3.27 can be alternatively expressed as

$$v_g = v_p - \lambda \frac{dv_p}{d\lambda}.$$

3.4 Find the Fourier series of the following periodic function, where the period is 4.

$$f(x) = \begin{cases} 0 & -2 < x < -1 \\ 2 & -1 < x < 1 \\ 0 & 1 < x < 2 \end{cases}$$

3.5 Find the Fourier integral of the following nonperiodic function:

$$f(x) = \begin{cases} -1 & -1 < x < 0 \\ 1 & 0 < x < 1 \\ 0 & \text{Otherwise} \end{cases}$$

3.6 A laser beam with a center wavelength of 632.8 nm has a coherent length of 10 m. Calculate its linewidth $\Delta\lambda$ and band width Δv.

3.7 White light is passed through a band-pass filter that transmits wavelengths in the range 550 ± 0.1 nm. What is the estimated coherent length of the transmitted light?

3.8 A continuous laser beam centered at 532 nm is chopped into 0.5 ns-pulses using an electric shutter. What are the linewidth $\Delta\lambda$, band width Δv, and coherent length of the pulse?

CHAPTER 4

Interference

4.1 INTRODUCTION

Interference is a phenomenon in which two or more waves are superimposed to give a resultant intensity different from the intensity summation of the individual waves. Interference effects can be observed in all types of waves, including electromagnetic wave, acoustic wave, and matter wave, though our discussion here is limited to the interference of light. In order for light waves to interfere, they should be correlated or coherent with one another. This requires that the waves involved in interference come from the same source or have the same frequency. The principle of interference is equal to the principle of superposition. If a crest of a wave meets a crest of another wave of the same frequency, the resultant amplitude becomes the vector sum of the individual amplitudes. Thus, the two waves must have the same polarization in order to maximize the interference effect. When waves of different polarization are added together, they result in a wave that has a polarization state different from those of the original waves. Interference can be easily observed in our daily life. For example, the colors observed in a soap bubble arise from the interference of light reflecting off the front and rear surfaces of the bubble film. The surfaces of many objects (butterfly wings, sea shells, and some minerals) appear to change color as the angle of illumination or the angle of observation changes. This is often created by the interference of light with surface microstructures. A simple one-dimensional (1D) interference pattern (or fringe) is obtained when two plane waves of the same frequency intersect at a non-zero angle. The resulting fringe consists of alternating bright (constructive interference) and dark (destructive interference) regions. In this case, interference is a kind of energy redistribution process. The energy lost at the destructive interference is regained at the constructive interference. Interference can also occur between multiple beams, once the phase differences between them remain constant over the observation time. Interferometry refers to a family of techniques in which waves, usually electromagnetic waves, are superimposed to retrieve certain information. Interferometers are widely used for the measurements of small displacements, refractive index changes, and surface irregularities. Interferometry has made a significant contribution to the advancement of astronomy, spectroscopy, and optical/engineering metrology. Interferometers are generally classified into amplitude-splitting and wavefront-splitting systems.[1-6] In the case of amplitude-splitting, a beam splitter divides a light wave into two beams traveling in different directions, which are eventually combined to produce an interference pattern. In the wavefront-splitting system, a light wave is divided in space through small apertures or narrow slits and the divided waves are allowed to recombine after traveling different paths. Young's double-slit interferometer is a classic example of the wavefront-splitting system.

4.2 YOUNG'S DOUBLE-SLIT EXPERIMENT

The first classic experiment that demonstrates interference of light was carried out by Thomas Young in 1802. His double-slit experiment played a crucial role in the general acceptance of the wave nature of light. In his original experiment, sunlight was passed through a small pinhole S to ultimately illuminate two narrow slits S_1 and S_2, as shown in Figure 4.1. After passing through the pinhole, the light spreads out in a spherical wave as if it emanated from a point source S. The spherical wave falls on the two closely spaced slits S_1 and S_2, from which coherent cylindrical waves are generated. Then, the slits become two coherent sources of light, whose interference fringe can be observed on a remote screen. A key to the experiment is the use of a small pinhole, since it makes the light waves coming from the slits mutually coherent. The generation of a spherical wave from a pinhole (and also a cylindrical wave from a narrow slit) is according to Huygens–Fresnel principle, which will be described in the following chapter. The Young's experiment can be analyzed by considering the physical parameters of the geometry shown in Figure 4.2. Here d is the slit separation and D is the distance from the slit aperture to the screen. The distance y represents the position of an observation point P on the screen, measured from the central axis. If the two slits are equal in shape and size, light emanating from the slits will have nearly identical amplitude. Then, the intensity at P is determined by the difference in phase between the two waves arriving at P. The phase difference depends on the difference in path lengths L_1 and L_2, which is given as

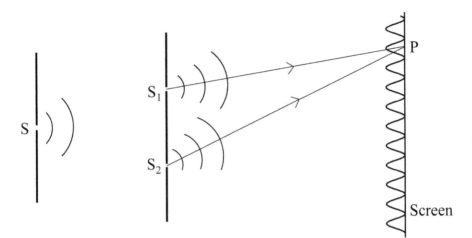

FIGURE 4.1   Young's double-slit experiment.

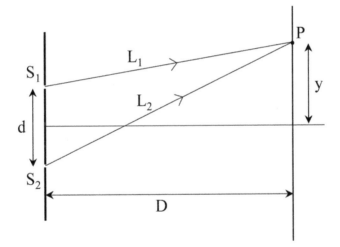

FIGURE 4.2   Geometry for analyzing the Young's double-slit experiment.

$$\left[D^2 + (y + d/2)^2 \right]^{1/2} - \left[D^2 + (y - d/2)^2 \right]^{1/2} \cong yd/D.$$

The above approximation is valid when d and y are both small compared to D. Constructive interference occurs when the path length difference, yd/D, is equal to an integral number of wavelengths. Thus, bright fringes occur at the points

$$y = 0, \pm \frac{\lambda D}{d}, \pm \frac{2\lambda D}{d}, \pm \frac{3\lambda D}{d}, \dots.$$

Dark fringes resulting from destructive interference are located midway between these points. The spacing of bright fringes is given by

$$\Delta y = \frac{\lambda D}{d}.$$

Thus, decreasing the slit separation d will expand the fringe pattern formed by each color. The wavelength of the light can be determined by measuring the fringe spacing. Young was able to estimate the wavelength of different colors in the spectrum from the spacing of the fringes. The small pinhole, used to provide a degree of spatial coherence, is unnecessary if a laser beam, both highly monochromatic and spatially coherent, is employed to illuminate the slits. If the slits are covered by optical elements such as phase retarders and polarizers, the fringe pattern will change. For instance, if a relative phase shift of 180° is introduced by placing a thin transparent plate over one slit, the resulting interference pattern will shift by half the fringe spacing from the pattern obtained in the absence of the plate. That is, bright fringes are observed in the regions where dark fringes were previously located. If polarizers are placed over the two slits in such a way that the waves passing through the slits are orthogonally polarized, no interference fringes occur.

There are many other arrangements for demonstrating interference, including Fresnel's double mirrors, Fresnel's biprism, and Lloyd's mirror. Since they make use of reflection or refraction to obtain two mutually coherent waves from a single source, the basic principle is the same as in Young's double-slit interferometer. In Lloyd's mirror, a single slit source S is placed near a flat mirror, as shown in Figure 4.3. A portion of the cylindrical wavefront coming from the source is reflected off the mirror. The reflected light appears to originate from a virtual source S' below the mirror. Another portion of the wavefront directly proceeds to the screen from the slit. The superposition of the direct and reflected light produces interference fringes on the screen. If we take the separation between the actual source and its virtual image to be d, the spacing of the fringes is again given by the above equation: $\Delta y = \lambda D/d$. It is to be noted, however, that when calculating the intensity at an observation point P, the phase change that occurs on reflection should be taken into account. If the reflected beam undergoes a phase shift of π, the interference pattern for Lloyd's mirror is complimentary to that of Young's double-slit interferometer; when one pattern has bright fringes at some y values, dark fringes occur at the same values in the other pattern. In a modern implementation of Lloyd's mirror, a diverging laser beam is made to strike the surface of a mirror at a grazing angle, so that the light reflecting off the mirror interferes with the light directly traveling to the screen. While the lower half of the interference pattern is blocked by the mirror itself, Lloyd's mirror has some advantages over Young's double-slit arrangement. In order to generate closely spaced interference fringes using a double-slit, the separation d between the slits should be increased. Increasing the slit separation, however, requires a very broad beam to cover both slits. This results in a large loss of beam energy. In contrast, the second slit in Lloyd's mirror is just the virtual image of the actual source. Therefore, the slit separation can be increased without a loss in the energy. The most common application of Lloyd's mirror lies in UV lithography for nanopatterning. Interference lithography (or holographic lithography) is a technique for patterning regular arrays of fine features, without the use of complex optical systems or photomasks. In this technique, an interference pattern formed by two or more coherent light waves is recorded in a medium to be patterned. Upon post-exposure processing such as etching, a fine structure corresponding to the interference pattern is generated in the medium. By using three-beam interference, two-dimensional arrays with hexagonal symmetry are generated, while three-dimensional arrays or photonic crystals can be produced with four beams. A detailed description of the topic is given in Chapter 9.

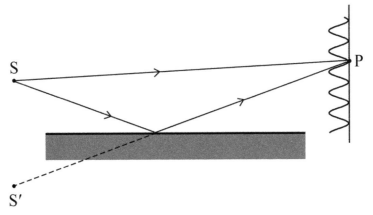

FIGURE 4.3   Lloyd's mirror.

Figure 4.4 shows the Fresnel's biprism arrangement, in which a cylindrical wavefront emanating from a slit source S impinges on the biprism.

The wavefront is refracted downward and upward at the upper and lower parts of the prism, respectively. Interference fringes are observed in the overlap region on the screen. This arrangement has two virtual sources S_1 and S_2, separated by a distance d. The value of d depends on the angle α and refractive index n of the prism, and the slit-to-prism distance. In practice, the prism angle α is very small so that the virtual sources are only slightly separated. The spacing of the resulting fringes is the same as before. However, the fringes are much brighter in comparison to those produced by Young's double-slit arrangement, because a much greater amount of light can pass through the prism than the double slits.

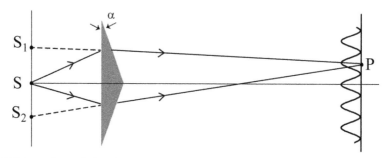

FIGURE 4.4   Fresnel's biprism.

4.3 AMPLITUDE-SPLITTING INTERFEROMETERS

An instrument used to exploit the interference of light and the resulting fringe pattern is called an optical interferometer. An interferometer splits an initial beam into two or more parts that travel different optical paths and then recombine to form an interference pattern. Interferometers are classified into wavefront-splitting and amplitude-splitting types, depending on how the initial beam is split. Young's double-slit, Lloyd's mirror, and Fresnel's biprism are all *wavefront-splitting interferometers*, in which a single pinhole or slit emits waves in different directions. The waves are eventually brought together by use of slits, mirrors, and prisms to produce interference fringes. In *amplitude-splitting interferometers*, a single beam of light is divided into two or more beams by partial reflection utilizing beam splitters and mirrors. The Michelson, Mach–Zehnder, and Fabry–Perot interferometers are of this type. In this section, we introduce the Michelson and Mach–Zehnder interferometers. As the Fabry–Perot interferometer is related to multi-beam interference, it is described in Section 4.4.

4.3.1 THE MICHELSON INTERFEROMETER

A schematic of the Michelson interferometer is illustrated in Figure 4.5. Light from an extended source S is incident onto a beam splitter A, which divides the beam into two parts. The beam splitter usually consists of a very thin metal film coated on the front surface of a glass plate. The divided beams are reflected by mirrors M_1 and M_2 and come back to A. The wave coming from M_1 is partially deflected by the beam splitter toward a detector C and part of the wave reflected from M_2 passes through the beam splitter. The two waves recombine at A to produce an interference pattern incident on the detector (or on the retina of a human eye). It is to be noted that one beam traverses the beam splitter three times, while the other passes through it once. This results in a difference in the optical path lengths of the two beams. For some applications, it is required to make the optical path lengths exactly identical. Thus, a compensating glass plate B is usually inserted in the path of one beam. This compensator is particularly necessary when white light is used. Although the optical path lengths may be equalized by adjusting the distance between the beam splitter and one of the mirrors, this correction is appropriate only for a specific wavelength but does not suffice for other wavelengths due to the dispersion of the glass. The insertion of a glass plate, which has the same thickness as the beam splitter, compensates for all wavelengths and invalidates the effect of dispersion. Without the compensator, only a monochromatic source can be used. The beam splitter and compensator are positioned at an angle of 45° with the incident beam. In practice, the surfaces of both mirrors are made to perpendicular to each other. One of the mirrors is movable along the beam direction so that the optical path difference can be varied. With the compensator inserted, any optical path difference arises from the actual path difference.

The interferometer shown in Figure 4.5 has two optical axes at right angle to one another. How the interferometer generates interference fringes can be better understood with the geometry shown in Figure 4.6, where the source and one of the mirrors are rotated counterclockwise by 90° so that all the elements are displayed on a single optic axis. Rays from the source S are divergent even if it is an extended source rather than a point source. Figure 4.6 shows the trajectory of a ray striking the mirror at an angle of θ. In reality, light from S is split at the beam splitter and its segments are reflected by M_1 and M_2. These two segments are denoted as ray 1 and ray 2, respectively, in the diagram of Figure 4.6. To an observer (or a detector), the two reflected rays will appear to come from two virtual sources S_1 and S_2. The corresponding virtual sources are mutually coherent. If one mirror is farther from the beam splitter than the other by a distance d, the optical path difference between the two rays will be $2d \cos \theta$, as indicated in the diagram. This corresponds to a phase difference of $4\pi d \cos \theta / \lambda$. There is an additional phase difference of π. In conjunction with the configuration of Figure 4.6, the ray 2 is externally reflected in the beam splitter, whereas the ray 1 is internally reflected there. Taking a relative phase shift of π into account, destructive interference will occur when

$$2d \cos \theta = m\lambda . \tag{4.1}$$

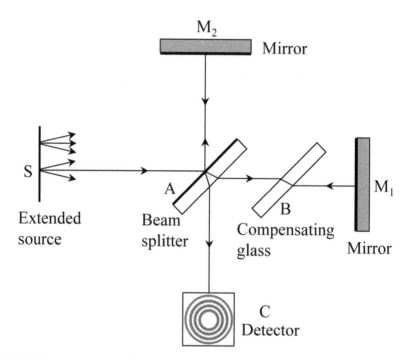

FIGURE 4.5   A schematic of the Michelson interferometer.

Here, m is an integer. If eq 4.1 is satisfied for the point S, it will be equally satisfied for any point lying on the circle of radius OS. The resulting fringes would thus be circles centered on the normal to M_1 and M_2. The interference pattern obtained by monochromatic light typically consists of a number of alternately dark and bright rings. A particular dark ring corresponds to a fixed value of m. A smaller ring has a larger m value. The central fringe ($\theta = 0$) is dark and has its order given by

$$m_{\text{max}} = 2d / \lambda. \tag{4.2}$$

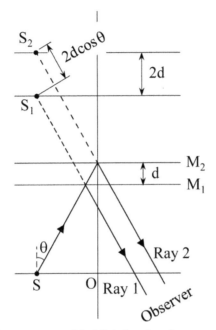

FIGURE 4.6 Equivalent geometry of the Michelson interferometer.

Here m_{max} is generally a very large integer. As d decreases, θ_m corresponding to a particular value of m also decreases. Therefore, the rings shrink toward the center with decreasing d. The fringe of the highest order, that is, the central dark fringe, disappears whenever d decreases by $\lambda/2$. At the same time, the remaining rings are broadened as more and more fringes vanish at the center. Rearrangement of eq 4.1 gives

$$\Delta\theta = \frac{\lambda\Delta m}{2d\sin\theta}. \tag{4.3}$$

The angular separation of two adjacent fringes (i.e., $\Delta m = 1$) is inversely proportional to d. This means that the fringes are more widely spaced as d gets smaller. For $d = \lambda/2$, we have $\cos\theta = m$ from eq 4.1. Since m is an integer, the allowed values of θ are 0° and 90°. So we will see only one central dark fringe. If the entire field of view is very wide, a bright ring may be observed at $\theta = 60°$ (note that constructive interference occurs under the condition of $\cos\theta = 1/2$). When $d = 0$, the central dark fringe spreads out, filling the entire field of view.

The Michelson interferometer configuration is used in a number of applications. Equation 4.2 shows that if d decreases by $\lambda/2$, the original central fringe disappears and its position is occupied by another fringe that was previously located nearest to the center. Thus, for a mirror translation of Δd, the number Δm of disappearing (or emerging) fringes is given by

$$\Delta m = \frac{2\Delta d}{\lambda}. \qquad (4.4)$$

This relation provides an experimental means of determining the unknown wavelength of light when Δd is known or making accurate length measurements using monochromatic light with a specific wavelength. In either case, we have only to count the number of changing fringes. A primary use of the Michelson interferometer is the measurement of the index of refraction of a gas. An evacuate cell is placed in one of the beam paths and the gas whose refractive index is to be determined is flowed into the cell. This is equivalent to changing the optical path length. If the length of the cell is L, the change (Δd) in optical path length is given by

$$\Delta d = nL - L = \frac{\lambda \Delta m}{2}. \qquad (4.5)$$

From this relation, the refractive index n can be determined.

When a light source containing a number of wavelength components, such as white light, is used, each component generates a fringe pattern of its own. Since the coherent length of white light is very short, the optical path difference should be made very small, nearly zero. Otherwise we would not see any fringes. In contrast, laser light has a fairly long coherent length and such a path difference of tens of centimeters has little influence. If the mirrors of the interferometer are tilted with respect to each other (i.e., if they are not mutually perpendicular), the interference fringes generally take the shape of conic sections. Various types of fringes (circular, straight, elliptical, parabolic, and hyperbolic) can appear depending on the orientation of the mirrors.

4.3.2 THE MACH–ZEHNDER INTERFEROMETER

Another amplitude-splitting interferometer is the Mach–Zehnder interferometer. It consists of two beam splitters and two mirrors, as shown in Figure 4.7. The incident beam, usually a collimated beam, is divided into two beams by a beam splitter. One beam is called the reference beam and

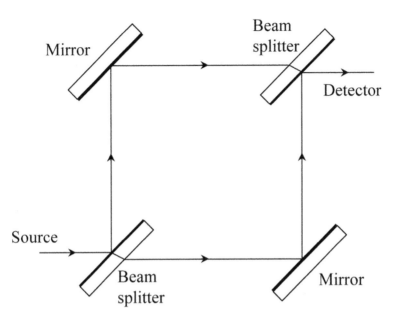

FIGURE 4.7 The Mach–Zehnder interferometer.

another, the object beam. The two beams are totally reflected by mirrors and then made coincident by another beam splitter before entering the detector. The Mach–Zehnder interferometer is a device employed to determine the relative phase shift variations between two collimated beams derived by splitting light from a single source. In the configuration of Figure 4.7, the reference and object beams travel an equal optical path length and produce a fringe of constructive interference. A collimated source results in a nonlocalized fringe pattern. Localized fringes are obtained when an extended source is used. An object placed in the path of one beam (i.e., the object beam) will give rise to an optical path length difference, thereby altering the fringe pattern. The Mach–Zehnder interferometer is an easily configurable instrument. In contrast to the Michelson interferometer, each of the two separated beams is traversed only once. It also has large and freely accessible working space. This makes the Mach–Zehnder interferometer suitable for visualizing gas flow in wind channels. For example, the geometry of air flow around an object in a windowed test chamber can be revealed if one beam is made to pass through this test chamber and an identical chamber is placed in the path of another beam. The beam within the test chamber will propagate through regions that have a spatially varying pressure and refractive index. The resulting distortion in the wavefront changes the fringe pattern. Thus, the Mach–Zehnder interferometer is frequently used in the fields of aerodynamics, plasma physics, and heat transfer to measure pressure, density, and temperature changes in gases. Mach–Zehnder-type interferometers are also widely used in electro-optic modulators, which are essential components for fiber-optic communications. Mach–Zehnder modulators can be incorporated in integrated circuits to offer high-bandwidth electro-optic amplitude and phase modulations.

4.4   MULTI-BEAM INTERFERENCE

4.4.1   MULTI-BEAM INTERFERENCE

Thus far, we have been concerned only with interference between two beams. However, there are many circumstances under which a large number of coherent beams are made to interfere. We found in Sections 2.4 and 2.5 that the reflectance of a material of infinite extent is determined by the refractive index of the material and the incident angle of light. This is the case in which the material extends semi-infinitely and thus the incident light undergoes a single external reflection. For a parallel plate, multiple reflections occur at the front and rear surfaces of the plate. When the thickness of the plate is small compared to the coherent length of the light, the waves reflecting off both surfaces interfere with one another. Consequently, the reflectance is influenced by the phase relationship between these multiply reflected waves. Before discussing the problem of multi-beam interference, let us examine the optical path length difference between two successive rays reflected from a parallel plate (Fig. 4.8a). Here the plate has a refractive index n_f and thickness d, and the surrounding medium has a refractive index of n. The optical path length difference for the two rays is given by

$$\Lambda = n_f (\overline{AB} + \overline{BC}) - n\overline{AD} \cdot \tag{4.6}$$

Using Snell's law, eq 4.6 can be rewritten as

$$\Lambda = 2n_f d \cos\theta_t. \tag{4.7}$$

The corresponding phase difference is then given by

$$\delta = \frac{4\pi}{\lambda_o} n_f d \cos\theta_t. \tag{4.8}$$

Here λ_o is the vacuum wavelength. Although the rays 1 and 2 of Figure 4.8a are spatially apart from each other, interference is determined by their phase relationship. Figure 4.8b shows two rays 2 and 3, both of which undergo a single internal reflection. These two rays maintain the same phase all the way along their paths because the angle of incidence is identical to that of reflection. When the incident beam is fairly wide compared to the plate thickness, actual interference occurs between the ray 1 of Figure 4.8a and the ray 3 of Figure 4.8b. Since the rays 2 and 3 are always in phase, we can instead consider the phase difference between the rays 1 and 2.

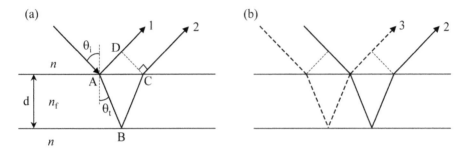

FIGURE 4.8 (a) Two successive rays reflected from a parallel plate. (b) Two rays undergoing a single internal reflection.

Multi-beam interference from a parallel dielectric plate is depicted in Figure 4.9. Suppose that a light beam of amplitude E_o is incident at an angle of θ_i. This primary beam is partially reflected and partially transmitted at the top surface. The transmitted segment is subsequently reflected back and forth between the top and bottom surfaces. Let r and t be the reflection and transmission amplitude coefficients, respectively. The corresponding coefficients for internal reflection are r' and t'. We already know from the Stokes relations of eq 2.51 that $r' = -r$ and $tt' = 1 - r^2$. The amplitudes of the successive reflected rays are rE_o, $tr't'E_o$, $tr'^3t'E_o$, ..., where E_o is the amplitude of the primary ray. Taking the above phase difference between two successive rays into account, the resulting reflected beam is given by

$$E_r = rE_o e^{i\omega t} + tr't'E_o e^{i(\omega t - \delta)} + tr'^3 t'E_o e^{i(\omega t - 2\delta)} + tr'^5 t'E_o e^{i(\omega t - 3\delta)} + \cdots. \qquad (4.9)$$

Figure 4.10 describes the addition of the successive reflected rays on a phasor diagram. Equation 4.9 can be rearranged into

$$E_r = E_o e^{i\omega t}\left\{ r + tr't'e^{-i\delta}\left[1 + r'^2 e^{-i\delta} + r'^4 e^{-i2\delta} + \cdots \right] \right\}. \qquad (4.10)$$

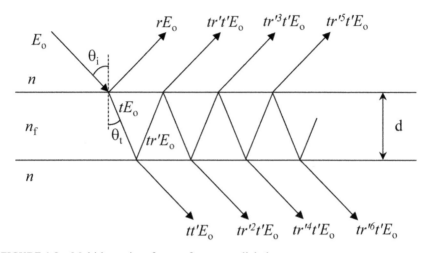

FIGURE 4.9   Multi-beam interference from a parallel plate.

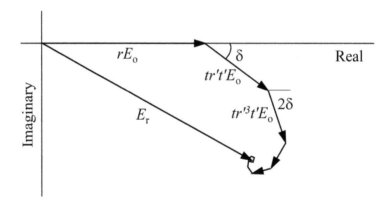

FIGURE 4.10   Addition of multiple reflected rays on a phasor diagram.

If the magnitude of $r'^2 e^{-i\delta}$ is less than 1, the series in the square brackets converges. Thus, we obtain

$$E_r = E_o e^{i\omega t}\left[r + \frac{tr't'e^{-i\delta}}{1-r'^2 e^{-i\delta}} \right].$$ (4.11)

Making use of the relations of $r' = -r$ and $tt' = 1 - r^2$, eq 4.11 becomes

$$E_r = E_o e^{i\omega t}\left[\frac{r(1-e^{-i\delta})}{1-r^2 e^{-i\delta}} \right].$$ (4.12)

The intensity I_r of the reflected beam is proportional to the square of the amplitude of E_r, which can be easily obtained by multiplying E_r by its complex conjugate. Then, the resulting intensity is given by

$$I_r = I_o \frac{2r^2(1-\cos\delta)}{(1+r^4)-2r^2\cos\delta}. \tag{4.13}$$

Here I_o is the intensity of the incident beam. A similar manipulation leads to the transmitted beam given by

$$E_t = E_o e^{i\omega t}\left[\frac{tt'}{1-r^2e^{-i\delta}}\right]. \tag{4.14}$$

The resulting transmitted intensity is then

$$I_t = I_o \frac{(1-r^2)^2}{(1+r^4)-2r^2\cos\delta}. \tag{4.15}$$

Indeed, we have $I_o = I_r + I_t$ satisfying the conservation of energy. The reflectance and transmittance, given by $R = I_r/I_o$ and $T = I_t/I_o$, respectively, are plotted in Figure 4.11 as a function of δ for some values of r^2. It is interesting to find that whenever δ is an integral multiple of 2π, R becomes zero regardless of the value of r. That is, the incident light is completely transmitted when $\delta = 2\pi m$, where m is an integer. It follows from eq 4.13 that I_r is zero when $\cos\delta = 1$, that is, $\delta = 2\pi m$. Meanwhile, $T = 0$ requires $r^2 = 1$. However, $|r| = 1$ is impractical in multi-beam interference. This value is achieved only when the surrounding medium has a higher refractive index than the plate and the incident angle is also larger than the critical angle. If this is the case, the primary beam will be totally internally reflected from the top surface of the plate.

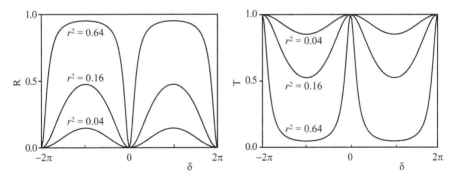

FIGURE 4.11 Reflectance and transmittance as a function of δ for some values of r^2.

It follows from eq 4.15 that a minimum in T exists when $\cos \delta = -1$, that is, when $\delta = (2m+1)\pi$. The minimum transmittance is

$$T_{\min} = \frac{(1-r^2)^2}{(1+r^2)^2}. \tag{4.15}$$

This values never becomes zero unless r^2 goes to unity, while $R_{\min} = 0$. How we obtain this result may be better understood with the below-mentioned two special cases of multi-beam interference. For simplicity, we assume that in Figure 4.9, the plate has a higher index than the surrounding medium, that is, $n_f > n$. It is to be noted, however, that the discussion to follow is equally applied to the opposite situation of $n_f < n$, as long as the incident light is multiply reflected. As the first case, consider the situation in which the optical path length difference between two successive reflected rays is equal to an integral multiple of λ_o, that is, $\Lambda = m\lambda_o$. From eq 4.8, the corresponding phase difference becomes $\delta = 2\pi m$. The total reflected light given by eq 4.10 is then

$$E_r = E_o e^{i\omega t} \left\{ r - trt'\left[1 + r^2 + r^4 + \cdots \right] \right\}. \tag{4.16}$$

Note that $e^{-i\delta} = 1$ and we replaced r' by $-r$. The phase difference given by eq 4.8, which is assumed to be $2\pi m$ in this case, arises from the path length difference. As discussed in Chapter 2, a phase shift may occur upon reflection. The total phase difference thus has two components: one originating from the optical path difference as described above and the other, from the phase change on reflection. However, the latter component is already embodied in the reflection amplitude coefficients. In the replacement of r' by $-r$, the minus sign indicates a phase shift of π. As long as r^2 is less than 1, the series in eq 4.16 converges, so that we have

$$E_r = \left[r - \frac{trt'}{1-r^2} \right] E_o e^{i\omega t}. \tag{4.17}$$

Since $tt' = 1 - r^2$, the total amplitude of the reflected waves is zero. As illustrated in the phasor diagram of Figure 4.12, the second reflected wave and all subsequent waves are in phase with one another but 180° out of phase with the first reflected wave. Since the reflectance is zero, the first reflected wave is exactly canceled by the sum of all the remaining waves. Although the path length difference between the first and second reflected waves is $m\lambda_o$, they are 180° out of phase with each other. This is because the second wave experiences an internal reflection, whereas the first wave undergoes an external reflection. As revealed by the relation of $r' = -r$, there is a relative phase shift of π between the externally and internally reflected waves. On the contrary, the second and all subsequent waves experience an odd number of internal reflections only. That is why they are all in phase with one another.

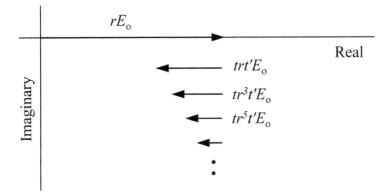

FIGURE 4.12   Phase-diagram representation of the reflected waves when $\Lambda = m\lambda_o$.

Let's consider another situation in which the optical path length difference between two successive reflected rays is $\Lambda = (m + 1/2)\lambda_o$. The phase difference corresponding to this path length difference is $\delta = 2\pi(m + 1/2) = \pi, 3\pi, 5\pi\ldots$ Equation 4.10 is then reduced to

$$E_r = E_o e^{i\omega t}\left\{r + trt'\left[1 - r^2 + r^4 - \cdots\right]\right\} = \frac{2r}{1 + r^2}E_o e^{i\omega t}. \tag{4.18}$$

Note that $e^{-i\delta} = -1$ in this case and we replaced r' by $-r$ again. The addition of the reflected waves is illustrated on a phasor diagram in Figure 4.13. The first and second reflected waves, which have relatively large amplitudes, are in phase. This should result in a large reflectance, as is already manifest from Figure 4.11a. The amplitude ratio of the first and second reflected waves is given by

$$\left|\frac{E_{2r}}{E_{1r}}\right| = 1 - r^2. \tag{4.19}$$

Meanwhile, the corresponding ratio of the second and third reflected waves is

$$\left|\frac{E_{3r}}{E_{2r}}\right| = r^2. \tag{4.20}$$

When r^2 is small, the amplitude ratio of the first and second waves is close to unity and as a consequence, the amplitude of the third reflected wave rapidly decreases. For normal incidence on a glass plate ($n_f = 1.5$) in air ($n = 1$), $r^2 = 0.04$. This means that the amplitude of the second reflected wave is 96% of that of the first reflected wave. Therefore, the overall reflectance is dominantly determined by the total phase difference between these two reflected waves because the amplitudes of the third and subsequent waves are very small. As discussed in Section 3.4, interference occurs between mutually coherent waves. Multi-beam interference is thus possible only when the incoming light has a coherent length larger than the optical path length difference between the reflected waves. For multi-beam interference to occur under white light, the parallel plate shown in Figure 4.9 should be very thin.

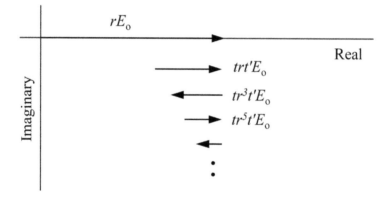

FIGURE 4.13   Phase-diagram representation of the reflected waves when $\Lambda = (m + 1/2)\lambda_o$.

4.4.2   *FABRY–PEROT INTERFEROMETER*

The Fabry–Perot interferometer is an interferometer employing multi-beam interference. It is used to measure the wavelengths of light and the refractive indices of gases with high precision and to study the fine structure of spectral lines. A Fabry–Perot interferometer consists of two parallel, partially reflecting plates or a transparent plate with two reflecting surfaces. In a strict sense, the former is an interferometer and the latter is called an *etalon*. However, the etalon is still an interferometer in a broad sense. Indeed, if the two surfaces of a quartz or glass plate are well polished and lightly silvered, it will serve as an interferometer. Etalons are widely used in telecommunications, lasers, and spectroscopy to control and measure the wavelengths of light and reflect out unwanted wavelengths. A typical arrangement of the

Fabry–Perot interferometer is shown in Figure 4.14. Two thick glass or quartz plates are separated by some distance to enclose an air gap between them. The incident beam is made to be multiply reflected within the enclosed air gap. The inner surfaces of the plates should be optically flat and parallel and are generally coated with a reflective layer of silver or aluminum. The outer surfaces of the plates are often made in a slight wedge shape (very small angle) to prevent the plate itself from producing interference fringes. The plate spacing (i.e., the thickness of the air gap), which can be mechanically varied, is an important parameter of the instrument. When the plates are held fixed by spacers, it is often referred to as an etalon. In a typical system, the interferometer is placed between a collimating lens and a focusing lens. It is illuminated by a broad source of light set at the focal plane of the collimating lens. If the plates were absent, the focusing lens would produce an inverted image of the source; all light emitted from a point on the source is focused to a single point in the image plane. In Figure 4.14, only one ray emitted from point S on the source is traced. As the ray passes through the paired plates, it is multiply reflected within the air gap to produce multiple transmitted rays. These rays are collected by the focusing lens and brought to point P on the screen placed at the focal plane of the converging lens, where they interfere to form either a bright or dark spot. In this system, every set of parallel rays entering the interferometer must arise from the same source point. In other words, there is a one-to-one correspondence between source and screen points. Thus if a broad source of light is used, interference fringes in the form of concentric circles appear on the screen because all rays incident on the interferometer at a given angle θ will result in a single circular ring. A given ring corresponds to a constant value of θ. For this reason, the circular interference fringes are called *fringes of equal inclination*. From eq 4.15, the transmittance of the interferometer is given by

$$T = \frac{(1-r^2)^2}{(1+r^4)-2r^2\cos\delta}.$$ (4.21)

This is known as *Airy function*. This equation can be more simplified to

$$T = \frac{1}{1+F\sin^2(\delta/2)}.$$ (4.22)

The quantity F, called the *coefficient of finesse*, is defined as

$$F = \frac{4r^2}{(1-r^2)^2}.$$ (4.23)

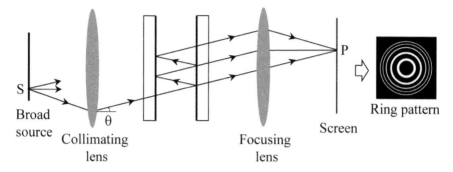

FIGURE 4.14   A typical arrangement of the Fabry–Perot interferometer.

This quantity is a measure of the sharpness of the interference fringes, that is, the fringe contrast. As r varies from 0 to 1, F varies from 0 to infinity. Note that $T_{max} = 1$ regardless of r and that T_{min} is never zero unless r is unity. Equation 4.21 has been derived under the assumption of no absorption. As long as the absorption by metal reflective layers is not zero, the actual transmittance will be somewhat less than that given by eq 4.21. The Airy function is plotted in Figure 4.11 for some values of F. Interferometers with high F show sharper transmission peaks with lower T_{min}, significantly enhancing the fringe contrast.

Suppose that the incident light has two slightly different wavelength components. The interferometer will give a double set of circular fringes, each belonging to one of the components. If a detector is scanned across two closely spaced rings, the intensity distribution will be a superposition of two peaks, as shown in Figure 4.15. If the two peaks are very close or severely overlapping, they may not be resolved into separate peaks. Instead, only the sum of the two peaks, given by a dashed line, will be detected. The intensity maxima occur at specific values of $\delta_{max} = 2\pi m$. The intensity will drop to half its maximum value at $\delta = 2\pi m \pm \delta_r$, where the Airy function of eq 4.22 becomes one-half. This gives

$$\sin\left(\frac{\delta_r}{2}\right) \approx \frac{\delta_r}{2} = \frac{1}{\sqrt{F}}.$$

Since δ_r is generally small, the approximation is valid. In order for the two peaks to be resolved, the phase difference between their intensity maxima should be at least twice the value of δ_r. In other words, the minimum phase difference between the two peaks is

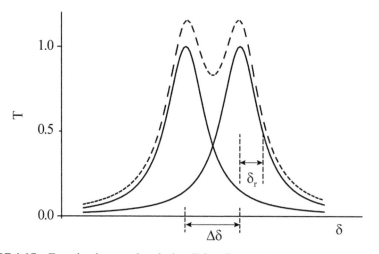

FIGURE 4.15 Two closely spaced peaks in a Fabry–Perot pattern.

$$(\Delta\delta)_{min} = \frac{4}{\sqrt{F}}. \tag{4.24}$$

The corresponding minimum wavelength difference can be found as follows. Based on the arrangement shown in Figure 4.14, the phase difference given by eq 4.8 is now

$$\delta = \frac{4\pi}{\lambda}d\cos\theta. \tag{4.25}$$

Here λ represents the wavelength in a vacuum. For small $\Delta\lambda$, the magnitude of $\Delta\delta$ is

$$\Delta\delta = \left(\frac{4\pi d\cos\theta}{\lambda^2}\right)\Delta\lambda. \tag{4.26}$$

At the intensity maxima, $2d\cos\theta = m\lambda$. Equation 4.26 is then reduced to

$$\Delta\delta = \left(\frac{2m\pi}{\lambda}\right)\Delta\lambda. \tag{4.27}$$

Combining this with eq 4.24 gives

$$(\Delta\lambda)_{min} = \frac{2\lambda}{m\pi\sqrt{F}}. \tag{4.28}$$

The *resolving power* is defined as

$$\mathcal{R} = \frac{\lambda}{(\Delta\lambda)_{min}} = \frac{m\pi\sqrt{F}}{2}. \tag{4.29}$$

Of course, a large resolving power is more desirable. For a given value of F, the resolving power can be made very large if the order of interference, m, is increased. This is simply accomplished by increasing the plate spacing d. However, it diminishes the separation between adjacent orders of interference. Thus, a proper compromise must be chosen.

4.5   INTERFERENCE IN THIN FILM

Interference in thin films is widely used in optics and optical components for control of light. These films are usually deposited on glass or metal substrates by evaporation. Many interesting and practical applications make use of multilayer thin films of different refractive index and thickness. It is possible to evaporate multiple films layer by layer to obtain any desired reflectance and transmittance characteristics. In this section, we only consider the case of normal incidence on a single dielectric film. The theory of multilayer films will be treated in Chapter 8. Suppose a thin dielectric film of refractive index n_1 and thickness d between two semi-infinite media of indices n_o and n_s, as shown in Figure 4.16. We assume that the film is sufficiently thin, so that the path difference between multiply reflected and transmitted waves remains small compared to the coherent length of incident light. Since the light is normally incident, both of the E- and H-fields are tangential to the interface. Here the E-field is chosen in a direction parallel to the plane of incidence, with the H-field perpendicular to it. When dealing with multi-beam interference in Section 4.4.1, we derived the amplitudes of all the individual reflected or transmitted waves and then added them to find the resultant reflectance or transmittance. This is possible because, in that case, the film was free-standing in a single surrounding medium and the reflection amplitude coefficient for internal reflection could be related to that for external reflection, that is, $r' = -r$. Since the film is now embedded between two different media, we develop a more general method of treating thin film interference, in which all the multiply reflected or transmitted waves are considered as already summed in the total electric and magnetic fields and their boundary conditions are applied at various interfaces. Figure 4.16 shows two boundaries denoted as I and II. The E-field amplitudes of the incident and transmitted beams are E_o and E_t, respectively. E_r is the amplitude

of the reflected beam. The incident beam undergoes multiple reflections between two boundaries. Thus, E_r represents the resultant amplitude of all the multiply reflected waves. While E_o, E_t, and E_r represent the amplitudes just outside of the film, E_{tI}, E_{iII}, E_{rII}, and E_{iI} are those present within the film. For instance, E_{tI} represents the sum of all the multiply transmitted waves at boundary I. The boundary conditions require that the tangential components of both the E- and H-fields be continuous at each interface (i.e., equal on both sides). At the boundary I, we have

$$E_{\mathrm{I}} = E_o + E_r = E_{tI} + E_{iI}. \tag{4.30}$$

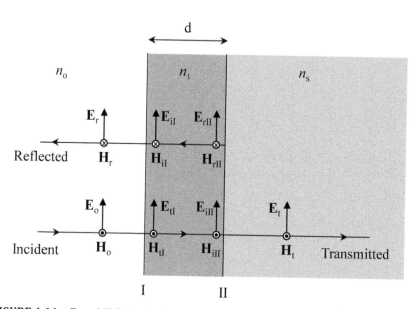

FIGURE 4.16 E- and H-fields for the case of normal incidence on a thin dielectric film.

The corresponding equation for the magnetic field is

$$H_{\mathrm{I}} = H_o - H_r = H_{tI} - H_{iI}. \tag{4.31}$$

At boundary II, the relations are

$$E_{\mathrm{II}} = E_{iII} + E_{rII} = E_t \tag{4.32}$$

and

$$H_{\mathrm{II}} = H_{iII} - H_{rII} = H_t. \tag{4.33}$$

Equations 4.31 and 4.33 can be alternatively expressed as

$$H_I = n_o \sqrt{\frac{\varepsilon_o}{\mu_o}}(E_o - E_r) = n_1 \sqrt{\frac{\varepsilon_o}{\mu_o}}(E_{tI} - E_{iI}) \qquad (4.34)$$

and

$$H_{II} = n_1 \sqrt{\frac{\varepsilon_o}{\mu_o}}(E_{iII} - E_{rII}) = n_s \sqrt{\frac{\varepsilon_o}{\mu_o}}E_t. \qquad (4.35)$$

The phase difference between E_{iI} and E_{iII} is

$$\delta = k_o n_1 d = \left(\frac{2\pi}{\lambda_o}\right)n_1 d. \qquad (4.36)$$

Thus we have

$$E_{iII} = E_{tI}e^{-i\delta}. \qquad (4.37)$$

Similarly

$$E_{rII} = E_{iI}e^{+i\delta}. \qquad (4.38)$$

The phase factors result from the fact that the wave travels a distance d from one interface to the other. If we introduce new parameters defined as follows:

$$Y_o = n_o \sqrt{\varepsilon_o / \mu_o}$$
$$Y_1 = n_1 \sqrt{\varepsilon_o / \mu_o}$$
$$Y_s = n_s \sqrt{\varepsilon_o / \mu_o}$$

Equations 4.32 and 4.35 can be rewritten as

$$E_{II} = E_{tI}e^{-i\delta} + E_{iI}e^{+i\delta} \qquad (4.39)$$

and

$$H_{II} = Y_1(E_{tI}e^{-i\delta} - E_{iI}e^{+i\delta}). \qquad (4.40)$$

These two equations can be solved for E_{tI} and E_{iI}, which can be substituted into eq 4.30 and eq 4.34 to yield

$$E_I = E_{II}\cos\delta + i(H_{II}/Y_1)\sin\delta \qquad (4.41)$$

and

$$H_I = iY_1 E_{II}\sin\delta + H_{II}\cos\delta. \qquad (4.42)$$

In matrix form, these relations are expressed as

$$\begin{bmatrix} E_I \\ H_I \end{bmatrix} = \begin{bmatrix} \cos\delta & i\sin\delta/Y_1 \\ iY_1\sin\delta & \cos\delta \end{bmatrix} \begin{bmatrix} E_{II} \\ H_{II} \end{bmatrix}. \tag{4.43}$$

The 2×2 matrix is called the *transfer or characteristic matrix* of the film. It relates the fields at one boundary to those at the other. Equation 4.43 can be reformatted into

$$\begin{bmatrix} E_o + E_r \\ Y_o(E_o - E_r) \end{bmatrix} = \begin{bmatrix} \cos\delta & i\sin\delta/Y_1 \\ iY_1\sin\delta & \cos\delta \end{bmatrix} \begin{bmatrix} E_t \\ Y_s E_t \end{bmatrix}. \tag{4.44}$$

By expanding the matrices, we can obtain the reflection and transmission amplitude coefficients as follows:

$$r = \frac{E_r}{E_o} = \frac{n_1(n_o - n_s)\cos\delta + i(n_o n_s - n_1^2)\sin\delta}{n_1(n_o + n_s)\cos\delta + i(n_o n_s + n_1^2)\sin\delta} \tag{4.45}$$

$$t = \frac{E_t}{E_o} = \frac{2n_o n_1}{n_1(n_o + n_s)\cos\delta + i(n_o n_s + n_1^2)\sin\delta} \tag{4.46}$$

Then, the reflectance and transmittance are

$$R = \left|\frac{E_r}{E_o}\right|^2 = \frac{n_1^2(n_o - n_s)^2\cos^2\delta + (n_o n_s - n_1^2)^2\sin^2\delta}{n_1^2(n_o + n_s)^2\cos^2\delta + (n_o n_s + n_1^2)^2\sin^2\delta} \tag{4.47}$$

$$T = \frac{n_s}{n_o}\left|\frac{E_t}{E_o}\right|^2 = \frac{4n_o n_s n_1^2}{n_1^2(n_o + n_s)^2\cos^2\delta + (n_o n_s + n_1^2)^2\sin^2\delta}. \tag{4.48}$$

As a specific example, let the incident and transmitting media be air ($n_o = 1.0$) and glass ($n_s = 1.52$), respectively. The reflectance for this system is plotted in Figure 4.17. Depending on the refractive index of the film, the reflectance is enhanced (when $n_1 > n_s$) or reduced (when $n_1 < n_s$) from that for uncoated glass. It is interesting to find that when δ is a multiple of π, the reflectance remains identical to the value ($R = 4.3\%$) of bare glass. If the optical thickness of the film is a quarter of the wavelength of incident light (i.e., $d = \lambda_o/4n_1$), we have $\cos\delta = 0$ and $\sin\delta = 1$. Equation 4.47 is then reduced to

$$R = \frac{(n_s - n_1^2)^2}{(n_s + n_1^2)^2}. \tag{4.49}$$

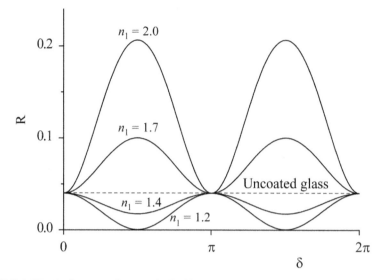

FIGURE 4.17 Reflectance from a single film of index n_1 versus phase difference in the system where the incident and transmitting media are air ($n_o = 1.0$) and glass ($n_s = 1.52$), respectively.

The reflectance is zero if

$$n_1 = \sqrt{n_s}. \tag{4.50}$$

However, few films have a refractive index of $n_1 = 1.23$. MgF_2, whose index is 1.38, is commonly used for coating glasses, although it does not exactly satisfy the requirement of eq 4.50. In general, the film thickness d is chosen so that the above quarter-wavelength condition is satisfied in the yellow-green region (e.g., $\lambda_o = 550$ nm) of the visible spectrum, where our eyes are most sensitive. Figure 4.18 plots the reflectance obtained when a 100 nm-MgF_2 film is coated onto glass. The reflectance is minimized at 550 nm, since the film is a quarter-wave layer at this wavelength. As shown, a single layer of MgF_2 can reduce the reflectance of glass from about 4% to less than 1.5%, over the visible spectrum. This enables a considerable saving of light energy in an optical system that has many reflecting surfaces.

Since the phase difference given by eq 4.36 is proportional to the film thickness, the ripples (maxima and minima in R) shown in Figure 4.17 appear more frequently as the film gets thicker. When a thin film of known refractive index is coated, its thickness can be determined from the wavelengths where the ripples appear. This is usually carried out with a measured transmission spectrum. As an example, Figure 4.19 shows the transmission spectrum of a

400 nm-ZrO$_2$ film (n_1 = 2.10) deposited on glass. The thickness of the film can be calculated from the following equation:

$$d = \frac{m\lambda_m}{4n_1}. \qquad (4.51)$$

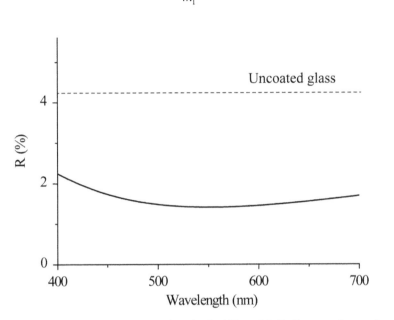

FIGURE 4.18   Reflectance versus wavelength of a 100 nm-MgF$_2$ film coated onto glass.

Here m is an integer and λ_m is the wavelengths where the transmission maxima and minima are observed. When $n_1 > n_s$, the transmission maxima and minima correspond to even and odd m values, respectively. Assume that the film thickness is unknown and to be determined. Since n_1 and d are constant, $m\lambda_m$ should also be a constant. If a particular value of m can be assigned to each λ_m so that their product $m\lambda_m$ remains constant, we can obtain the thickness. A simple way of checking the thickness uniformity is to measure transmission spectra at different locations and compare the results. As d varies, R at a given wavelength also changes. A real-time monitoring of the thickness during a film deposition is thus possible if the surface reflectance is measured using a monochromatic light source (usually a weak laser beam). Most dielectric materials are dispersive and their refractive indices slightly increase with decreasing wavelength in the visible range. Equation 4.51 can also be used to derive the dispersion behavior of a thin dielectric film. Since both the thickness and the refractive index cannot be simultaneously determined from the transmission spectrum alone, the derivation of the dispersion relation requires that the film thickness be already known by other methods such as scanning or transmission electron microscopy. In this case, $m\lambda_m/n_1$ instead of $m\lambda_m$ is constant. Note that m is an integer in any case and that n_1 is a weak function of the wavelength. If a suitable set of m values can be fit into the ripple structure, we can obtain n_1 at various values of λ_m, that is, at various wavelengths.

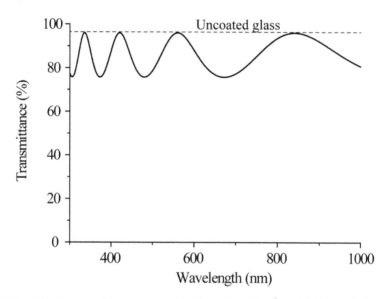

FIGURE 4.19   Transmission spectrum of a 400 nm-ZrO$_2$ film ($n_1 = 2.10$) deposited on glass.

PROBLEMS

4.1 A 1D interference pattern is generated if two plane waves of the same wavelength intersect at a non-zero angle, as shown in Figure 4.20. The fringe pattern produced by these waves is a series of straight lines and the fringes are observed wherever the two waves spatially overlap. This type of interference can be more easily understood in reciprocal space. When two nonparallel beams of wave vectors $\mathbf{k}_1$ and $\mathbf{k}_2$ are superimposed, an interference pattern is generated with a grating vector $\mathbf{g}$. The grating vector $\mathbf{g}$ is defined by the difference of the incident wave vectors, that is, $\mathbf{g} = \mathbf{k}_2 - \mathbf{k}_1$. This vector is perpendicular to the fringes and has a magnitude of g = $2\pi/d$, where d is the fringe spacing.

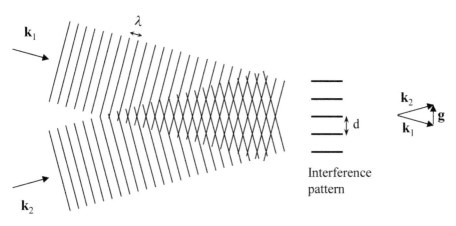

FIGURE 4.20   Interference of two plane waves.

(a) Show that when the intersecting angle of the two waves is θ, the fringe spacing d is given by

$$d = \frac{\lambda}{2\sin(\theta/2)}$$

(b) Two-beam interference pattern can be easily generated by using a biprism. Using the geometry shown in Figure 4.21a calculate the period "d" of the interference pattern. The prism has a refractive index of 1.5.

(c) Interference pattern can be recorded inside a photosensitive material. Suppose that a photosensitive film is coated on a glass substrate and that two interfering beams are incident from the backside of the substrate, as shown in Figure 4.21b. The beams will be refracted twice: one at the air–glass interface and another at the glass–film interface. Nevertheless, the interference pattern formed in the film has the same period as in air. Explain why?

4.2 A thin liquid film ($n = 1.3$) spread on a flat glass substrate ($n = 1.5$) was normally illuminated with white light to show a color pattern in reflection. This is probably due to local fluctuations in the film thickness. If a region of the film reflected green light (550 nm) strongly, how thick will this region be?

4.3 Consider a 100 nm-thick dielectric film ($n = 1.5$) free standing in air.

(a) When monochromatic light of $\lambda_o = 600$ nm is normally incident onto the film, calculate the reflectance and transmittance of the light.

(a)   (b)

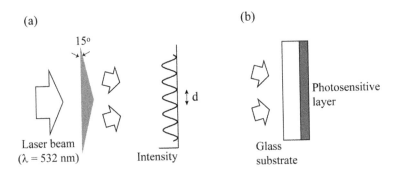

FIGURE 4.21   Interference using a prism

(b) If TE-polarized light of the same wavelength is incident at an angle of 45°, how will the reflectance and transmittance change?

4.4 A dielectric film of high refractive index was deposited on a glass substrate. When the film was very thin, it appeared weakly colored. However, the color disappeared as the film thickness increased. How can this effect be explained?

4.5 The following transmission curve (Fig. 4.22) was obtained from a thin film of unknown index, which was deposited on glass to a thickness of 200 nm. Transmission maxima were observed at 333 and 500 nm, while minima occurred at 400 and 667 nm. Derive the refractive index of the film.

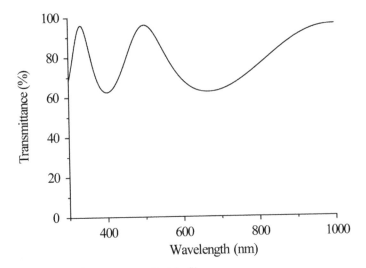

FIGURE 4.22   Transmittance curve of a thin film

REFERENCES

1. Hecht, E. *Optics,* 5th ed; Pearson: Boston, 2016.
2. Pedrotti, F.; Pedrotti, L. M.; Pedrotti, L. S. *Introduction to Optics,* 3rd ed; Addison-Wesley: Boston, 2006.
3. Fowles, G. *Introduction to Modern Optics,* 2nd ed; Dover: New York, 1989.
4. Guenther, B. *Modern Optics,* 2nd ed; Oxford University Press: New York, 2015.
5. Born, M.; Wolf, E. *Principles of Optics,* 7th ed.; Cambridge University Press: Cambridge, UK, 1999.
6. Steel, W. *Interferometry*; Cambridge University Press: New York, 1986.

CHAPTER 5

Diffraction

5.1 HUYGENS' PRINCIPLE

The Dutch physicist Christiaan Huygens showed an important insight into the nature of a propagating light wave, which is nowadays called *Huygens' principle*. This principle states that *every point on a primary wavefront serves as a source of secondary spherical waves (or wavelets) and the new wavefront is the surface tangential to all of these secondary waves.* In 1678, Huygens proposed that every point on the primary wavefront becomes a source of the secondary spherical waves and the sum of these secondary waves determines the form of the wave at any subsequent time. Simple illustrations of the principle are shown in Figure 5.1 for plane and spherical waves. According to Huygens' principle, the propagation of a light wave can be predicted if each point of the wavefront is regarded as the source of a secondary wave spreading out in all directions. Thus, we can suppose that a planar wavefront consists of a lot of imaginary oscillators and each oscillator generates a spherical wave. Since every point on the wavefront serves as the source, the oscillators are considered very closely spaced. Then, the new wavefront formed by these secondary waves will also be planar. A plane wave propagates through a medium in this way. Similarly, a spherical wave produces a spherical wavefront that will act as the source of a subsequently formed wavefront. According to Huygens' principle, there should also be a backward wave propagating toward the source. This backward wave is not observed in reality. He simply assumed that the secondary waves travel only in the forward direction. The atomic scattering of radiant energy can be rather nicely discussed with this principle. When a light wave propagates inside a material, each atom of the material that interacts with an incoming primary wavefront can be regarded as a point source of scattered secondary wavelets. However, things are not quite clear when the principle is applied to the propagation of light through a vacuum. Despite these weaknesses, the imaginary oscillator model proposed by Huygens is very useful for explaining some optical phenomena, including reflection and refraction. As an example, the application of Huygens' principle to the law of refraction is illustrated in Figure 5.2, where a plane wave incident from a medium of higher index (n_2) propagates into a medium of lower index (n_1). The incident primary wave generates secondary wavelets and these wavelets overlap to form a new wavefront in the lower-index medium. Since $n_2 > n_1$, the wavelet has a higher speed in the lower-index medium, that is, $v_1 > v_2$. This means that for the same time t, the wavelet travels a longer distance in the lower-index medium than in the higher-index medium. Thus, the wavefront formed in the lower index medium is more inclined toward the interface normal, making the angle of refraction larger than the angle of incidence. Snell's law of refraction is easily derived from this Huygens' construction. A similar geometry can be constructed to prove the law of reflection.

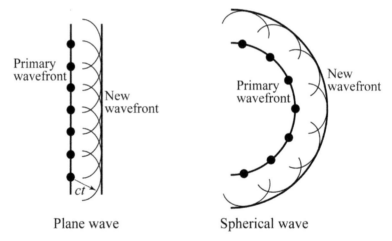

Plane wave   Spherical wave

FIGURE 5.1   Huygens' principle for plane and spherical waves.

Huygens was able to provide a qualitative explanation of the propagation of linear and spherical waves, and also to derive the laws of reflection and refraction. However, he could not explain the deviations from rectilinear propagation that occurs when light encounters apertures, obstacles, and edges, commonly known as *diffraction* effects. According to Huygens' principle, a plane wave rectilinearly propagates at all times. This is not observed in reality. A real wave has a finite width. Consider the plane wave shown in Figure 5.1. The extreme edges of the wave will propagate in different directions than the remainder central region because the new wavefront actually has curved edges. Huygens ignored the edge effect and thought that the new wavefront abruptly ends. As a consequence, Huygens' principle by itself is not able to account for the details of a diffraction process. In addition, any wavelength considerations are not included in the principle. This inadequacy was solved by Fresnel who employed his own concept of interference. He included additional assumptions on the phase and amplitude of the secondary waves to obtain agreement with experimental results. The corresponding *Huygens–Fresnel principle* states that *every unobstructed point on a wavefront, at a given instant, acts as a source of secondary spherical wavelets with the same frequency as that of the primary wave. The optical field at any point beyond the wavefront is a superposition of all these wavelets, taking into account their amplitudes and relative phases.* The Huygens–Fresnel principle provides a good basis for understanding and predicting the propagation of light. Although there are some limitations and different views as to whether it accurately represents reality, this principle is consistent with many experimental observations and can explain diffraction effects fairly well. A common example of diffraction is the spreading out of a wave passing through a small aperture. This is explained by the Huygens–Fresnel principle in Figure 5.3. While there are numerous point sources on the incident wavefront, secondary waves only from some points ahead of the aperture can pass through it and those from the others are blocked. As the aperture gets smaller, the number of unobstructed points decreases. Thus, the incoming wave is more widely spread on passing through a smaller aperture.

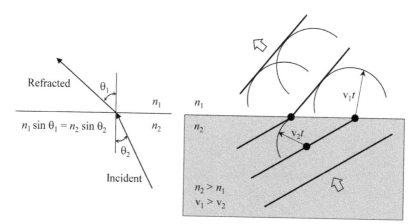

FIGURE 5.2 Application of Huygens' principle to the law of refraction.

If the aperture is a very narrow, long slit, a cylindrical wave will emerge (Fig. 5.4a). Young's double-slit experiment is slightly modified in

Figure 5.4b, where a plane wave falls on two narrow, parallel, and closely spaced slits. Two cylindrical waves coming from the slits are always in phase along certain directions and 180° out of phase along some others. If a screen is placed far away from the slits, alternating bright and dark fringes will appear on the screen. We here see that there is no physical difference between interference and diffraction. When a beam of light is focused by a lens, the beam size at focus is decreased as the focal length of the lens decreases. Since the beam shape is symmetric with respect to the focal point (refer to Fig. 1.7), a more tightly focused beam is more widely spread. Assuming that the areal density of point sources is constant, a small beam containing a limited number of sources will be highly divergent. As the beam size increases, the number of point sources increases as much, thus producing more secondary waves. This makes the wavefront more planar, as depicted in Figure 5.5.

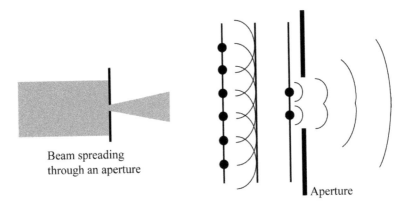

Beam spreading
through an aperture

Aperture

FIGURE 5.3   Diffraction (spreading out) of a light beam by a small aperture, explained by the Huygens–Fresnel principle.

5.2   FRAUNHOFER DIFFRACTION

Diffraction effects are a consequence of the wave nature of light. Any deviation from geometrical optics is referred to as diffraction. This deviation results from the obstruction of a wavefront of light. When an opaque body is placed between a point light source and a screen, the shadow cast by the body deviates from the perfect sharpness predicted by geometrical optics and some light goes over the dark region of the geometrical shadow. If the body is a thin opaque object with a circular hole, the circle of light appearing on the screen also shows smeared edges. In either case, a region of the wavefront is obstructed. Even though the obstacle is not opaque, such effects are observed once the wavefront is locally altered in amplitude or phase. Diffraction also occurs when a light wave travels through a transparent medium that has fluctuations in refractive index. Many optical elements such as lens, mirror, and pinhole make use of only a portion of the whole incident wavefront. Tiny imperfections in a lens produce undesirably distorted images. If all defects in a lens system were eliminated, the ultimate sharpness of an optical image would be limited by diffraction. In this sense, diffraction-limited optics is good optics.

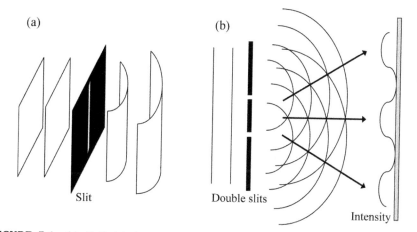

FIGURE 5.4 (a) Cylindrical waves emerging from a long, narrow slit. (b) Young's experiment with two narrow, parallel, and closely spaced slits.

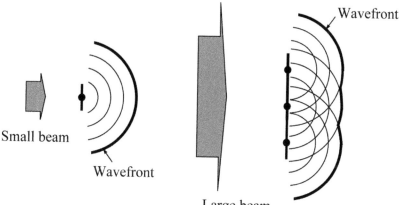

FIGURE 5.5 Beam size versus wavefront shape.

In the detailed treatment of diffraction, it is customary to distinguish between two general cases. Suppose that a small aperture is illuminated by a plane wave coming from a distant point source, as shown in Figure 5.6. Note that when the point source is located far away from the aperture, the incoming wavefront is nearly planar. If the screen of observation is far away from the aperture, the obtained diffraction pattern (i.e., the intensity distribution on the screen) bears little resemblance to the actual aperture. In this far-field region, moving the screen position changes only the size of the pattern without altering its shape. This is known as *Fraunhofer* or *far-field* diffraction. Fraunhofer diffraction occurs when the wavefronts arriving at the aperture and observation screen are effectively planar. This corresponds to the case where the distances from the source to the aperture and from the aperture to the screen are both large enough. In contrast, the diffraction pattern observed near the aperture is governed by *Fresnel* or *near-field* diffraction. In the near-field where Fresnel diffraction prevails, the curvature of the wavefront is significant. Therefore, both of the pattern size and shape vary with distance from the aperture. The fringe effects observed around shadows are examples of Fresnel diffraction. It is to be noted that Fresnel diffraction occurs when either the point source or the observation screen is close to the aperture. If the point source is moved toward the aperture, a spherical wave will impinge on the aperture. A Fresnel pattern then exists, even on a distant screen. Of course, there is no sharp boundary separating the near- and far-field regions. When the point source is far away from the aperture or the incident wave is a plane wave, the far-field approximation is valid if the distance R between the aperture and observation screen is sufficiently large so that

$$R \gg a^2/\lambda. \tag{5.1}$$

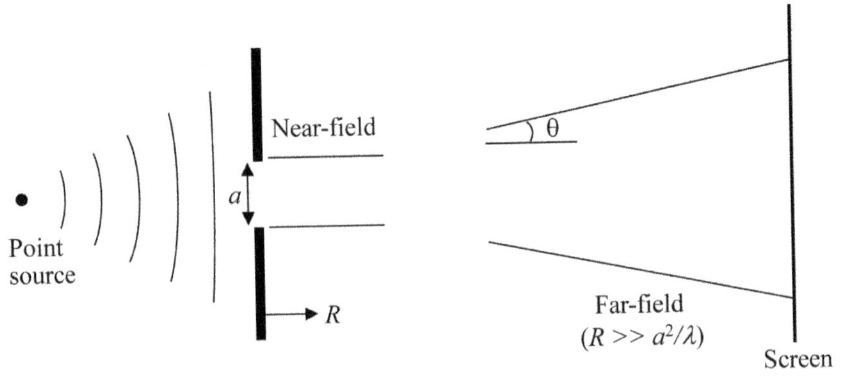

FIGURE 5.6   Far-field (Fraunhofer) diffraction.

Here, a is the largest dimension of the aperture. In terms of optical applications, Fraunhofer diffraction is far more important than Fresnel diffraction. Fresnel diffraction is mathematically more difficult to treat than Fraunhofer diffraction.[1-8] Fresnel diffraction is actually simpler to observe experimentally. However, its mathematical treatment is much more complex and should be handled by many approximation techniques. Our discussion in this book is limited to Fraunhofer diffraction.

5.2.1 DIFFRACTION FROM A SINGLE SLIT

We first consider the Fraunhofer diffraction from a single slit whose height is much larger than its width. Suppose that a slit of width b is uniformly illuminated by a plane wave of wavelength λ, as shown in Figure 5.7. Here, the slit extends from $y = -b/2$ to $+b/2$ in the width direction. According to the Huygens–Fresnel principle, there are numerous coherent point sources of spherical waves within the unobstructed width b of the slit. In the forward direction ($\theta = 0$), the waves produced from the individual sources would all be in phase, constituting a forward wave along the x-direction. If the total number of sources is N and each source produces a spherical wave of amplitude E_o, a small segment dy contains $dy(N/b)$ sources and the strength (amplitude) of that segment will be $dy(N/b)E_o$. N is a very large quantity. As N approaches infinity, the strength of the individual source, E_o, should diminish to almost zero, if the total output remains finite. Then we can define a constant $E_L = NE_o/b$ as the source strength (or amplitude) per unit length. Let us derive the optical field at a point P on the screen, which is a consequence of the diffraction occurring at an angle of θ. The total field at P can be viewed as the sum of all waves emanating from all point sources in the slit. Since the screen is far away from the slit, the waves arriving at P (i.e., light rays from different point sources to P) are almost parallel. Alternatively, we can use a lens to focus the diffracted parallel rays to P. The field at P resulting from the differential segment dy is

$$dE = \frac{E_L}{r} e^{i(kr - \omega t)} dy. \tag{5.2}$$

Here r is the distance from the segment to P. It appears because the amplitude of a spherical wave scales with $1/r$. Both of the phase and amplitude depend on r, which is ultimately a function of y. Note that the phase is much more sensitive to $r(y)$ than is the amplitude. Then eq 5.2 can be rewritten as

$$dE = \frac{E_L}{r_o} e^{i(kr_o - ky\sin\theta - \omega t)} dy. \qquad (5.3)$$

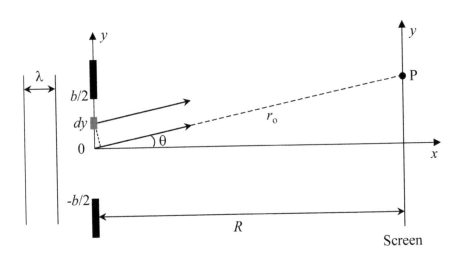

FIGURE 5.7   Fraunhofer diffraction by a single slit.

Here r_o is the distance from the center of the slit to P $(r \cong r_o)$. Since dy is sufficiently small, the phase difference between the sources within dy can be neglected. Integration over the whole width of the slit leads to

$$E = \frac{E_L}{r_o} e^{i(kr_o - \omega t)} \int_{-b/2}^{b/2} e^{-iky\sin\theta} dy = \frac{E_L b}{r_o} \frac{\sin\left[(kb/2)\sin\theta\right]}{(kb/2)\sin\theta} e^{i(kr_o - \omega t)}. \qquad (5.4)$$

By introducing a new term defined as

$$\beta = (kb/2)\sin\theta, \qquad (5.5)$$

we can get a more simplified expression of the field.

$$E = \frac{E_L b}{r_o}\left(\frac{\sin\beta}{\beta}\right) e^{i(kr_o - \omega t)} \qquad (5.6)$$

The amplitude has the form of a sinc function: sinc $(\beta) = \sin(\beta)/\beta$. The intensity is proportional to the square of the amplitude. Thus

$$I(\theta) = I(0)\,\text{sinc}^2(\beta). \tag{5.7}$$

We here assumed that the slit is very long compared to its width. Therefore, the expression given by eq 5.7 holds at any vertical position, that is, in any plane perpendicular to the length direction of the slit. When $\theta = 0$, sinc $(\beta) = 1$. $I(0)$ represents the intensity at $\theta = 0$, which corresponds to the principal maximum. The intensity drops to zero when sin $\beta = 0$, that is, when $\beta = (kb/2)\sin\theta = \pm\pi, \pm 2\pi, \pm 3$, and so on. Since θ is a small angle in the far-field approximation where R is very large, the positions (y values) of zero-intensity on the screen can be obtained from the relation of sin $\theta \cong \theta \cong y/R$. The intensity variation on the screen is schematically depicted in Figure 5.8. The diffraction pattern consists of bright and dark regions, which is certainly a consequence of the interference of waves from the slit. The width (D) of the central bright peak is approximately $D = 4\pi R/kb = 2\lambda R/b$. In Fraunhofer diffraction, the central bright region is wider than the width of the aperture. As the slit width b decreases, D increases. In other words, a wave is more widely spread when it is diffracted by a narrower slit. By the same token, diffraction becomes more significant at longer wavelengths. For fixed λ and b, D is proportional to R. The size of the diffraction pattern thus depends on the position of the observation screen, as shown in Figure 5.9. However, its shape remains almost unaltered.

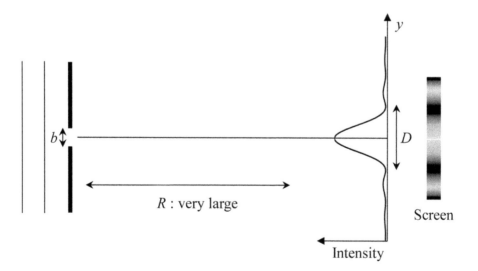

FIGURE 5.8   Fraunhofer (far-field) diffraction pattern from a single slit.

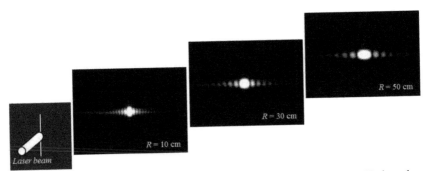

FIGURE 5.9   Diffraction patterns obtained at different positions when a He-Ne laser beam at 633 nm illuminates a slit of 25 μm width.

5.2.2   DIFFRACTION FROM MANY SLITS

Before discussing diffraction from multiple slits, let us first consider a linear array of N coherent point sources illustrated in Figure 5.10. These point sources of spherical waves are assumed to have the same initial phase angle and we are concerned with measuring the optical field and intensity at some distant point P. If the spatial extent of the array is small compared to the distance to P, the waves arriving at P will be essentially of equal amplitude. In addition, the rays from individual sources to P are also nearly parallel to one another. Then, the total electric field at P may be written in its complex exponential form as follows.

$$E = E_o e^{i\{kr_1 - \omega t\}} + E_o e^{i\{kr_1 + ka\sin\theta - \omega t\}} + \cdots + E_o e^{i\{kr_1 + (N-1)ka\sin\theta - \omega t\}}, \qquad (5.8)$$

where E_o is the amplitude and a is the separation between two adjacent sources. Equation 5.8 can be rewritten as

$$E = E_o e^{i(kr_1 - \omega t)} \left[1 + e^{i\delta} + e^{i2\delta} + \cdots + e^{i(N-1)\delta} \right]. \qquad (5.9)$$

Here $\delta = ka \sin \theta$ is the phase difference between the waves generated from two adjacent sources. The parenthesized geometric series has the value of $(1 - e^{iN\delta})/(1 - e^{i\delta})$, which is equivalently

$$e^{i(N-1)\delta/2} \frac{\sin(N\delta/2)}{\sin(\delta/2)}. \qquad (5.10)$$

The total field then becomes

$$E = E_o \frac{\sin\left(N\delta/2\right)}{\sin\left(\delta/2\right)} e^{i\{kr_1 - \omega t + (N-1)\delta/2\}}. \qquad (5.11)$$

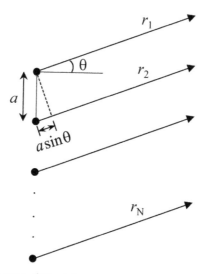

FIGURE 5.10 A linear array of N point sources.

The total intensity is

$$I = E_o^2 \frac{\sin^2{(N\delta/2)}}{\sin^2{(\delta/2)}} = E_o^2 \frac{\sin^2{(N\alpha)}}{\sin^2{(\alpha)}}. \tag{5.12}$$

Here α is defined as $\alpha = \delta/2 = (ka/2)\sin\theta$. The numerator undergoes rapid fluctuations, while the denominator varies rather slowly. This gives rise to a series of sharp peaks at $\alpha = (ka/2)\sin\theta = m\pi$, where m is an integer. Constructive interference occurs in certain directions that satisfy this condition. There are $(N-2)$ subsidiary intensity maxima between consecutive principal maxima. As the number of sources increases, the principal peaks become narrower and stronger, and the number of subsidiary peaks increases (Fig. 5.11).

We now turn to the Fraunhofer diffraction from multiple slits. Consider N parallel, narrow slits, each of width b and center-to-center separation a, as shown in Figure 5.12. Here N is assumed to be an even number and the center of the slits is taken as the origin of the coordinate system. All the slits are close to the origin so that the approximation given by eq 5.3 applies over the entire array. Since the slits are symmetrically located with respect to the origin (i.e., $y = 0$), two symmetric slits can be assigned to a pair. The first pair is those just above and below the origin. Based on eq 5.3, the contribution from the jth pair is

$$E_j = \frac{E_L}{r_o} e^{i(kr_o - \omega t)} \left[\int_{-(2j-1)a/2-b/2}^{-(2j-1)a/2+b/2} e^{-iky\sin\theta} dy + \int_{(2j-1)a/2-b/2}^{(2j-1)a/2+b/2} e^{-iky\sin\theta} dy \right]. \tag{5.13}$$

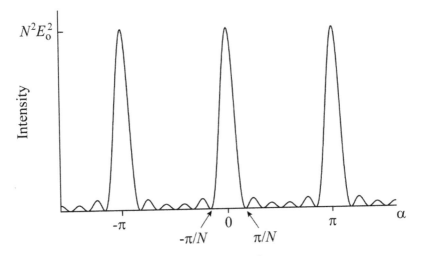

FIGURE 5.11   Dependence of intensity on the value of α.

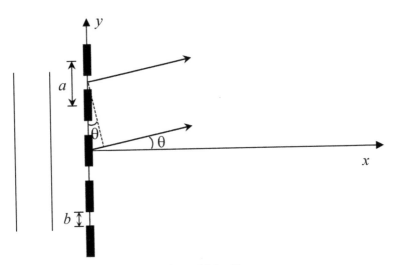

FIGURE 5.12   Fraunhofer diffraction by multiple slits.

Integration leads to

$$E_j = \frac{E_L}{r_o} e^{i(kr_o - \omega t)} \frac{2b\sin\beta}{\beta} \cos\{(2j-1)\alpha\}, \tag{5.14}$$

where $\alpha = (ka/2)\sin\theta$ and $\beta = (kb/2)\sin\theta$. The total field is simply the sum of the contributions from each pair of the slits, as follows.

$$E = \sum_{j=1}^{N/2} E_j = \frac{E_L}{r_o} e^{i(kr_o - \omega t)} \frac{2b \sin \beta}{\beta} \sum_{j=1}^{N/2} \cos\{(2j-1)\alpha\} \quad (5.15)$$

The cosine function in the summation can be expressed as the real part of a complex exponential so that

$$\sum_{j=1}^{N/2} \cos\{(2j-1)\alpha\} = \text{Re}\left[e^{i\alpha} + e^{i3\alpha} + e^{i5\alpha} + \ldots + e^{i(N-1)\alpha} \right]. \quad (5.16)$$

The terms in the square brackets are a geometric series and the sum of the series leads to

$$\text{Re}\left[e^{i\alpha} + e^{i3\alpha} + e^{i5\alpha} + \ldots + e^{i(N-1)\alpha} \right] = \text{Re}\left[\frac{e^{i\alpha} \left(1 - e^{iN\alpha}\right)}{1 - e^{i2\alpha}} \right]$$

$$= \text{Re}\left[\frac{1 - e^{iN\alpha}}{e^{-i\alpha} - e^{i\alpha}} \right] = \frac{\sin N\alpha}{2\sin \alpha}. \quad (5.17)$$

The total field is then

$$E = \frac{bE_L}{r_o} \frac{\sin \beta}{\beta} \frac{\sin N\alpha}{\sin \alpha} e^{i(kr_o - \omega t)}. \quad (5.18)$$

The resulting intensity is

$$I(\theta) = I_o \left(\frac{\sin \beta}{\beta}\right)^2 \left(\frac{\sin N\alpha}{\sin \alpha}\right)^2. \quad (5.19)$$

To normalize the expression, eq 5.19 is alternatively expressed as

$$I(\theta) = \frac{I(0)}{N^2} \left(\frac{\sin \beta}{\beta}\right)^2 \left(\frac{\sin N\alpha}{\sin \alpha}\right)^2. \quad (5.20)$$

This makes $I = I(0)$ at $\theta = 0$. Although eq 5.20 was derived for an even number of slits, the result is also valid for odd N. For an odd number of slits, the same type of expression is obtained if the center of the central slit is taken as the origin. When $N = 1$, eq 5.20 reduces to the result of a single slit given by eq 5.7.

For multiple slits, the diffraction pattern has multiple principal maxima. The general behavior of diffraction angle versus intensity is illustrated in Figure 5.13. When $\alpha = (ka/2) \sin \theta$ is equal to 0, $\pm\pi$, $\pm2\pi$,..., $(\sin N\alpha/\sin \alpha)^2 = N^2$. The principal maxima observed at this condition (i.e., at the corresponding angle) are a consequence of the interference between slits. There are $(N-1)$ minima between consecutive principal maxima. As θ (or sin θ) increases, the overall intensity decreases due to the diffraction envelope factor, (sin β/β)², which describes the Fraunhofer diffraction of each slit. If the slit width were reduced to zero, this envelope factor would vanish and then all the principal maxima would have the same intensity, as is the case of a linear array of coherent point sources. When light is diffracted from a multi-slit system of finite slit width, its diffraction pattern is a result of the interference between slits modulated by the interference characteristics occurring in a single slit. Figure 5.14 shows how the diffraction pattern changes with N for the case of $a = 4b$. As N increases, the principal maxima become narrower. Although there are $(N-2)$ subsidiary maxima between consecutive principal maxima, these subsidiary peaks may be hard to observe in real diffraction patterns. It is apparent, however, that the dark regions between principal maxima are broadened as the number of slits increases. As the width b of each slit decreases, the envelope factor (sin β/β)² becomes more slowly varying with respect to sin θ (also θ). This will make the first few principal peaks have nearly the same intensity, as illustrated in Figure 5.15. It is to be noted that the angular positions of the principal maxima are determined by the inter-slit separation a, being independent of the slit width b. For large N, the principal peaks are bright, distinct, and well spatially separated. The principal diffraction peaks are obtained when

$$a\sin\theta_m = m\lambda. \tag{5.21}$$

This is known as the *diffraction-grating equation*, where $m = 0, \pm1, \pm2,...$ is called the *order* of the diffraction. A diffraction grating diffracts light into several beams traveling in different directions.

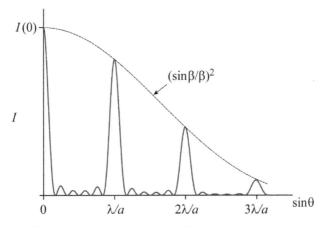

FIGURE 5.13   Diffraction pattern from multiple slits.

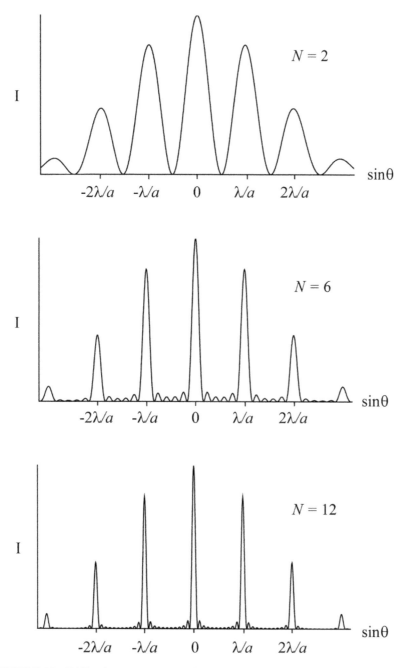

FIGURE 5.14   Diffraction pattern versus the number of slits ($a = 4b$).

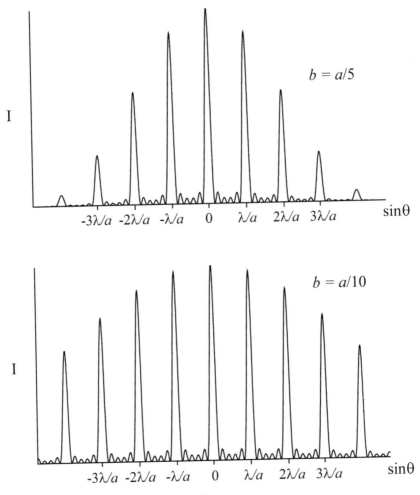

FIGURE 5.15   Effect of slit width b ($N = 6$).

5.2.3   DIFFRACTION FROM RECTANGULAR AND CIRCULAR APERTURES

In Section 5.2.1, we described diffraction from a single slit whose length is much larger than its width. When the length is made small and comparable to the width, each dimension would result in appreciable light spreading, as shown in Figure 5.16. From eq 5.7, a rectangular aperture with dimensions a and b will give the following intensity distribution.

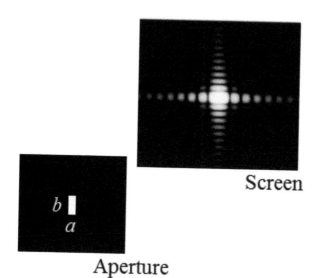

Screen

b

a

Aperture

FIGURE 5.16   Diffraction pattern from a rectangular aperture.

$$I(\theta) = I(0)\left(\frac{\sin \alpha}{\alpha}\right)^2 \left(\frac{\sin \beta}{\beta}\right)^2, \tag{5.22}$$

where $\alpha = (ka/2)\sin \theta$ and $\beta = (kb/2)\sin \theta$. It involves the multiplication of two individual sinc functions. When θ is a small angle, $\sin \theta \cong \theta$. The two-dimensional pattern can then be expressed in term of x and y values on the screen.

$$I(x, y) = I(0)\left[\frac{\sin(\pi ax / R\lambda)}{(\pi ax / R\lambda)}\right]^2 \left[\frac{\sin(\pi by / R\lambda)}{(\pi by / R\lambda)}\right]^2. \tag{5.23}$$

Here R is the distance from the aperture to the screen and $I(0)$, the intensity at the center of the screen.

Fraunhofer diffraction by a circular aperture is of great practical significance in many optical instruments. To calculate the diffraction pattern of a circular aperture, we need to use polar coordinates, as shown in Figure 5.17. If we consider an element of area dA on the aperture, the total field at an observation point P will be

$$E = \frac{E_A}{r_0} \int_{Aperture} e^{i(kr - \omega t)} dA, \tag{5.24}$$

where E_A now represents the amplitude per unit area. The differential element of area is $dA = \rho d\rho d\varphi$. When this element has y and z values on the aperture, and *P* has *Y* and *Z* values on the screen, the distance from the element to *P* is given by

$$r \cong r_0 - (yY + zZ)/r_0.$$ (5.25)

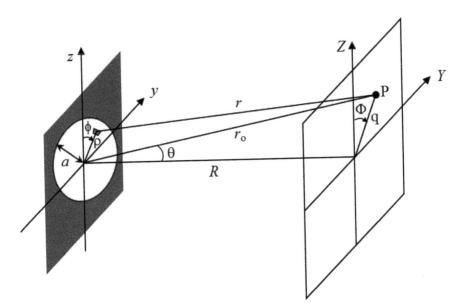

FIGURE 5.17   Fraunhofer diffraction by a circular aperture.

Then, eq 5.24 is rewritten as

$$E = \frac{E_A}{r_0} e^{i(kr_0 - \omega t)} \int_{Aperture} e^{-ik(yY + zZ)/r_0} dA.$$ (5.26)

Using the relations of $y = \rho \sin \phi$, $z = \rho \cos \phi$ and $Y = q \sin \Phi$, $Z = q \cos \Phi$, this equation becomes

$$E = \frac{E_A}{r_0} e^{i(kr_0 - \omega t)} \int_{\rho=0}^{a} \int_{\phi=0}^{2\pi} e^{-i(kq\rho/r_0)\cos(\phi - \Phi)} \rho d\phi d\rho.$$ (5.27)

Here *a* is the radius of the aperture. Since the optical field will be independent of Φ, we can simply set $\Phi = 0$. The inner integral is the *Bessel function* of order zero, denoted as J_0. Equation 5.27 can be rewritten as

$$E = \frac{E_A}{r_0} e^{i(kr_0 - \omega t)} 2\pi \int_{\rho=0}^{a} J_0(kq\rho/r_0) \rho d\rho.$$ (5.28)

The integration of the Bessel function of order zero can be converted into the Bessel function of order one, J_1, so that the intensity at P is expressed by

$$I = I(0)\left[\frac{2J_1\left(kaq/r_o\right)}{kaq/r_o}\right]^2.$$
(5.29)

This is plotted in Figure 5.18. The central bright region is known as the *Airy disk*. Since $\sin\theta = q/r_o$, it can be equivalently written as

$$I(\theta) = I(0)\left[\frac{2J_1\left(ka\sin\theta\right)}{ka\sin\theta}\right]^2.$$
(5.30)

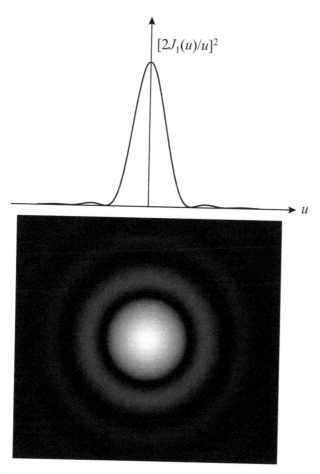

FIGURE 5.18   Airy pattern.

The Airy disk is surrounded by a dark ring whose size is given by the first zero of the Bessel function: $J_1(u) = 0$ at $u = 3.83$ from the mathematical table. Then, the radius of the Airy disk, q_1, is

$$q_1 = 1.22 \frac{\lambda r_0}{2a} \cong 1.22 \frac{\lambda R}{2a}. \tag{5.31}$$

By using a focusing lens, we can bring the observation screen in close proximity to the aperture without altering the pattern shape. If a lens of focal length f is used to focus the diffracted light, the pattern observed in the focal plane is therefore identical to eq 5.29, with the distance r_0 replaced by f. Then the radius of the focused spot is

$$q_1 \cong 1.22 \frac{\lambda f}{D}. \tag{5.32}$$

Here, D is the diameter of the aperture. q_1 is inversely proportional to D.

5.3   DIFFRACTION GRATING

A diffraction grating is an optical element that splits and diffracts light into several beams traveling in different directions. A repetitive array of diffracting elements (apertures, obstacles, refractive index variations, etc.) has the effect of producing periodic alterations in the amplitude, phase, or both of the light. This makes constructive interference occur along certain well-defined directions. The multi-slit system described in Section 5.2.2 is a simple example showing the effect. Suppose that the width of each slit is much smaller than the inter-slit separation a. When a plane wave is incident, the slits become coherent sources as illustrated in Figure 5.19. Wavelets emanating at an angle θ from two adjacent slits have a path length difference of $a \sin \theta$. Obviously, constructive interference occurs at this angle when the path length difference equals an integral multiple of wavelengths, that is,

$$a \sin \theta_m = m\lambda. \tag{5.33}$$

This is the *grating equation* for normal incidence. This equation, also known as the *Bragg diffraction condition*, is valid for any grating whose period is a. The diffraction directions depend on the grating period and the wavelength of the light. The value of m defines the diffraction order; $m = 0$ is zero order, $m = \pm 1$ being first order, and so on. For a finite slit width, the diffraction intensity decreases as the magnitude of m, that is, the diffraction angle increases. If the slit width decreases essentially to zero, all the diffracted beams have the same intensity, as discussed in Section 5.2.2. The wavelength dependence in eq 5.33 shows that the diffraction grating can separate an incident white beam into its constituent wavelength components. Each component is diffracted into a different direction, producing a rainbow of colors. This is visually similar to the operation of a prism, although the underlying mechanism is much different. Note that for a white source, the 0th-order (i.e., $m = 0$) beam corresponds to the undeflected white-light view of the source. Since $\sin \theta \leq 1$, the maximum wavelength of light that a grating can diffract is equal to the grating period. In other words, a diffraction grating works for wavelengths smaller than its period. The smaller the grating period, the fewer is the number of visible diffraction orders.

In the multi-slit system shown in Figure 5.19, the incident wavefront is confronted by alternate opaque and transparent regions, thus undergoing a modulation in amplitude. Accordingly, a multi-slit grating is a kind of *amplitude grating*. More precisely, it is a *transmission amplitude grating* because the incident and diffracted beams are on opposite sides of the grating. For practical applications, a transmission grating is generally made by forming parallel grooves or ridges on the surface of a flat glass plate (Fig. 5.20). Each of the grooves serves as a scattering source of light and they altogether form a regular array of sources. When the grating is highly transparent, there is negligible amplitude modulation. Instead, the phase of the incident wave is modulated in a regular periodic fashion due to the varying optical thickness of the grating. This is known as a transmission *phase grating*. In a phase grating, the wavelets can be envisioned as emanated with different phases over the grating surface. Diffraction of a laser beam by a transmission phase grating is illustrated in Figure 5.21. Note that only the principal diffraction peaks expected from eq 5.33 are clearly visible, while the subsidiary peaks are hardly observable in real diffraction patterns. In reflection phase gratings, the grating faces are made highly reflecting. Periodic reflecting structures are generally formed by etching grooves in metal thin films evaporated onto optically flat glass plates (Fig. 5.22). The interference pattern generated after reflection is quite similar to that resulting from transmission. When the incident beam is not normal to the diffraction beam, the grating equation becomes

$$a\left(\sin \theta_m - \sin \theta_i\right) = m\lambda. \tag{5.34}$$

Here θ_i is the angle of incidence measured from the surface normal and θ_m represents the diffraction angle for the mth mode. This expression holds

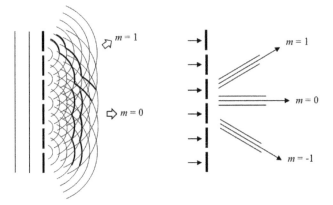

FIGURE 5.19 Split of light into several beams by a diffraction grating (wave optics and ray optics representations).

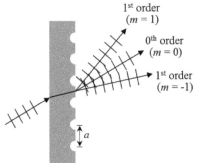

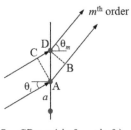

$$AB - CD = a\,(\sin\theta_m - \sin\theta_i)$$

FIGURE 5.20 A transmission grating.

FIGURE 5.21 Diffraction of a laser beam by a transmission phase grating. The grating period is 2 μm and the laser wavelength is 532 nm.

for both transmission and reflection. For normal incidence ($\theta_i = 0$), eq 5.34 reduces to eq 5.33. The diffraction equation presented here assumes that both sides of the grating are in contact with the same medium (e.g., air). The light that corresponds to direct transmission (or specular reflection in the case of a reflection grating) is called the 0th-order beam and is denoted by $m = 0$. Non-zero integers ($m \neq 0$), which can be positive or negative, result in diffracted beams on both sides of the 0th-order beam. Although the spatial intensity distribution of the diffracted light depends on the structure (i.e., shape and depth) of the grating, the diffraction directions are always given by the above diffraction equation and are independent of the detailed structure of the grating elements. This is because once the grating period is fixed, the phase relationship between light scattered from adjacent elements of the grating remains the same.

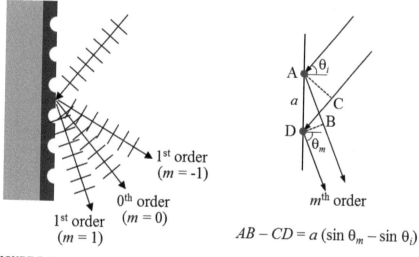

FIGURE 5.22 A reflection grating.

Diffraction gratings are widely used in spectroscopic applications because they enable wavelength-dependent light deflection. In such applications, the undeflected light corresponding to $m = 0$ is clearly undesirable because it simply represents a waste of light energy. Note that the 0th-order beam provides no information about the wavelength of the illuminating light. Information about the wavelength of the illumination can only be obtained by measuring the angular positions of the first- or higher-order diffraction beams. For gratings like those shown in Figures 5.20 and 5.22, a considerable portion of the incident light energy is essentially wasted through direct transmission or specular reflection. An effective way of reducing the energy waste is to blaze the grating. The effect of blazing is explained with a blazed reflection grating in Figure 5.23. The underlying principle is to shift the wasted energy to one of the higher-order beams (usually $m = -1$ or 1) by making specular reflection occur in the direction of this order. Consider a beam of light incident on a groove face at θ_i relative to the normal N to the grating plane. The first-order diffracted beam ($m = -1$) will make an angle of θ_m with N. When the blazing angle is γ, the normal N' to the groove face makes an angle γ with N. Now, the angle between the incident beam and N' is $\theta_i - \gamma$. Then the incident beam undergoes specular reflection at $(\theta_i - \gamma)$ with respect to N'. The first-order diffraction occurs at $(\theta_m + \gamma)$ with respect to N'. If these two angles are identical, the diffracted beam will be very strong because it is diffracted in the direction of specular reflection. This condition is satisfied when

$$\gamma = \frac{\theta_i - \theta_m}{2}. \tag{5.35}$$

Substitution of eq 5.35 into eq 5.34 leads to

$$a\{\sin\theta_i - \sin(\theta_i - 2\gamma)\} = \lambda. \tag{5.36}$$

For a given blazing angle γ, an incident beam that satisfies this equation will be strongly diffracted. As a special case, consider the configuration depicted in Figure 5.24, where the beam is incident along the normal N to the grating plane. Then $\theta_i = 0$ and $\theta_m = -2\gamma$. Note that θ_m is negative in this case. Equation 5.36 reduces to

$$a\sin(2\gamma) = \lambda. \tag{5.37}$$

When this condition is satisfied, one of the diffraction orders (i.e., $m = -1$) corresponds to specular reflection off the individual groove faces. Then, most of the diffracted light is concentrated about $\theta_m = -2\gamma$. This also means that a substantial portion of the incident energy is diffracted along the direction of $\theta_m = -2\gamma$. If the incident beam is white light, the wavelength component that satisfies eq 5.37 will be deflected into the same direction. In the symmetric grating structure shown in Figure 5.22, the 0th-order diffraction has the highest intensity. For normal incidence, a great portion of the incoming energy is thus reflected back into the incident direction. Due to this reason, most commercial diffraction gratings are of the blaze-type.

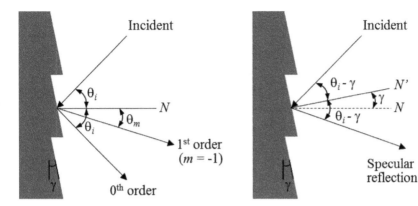

FIGURE 5.23   A blazed reflection grating.

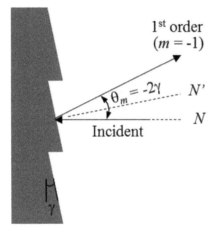

FIGURE 5.24   Strong diffraction by a blazed reflection grating, where the first-order diffraction ($m = -1$) corresponds to specular reflection.

5.4   X-RAY DIFFRACTION

5.4.1   INTRODUCTION

X-ray diffraction is a very powerful, nondestructive tool for analyzing crystalline materials. The diffraction pattern, characterized by the direction and intensity of the diffracted beams, is characteristic of the matter and its internal structure. Therefore, a variety of information about the matter (crystal structure, phase, lattice parameter, etc.) can be deduced from the obtained diffraction pattern. X-rays refer to the electromagnetic radiations that have a wavelength range of 10^{-3}–10 nm (see Fig. 1.14). X-rays deeply penetrate into all substances (even into metals), although the penetration depth depends on the substance. As shown in the previous section, when a light wave encounters a periodic grating structure, it is diffracted into several beams traveling in different directions. The diffraction effect becomes profound when the wavelength of the incident light is comparable to the grating period. Most materials are crystalline, consisting of regularly arranged atoms. X-rays have wavelengths similar to the interatomic distances and a crystalline material contains many atomic planes of different orientation and spacing. Therefore, each atomic plane can act as an efficient diffraction grating when an X-ray beam is incident. X-ray diffraction is a scattering phenomenon in which a large number of atoms are involved. Since the atoms are periodically arranged in a crystalline matter, X-rays scattered by these atoms may be in phase with one another and constructively interfere in some directions. The analysis of materials by X-rays is much facilitated by the fact that all materials have a refractive index very close to one at X-ray wavelengths. Since there is essentially no refraction at the air-material interface, special sample preparation such as polishing is unnecessary. Of course, we do not have to calculate the propagation direction within the material because it is basically the same as the incident direction. X-ray diffraction is also an indispensable technique for nondestructive in situ characterization of thin films. It can provide information on their phase, lattice parameter, thickness, internal stress and strain, and orientation with respect to the underlying substrate. This section describes the basic principle of X-ray diffraction.

5.4.2   BRAGG LAW

In the diffraction grating depicted in Figure 5.20, diffraction occurs by the periodic surface structure where each of the grooves serves as a source of scattered light. In X-ray diffraction, however, all atoms residing in the beam path take part in scattering and diffraction. Suppose that a plane-wave X-ray beam is incident into a crystal with atoms arranged on a set of parallel planes and that the incident beam makes an angle of θ with the crystal plane, as shown in Figure 5.25. Although the X-ray wave is scattered in all directions, we here consider scattering that occurs at an angle of 2θ with respect to the incident beam. The ray *1* scattered by the atom located at point *A* always has the same phase as the ray *2* scattered by the atom at *B* because the length *AD* is equal to the length *CB*. In fact, there is no path length difference between the rays scattered by the atoms of a plane when the angle of scattering equals the angle of incidence with respect to the plane. The scattered rays *1* and *2* in Figure 5.25 are spatially apart from each other. In Figure 5.26, the ray *1S*, scattered from the first plane, spatially overlaps the ray *3S* scattered from the second plane. The diffraction intensity is ultimately decided by the phase relationship between these rays. The rays *3S* and *2S*, both scattered from the second plane, have no path length difference. Therefore, the phase difference between the rays *1S* and *3S* is identical to that between the rays *1S* and *2S*. It is easier to calculate the phase difference with the rays *1S* and *2S*, rather than with the rays *1S* and *3S*. Figure 5.27a depicts how the condition for diffraction is derived. The difference in path length between the rays *1S* and *2S* is $CA + AD = 2d\sin\theta$, where d is the interplanar spacing. When this path length difference is equal to an integral multiple of wavelengths, the two rays are completely in phase and an arbitrary plane *XY* normal to the scattering direction becomes a planar wavefront. The condition for diffraction is then

$$n\lambda = 2d\sin\theta. \tag{5.38}$$

Here n is an integer. This relation was first derived by W. L. Bragg and is known as the *Bragg law* or equation. It states that the incident and diffracted beams are coplanar with the normal to the lattice planes and equally inclined to the plane normal. The angle θ (called the Bragg angle) is related to the X-ray wavelength and to the interplanar spacing. A number of planes are involved in scattering because the beam size used for X-ray diffraction is much larger than the interplanar spacing.

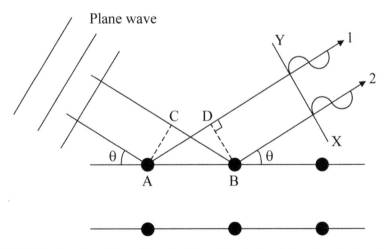

FIGURE 5.25   Rays 1 and 2 are always in phase when the incident and scattered angles are identical.

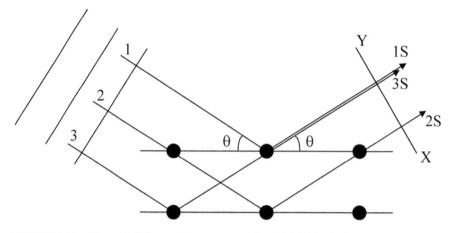

FIGURE 5.26 The path difference between rays 1S and 3S is identical to that between rays 1S and 2S.

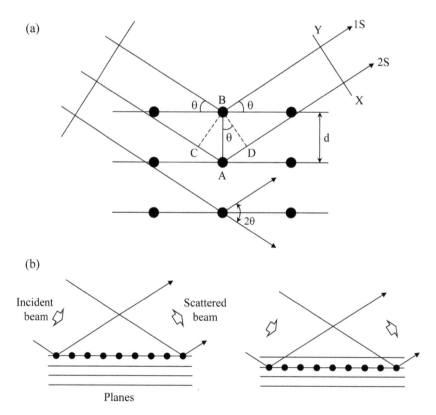

FIGURE 5.27 (a) X-ray diffraction by a crystal. (b) Reflection from successive lattice planes.

An alternative way of interpreting the Bragg equation is that the diffracted beam can be regarded as a series of reflection of the incident beam by a set of lattice planes. Bragg considered first how the X-rays scattered by all the lattice points (or atoms) in a single plane might be in phase. The condition for optical reflection is that the angle of incidence is equal to the angle of reflection (see Fig. 2.3). This ensures that the waves scattered by all points in that plane are in phase with one another, as illustrated in Figure 5.25. In general, the waves reflected from successive lattice planes will not be in phase. Figure 5.27b shows reflection from two adjacent planes. The waves reflected from the upper plane have a shorter path length than those reflected from the lower plane. When the path length difference is equal to an integral multiple of wavelengths, the waves reflected from successive planes are in phase and reinforce one another. In other words, the X-rays scattered by all the atoms in all the planes constructively interfere to form a diffracted beam in the given direction. The waves scattered in other directions are out of phase and cancel out with one another. In both of the optical reflection and Bragg reflection (i.e., X-ray diffraction), the angle of incidence is equal to the angle of reflection. However, two phenomena are fundamentally different. The optical reflection occurs in a very thin surface layer. On the contrary, the diffracted beam from a crystal is built up of waves scattered by all the atoms of the crystal irradiated by the incident X-ray beam. While the reflection of visible light can take place at any angles of incidence, the diffraction of a monochromatic X-ray beam is possible only at specific angles of incidence that satisfy the Bragg law. The integer n of eq 5.38, which is called the order of reflection, can take on any value unless $\sin \theta$ exceeds unity. Therefore, for fixed λ and d, there may be several angles of incidence θ_1, θ_2..., which correspond to $n = 1, 2,...$ For the first-order reflection ($n = 1$), the scattered rays $1S$ and $2S$ of Figure 5.27a would differ in path length by λ. For $n = 2$, their path length difference would be twice, that is, 2λ. Consider the first- and second-order reflections from a set of (001) planes, as illustrated in Figure 5.28. If the angles of incidence for the two reflections are θ_1 and θ_2, respectively, we obtain the following relations.

$$\lambda = 2d_{001} \sin \theta_1$$
$$2\lambda = 2d_{001} \sin \theta_2$$

The second-order reflection can be alternatively expressed as $\lambda = 2(d_{001}/2)$ $\sin \theta_2 = 2d_{002} \sin \theta_2$. This means that the second-order reflection from the (001) planes is equivalent to the first-order reflection from (002) planes. In Figure 5.28, the dotted plane midway between the (001) planes corresponds

to part of the (002) set of planes. An *n*th-order reflection from (*hkl*) planes
may be regarded as a first-order reflection from the (*nh nk nl*) planes with 1/
*n*th the spacing of the former. It is conventional to represent the reflection
from the (*hkl*) planes by *hkl* without parentheses. Although 002, 003, and
004 reflections are equivalent to the second-, third-, and fourth-order reflections from the (001) planes, it is more general to view them as the first-order
reflections from the (002), (003), and (004) planes. This allows us to write
the Bragg law of eq 5.38 simply as

$$\lambda = 2d \sin \theta_B. \tag{5.39}$$

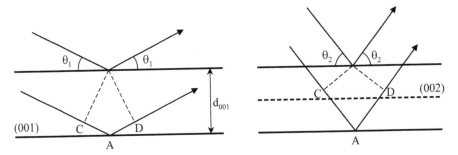

FIGURE 5.28   First-order and second-order reflections from (001) planes. The second-
order 001 reflection is equivalent to the first-order 002 reflection.

Here θ_B is the Bragg angle. This new form of *Bragg law* is more dominantly used in X-ray diffraction. The Bragg law can be experimentally
applied in two different ways. By using a monochromatic X-ray beam of
known wavelength λ and measuring θ, we can determine the spacing d of
various planes. Alternatively, a crystal with planes of known spacing d can
be used to determine the radiation wavelength λ. Most applications of X-ray
diffraction are associated with the measurement of the diffraction angle (2θ
rather than θ is experimentally measured). The essential components of
typical X-ray diffractometer are shown in Figure 5.29. A collimated beam
from the X-ray source is incident onto a sample stationed on the sample
holder, which may be set at any desired angle to the incident beam. A
detector measures the intensity of the diffracted beam; it can be rotated
around the sample and set at any desired angular position. The sample
holder can also be rotated around its center independently of or in conjunction with the detector. The diffractometer measures the angle 2θ between
the incident and detected beams. In the symmetric scan (often called θ–2θ
scan or 2θ–θ scan), the angle θ between the incident beam and the sample

holder is maintained at half the measured diffraction angle 2θ, as depicted in Figure 5.29.

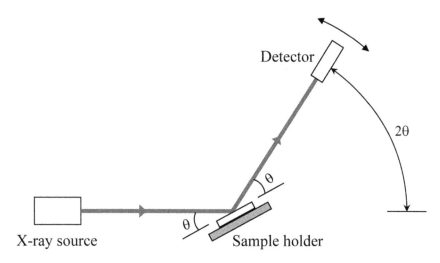

FIGURE 5.29 Typical X-ray diffractometer.

The Bragg law of eq 5.39 is given in a simple scalar form. The Bragg equation for (*hkl*) planes can be reformatted as

$$(2\sin\theta_B)(2\pi/\lambda) = 2\pi/d_{hkl}. \qquad (5.40)$$

Figure 5.30 shows a graphical configuration for the Bragg law in which the incident and diffracted beams are represented by their wave vectors $\mathbf{K}_i$ and $\mathbf{K}_d$, respectively. The magnitude of $(\mathbf{K}_d - \mathbf{K}_i)$ is equal to the left-hand side of eq 5.40. If we introduce a reciprocal lattice vector $\mathbf{H}_{hkl}$, which is normal to the (*hkl*) plane and has a magnitude of $2\pi/d_{hkl}$, the relation between these three vectors is given by

$$\mathbf{K}_d - \mathbf{K}_i = \mathbf{H}_{hkl}. \qquad (5.41)$$

Equation 5.41 is the Bragg law in vector form or in reciprocal space. It states that the difference between incoming and diffracted wave vectors is a reciprocal lattice vector. The concept of *reciprocal lattice* plays an important role in most analytic studies of periodic structures, particularly in the theory of diffraction. There are many different sets of atomic planes in a crystal. Each set is given a reciprocal lattice point. The reciprocal lattice point corresponding to the (*hkl*) plane is labeled with *hkl*. The reciprocal lattice vector

$\mathbf{H}_{hkl}$, which connects the origin 000 of the reciprocal lattice to the point *hkl*, is perpendicular to the (*hkl*) plane and has a magnitude of $2\pi/d_{hkl}$. While $\mathbf{H}_{001}$ and $\mathbf{H}_{002}$ have the same direction, the latter is twice as long as the former. The reciprocal lattice can provide a simple graphical view of the Bragg law, as illustrated in Figure 5.31. The incident beam vector $\mathbf{K}_i$ is first drawn to the origin of the reciprocal lattice. A sphere of radius $2\pi/\lambda$ is then drawn centered on the initial point of the incident beam vector. This sphere is known as the sphere of reflection or *Ewald sphere*. The condition for diffraction is satisfied when any reciprocal lattice point *hkl* falls on the surface of this sphere. The direction of the diffracted beam is given by the vector $\mathbf{K}_d$ drawn from the origin of the sphere to the point *hkl*. This construction is known as the Ewald construction. It is evident that the relationship between $\mathbf{K}_i$, $\mathbf{K}_d$, and $\mathbf{H}_{hkl}$ is that of eq 5.41. When the reflecting (*hkl*) plane and the X-ray wavelength are fixed, the beam should be incident at a specific angle in order for diffraction to occur. This angle is the Bragg angle (θ_B). For other angles of incidence, the $\mathbf{H}_{hkl}$ vector cannot terminate on the surface of the Ewald sphere. A real crystal has a number of atomic planes of different orientation and spacing. There are also various $\mathbf{H}$ vectors corresponding to these planes. When an X-ray beam of fixed wavelength is incident onto the crystal at an arbitrary angle, diffraction occurs by the $\mathbf{H}$ vector that terminates on the surface of the Ewald sphere. The diffraction condition for this case, both in real and reciprocal space, is illustrated in Figure 5.32.

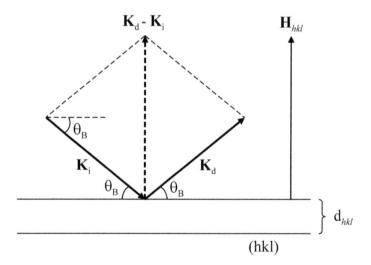

FIGURE 5.30 A graphical configuration for the Bragg law.

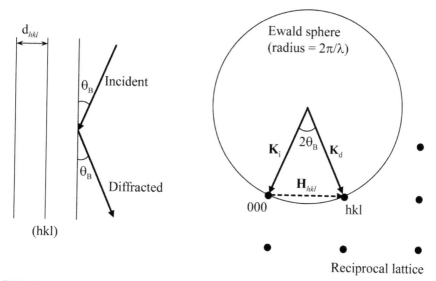

FIGURE 5.31   Ewald construction.

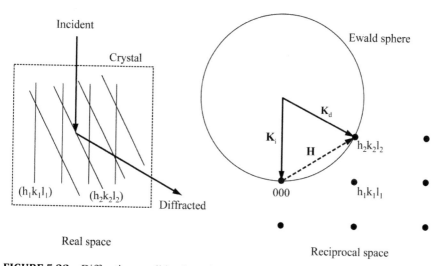

FIGURE 5.32   Diffraction condition in real and reciprocal space.

Although the Bragg law has been derived from the interaction of an X-ray beam and a set of atomic planes, it applies to the diffraction of visible light by a one-dimensional periodic structure (e.g., refractive-index variation) formed inside a material. Of course, the period of the structure should be comparable to or larger than visible wavelengths in this case. As will be

Optics for Materials Scientists

discussed in Chapter 9, the vector relation for two-beam interference is of the same form as eq 5.41. When two plane waves of wave vectors $\mathbf{K}_1$ and $\mathbf{K}_2$ intersect at a non-zero angle, they generate a one-dimensional interference pattern characterized by the difference vector $\mathbf{K} = \mathbf{K}_2 - \mathbf{K}_1$. Just as $\mathbf{H}_{hkl}$ is normal to the reflecting planes, $\Delta\mathbf{K}$ is also perpendicular to the generated interference fringes and has a magnitude of $2/d$, where d is the period of the interference fringes. The difference between two-beam interference and X-ray diffraction is that the former involves the formation of a new grating with two incident beams, while the latter is to generate a secondary beam from an already existing grating (i.e., a set of atomic planes).

5.4.3 INTENSITY OF DIFFRACTED BEAM

The Bragg law states that an X-ray beam of a given wavelength can be diffracted from a given set of reflecting planes when it is incident at a specific angle called the Bragg angle. For a set of (hkl) planes, the diffraction angle $2\theta_{hkl}$ can be obtained from the relation of $\sin\theta_{hkl} = \lambda/2d_{hkl}$. For any set of planes, the diffraction direction (i.e., the diffraction angle) is influenced only by the interplanar spacing. Figure 5.33 shows an example where the Bragg angle for (001) decreases as a result of the increase in d_{001}. Since the interplanar spacing d_{hkl} is dependent on the crystal system to which the crystal belongs and its lattice constants, the diffraction direction is solely determined by the shape and size of the unit cell. The crystal structures shown in Figure 5.33 are rather simple, with atoms only on the corners of the unit cell. Even though the crystal has a complex structure with many atoms within the unit cell, the diffraction angle for a specific set of (hkl) planes is not influenced by the presence of these atoms once the unit cell dimensions remain unaltered. It is important to note that *the Bragg law mentions nothing about the intensity of a diffracted beam and suggests only the possible directions of diffraction*. No diffraction signal may be observed in the direction predicted by the Bragg law. The zero intensity means that diffraction does not take place. The Bragg law is a necessary condition for diffraction to occur, but it is not a sufficient condition. Then, what determines the intensity of a diffracted beam? Although there are many variables involved, the diffraction intensity is dominantly determined by the arrangement of atoms within the unit cell. As an example, consider 001 reflection from three different cubic crystals of lattice constant "a." Figure 5.34 shows simple cubic (top), body-centered cubic (middle), and CsCl (bottom) structures. Since all these structures are assumed to have the same lattice constant (also

d_{001} value), their Bragg angles for 001 reflection, θ_{001}, are also identical. Thus, an X-ray beam should be incident at this angle in all cases. It should be noted that this is the minimum requirement for diffraction and that the diffraction intensity depends on the actual crystal structure. When the crystal has a simple cubic structure, the path difference between rays *1* and *2* is one wavelength and diffraction occurs in the direction shown. In the crystal of body-centered cubic structure, rays *1* and *2* are also in phase with each other. In this case, however, there is another plane of atoms midway between two successive (001) planes. Since the path difference between rays *1* and *3* is one-half wavelength, they are completely out of phase and annul each other. Similarly, ray *4* from the next plane (not shown) annuls ray *2*. It means that the diffraction intensity from the set of (001) planes is zero, even though the Bragg condition is satisfied. In other words, there is no 001 reflection from the body-centered cubic structure. In the CsCl structure, rays *1* and *3* are also 180° out of phase. However, they are scattered from different types of atoms. Different atoms have different scattering strengths. Since the rays *1* and *3* have unequal amplitudes, no complete destructive interference will occur. Thus, we have non-zero diffraction intensity in the given direction. Of course, the diffracted beam will be much weaker in comparison to that from the simple cubic structure.

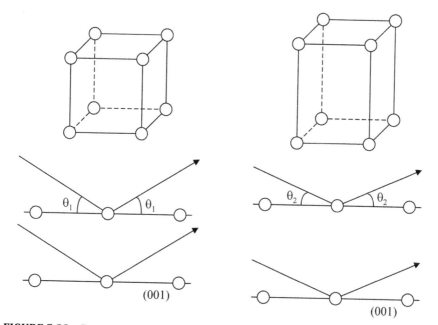

FIGURE 5.33 Decrease of the Bragg angle for (001) as a result of the increase in d_{001}.

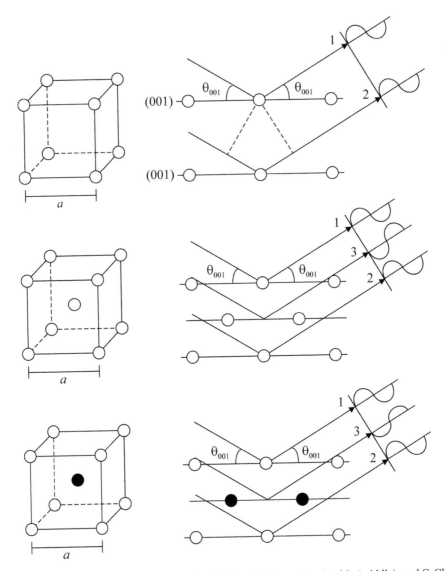

FIGURE 5.34 Diffraction from simple cubic (top), body-centered cubic (middle), and CsCl (bottom) structures.

The above example shows that the intensity of a diffraction signal is significantly affected by the actual arrangement of atoms within the unit cell. Conversely, it indicates that we can infer the atomic arrangement of a crystal by measuring its diffraction pattern. The Bragg condition in reciprocal space (Fig. 5.31) states that in order for a monochromic X-ray beam to reflect from a set of (*hkl*) planes, the difference between the incident and scattered wave vectors should be equal to the reciprocal lattice vector $\mathbf{H}_{hkl}$. As shown in the above example, we may have no diffracted signal in the expected direction even if the Bragg condition is satisfied. This means that the corresponding reciprocal lattice point is missing, which is known as *systematic absence*. The reciprocal lattice of a crystal should then be constructed taking this systematic absence into account. The systematic absence (i.e., missing reflections) forms a basis for the analysis of crystal structure and phase by X-ray diffraction. If a unit cell contains a total of N atoms, the resultant wave scattered by all the atoms of the unit cell is given by the following *structure factor, F.*

$$F_{hkl} = \sum_{n=1}^{N} f_n e^{2\pi i \left(hu_n + kv_n + lw_n \right)} \tag{5.42}$$

Here, f_n is the atomic scattering factor of the nth atom and u_n, v_n, and w_n represent its coordinates within the unit cell. The magnitude $|F|$ of the structure factor represents the amplitude of the resultant wave. The derivation of eq 5.42 can be referred to more specialized books on X-ray crystallography and diffraction.[9–14] When the Bragg law is satisfied for a set of (*hkl*) planes, the diffraction intensity from these planes is proportional to $|F_{hkl}|^2$. The structure factor may be real or a complex number. For a complex structure factor, the squared magnitude can be obtained with multiplication by its complex conjugate F^*: $|F|^2 = FF^*$. As an example, we consider the structure factor of NaCl structure (Fig. 5.35). A unit cell of the NaCl structure contains four Na and four Cl atoms and their coordinates are Na: (0,0,0), (1/2,1/2,0), (1/2,0,1/2), (0,1/2,1/2), and Cl: (1/2,1/2,1/2), (0,0,1/2), (0,1/2,0), (1/2,0,0). The structure factor is then

$$F = \left[1 + e^{\pi i (h+k)} + e^{\pi i (h+l)} + e^{\pi i (k+l)} \right] \left[f_{Na} + f_{Cl} e^{\pi i (h+k+l)} \right]. \tag{5.43}$$

When the plane indices h, k, and l are all even, we have $F^2 = 16(f_{Na} + f_{Cl})^2$. If h, k, and l are all odd, we have $F^2 = 16(f_{Na} - f_{Cl})^2$. This indicates that (200) planes will exhibit higher diffraction intensity than (311) planes. When the indices are mixed, the structure factor becomes zero, that is, $F = 0$. No reflection occurs from the corresponding planes due to the systematic absence. Figure 5.36 shows a diffraction pattern of NaCl powder where an X-ray beam of $\lambda = 1.54$ Å was used as the monochromatic source. Note that such diffraction peaks as 100 and 110 are absent from the pattern. We can also find that reflections from (200) and (220) planes are stronger than those from (111) and (311) planes, consistent with the structure factor.

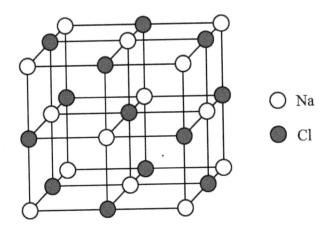

FIGURE 5.35   NaCl structure.

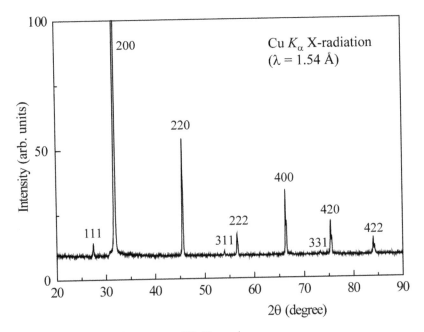

FIGURE 5.36   A diffraction pattern of NaCl powder.

5.5   ELECTRON DIFFRACTION

We take a brief look at electron diffraction in this section. The diffraction condition in reciprocal space is also very useful for thin film X-ray diffraction and transmission electron microscopy, both of which deal with materials of small dimensions. If the dimension of a feature is small along a certain direction in real space, the feature size along that direction is large in reciprocal space. This also holds for the opposite case. A real crystal with infinitely large dimensions has very tiny reciprocal lattice points that have essentially no volume. When the crystal dimension is reduced along a certain direction, it reciprocal lattice points are elongated along that direction. Thus, a thin plate crystal will exhibit reciprocal lattice points of rod shape, while a large bulk crystal has small spherical lattice points (Fig. 5.37). The rod shape is a consequence of the elongation of lattice points in the direction along which the crystal dimension is reduced. As the plate gets thinner, its reciprocal lattice points (actually lattice rods) become longer. Consider X-ray diffraction from a thin film whose surface is parallel to (001) planes, as depicted in Figure 5.38. This thin crystal has reciprocal lattice rods aligned perpendicular to the film surface. Then, the length of the $\mathbf{H}_{001}$ vector is not fixed but has a certain range. This enables diffraction to take place at other incidence angles near the Bragg angle θ_B, resulting in peak broadening. That is why a thin film exhibits weak and broad diffraction peaks. As the film gets thinner, the peak will be more broadened. When it is too thin, however, the overall diffraction intensity may be very low, often being undetectable. We have described diffraction from a linear array of coherent point sources in Section 5.2.2. As shown in Figure 5.11, the principal peak becomes weaker and broader as the number of sources decreases. When the number of sources is small, the total field has non-zero intensity even if the waves emanating from successive sources are not perfectly in phase. Similarly, a thin film, which has a small number of reflecting planes, will exhibit non-zero diffraction intensity at incident angles slightly deviating from the Bragg angle. In reciprocal space, this off-Bragg angle diffraction is explained with elongated lattice points. It is important to note that the reciprocal lattice vector $\mathbf{H}$ may have a certain range in its direction as well as in the magnitude. Consider a thin crystal layer of cubic structure. If the layer has a small dimension in the z-direction, its reciprocal lattice points are elongated along the z-direction, as illustrated in Figure 5.39. Then, the $\mathbf{H}_{010}$ vector (also other vectors such as $\mathbf{H}_{020}$ and $\mathbf{H}_{100}$) has divergent directions. In this case, the $\mathbf{H}_{001}$ vector has a range of lengths while its direction is nearly fixed.

Electron diffraction is most frequently used to study the crystal structure and phase of solids. It makes use of the wave nature of electrons. The periodic structure of a crystalline solid acts as a diffraction grating, scattering the electrons in a predictable manner. The de Broglie postulate, formulated in 1924, predicts that particles should also behave like waves. De Broglie's hypothesis was confirmed some years later with the observation of electron diffraction in two independent experiments by G. Thomson, C. Davisson, and L. Germer. Electron diffraction is usually carried out in a transmission electron microscope (TEM). The electrons in a TEM are accelerated to a velocity comparable to the speed of light. With a relativistic modification made, the electron wavelength is given by

$$\lambda = h \left[2m_o eV \left(1 + \frac{eV}{2m_o c^2} \right) \right]^{-1/2}. \tag{5.44}$$

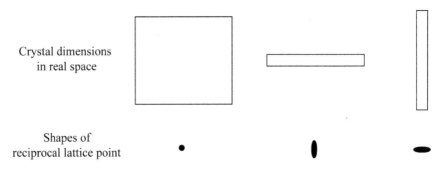

Crystal dimensions
in real space

Shapes of
reciprocal lattice point

FIGURE 5.37 Crystal dimensions in real space versus shapes of reciprocal lattice point.

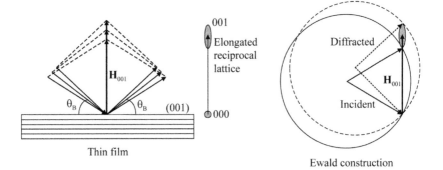

Thin film

Ewald construction

FIGURE 5.38 Peak broadening related to the diffraction condition in reciprocal space.

Thin layer in real space

Reciprocal lattice

FIGURE 5.39 Origin of divergent **H** vector directions.

Here, h is Plank's constant, c is the speed of light, and m_o is the rest mass of the electron. The electron wavelength in a 200 kV TEM is 0.025 Å (λ = 1.54 Å for Cu K_α X-ray). In a TEM analysis, the electrons pass through a thin sample to be investigated. When an electron beam is normally incident into the sample, it has reciprocal lattice points elongated along the incident direction of the beam (elongated reciprocal lattice points are hereafter referred to as *reciprocal lattice rods*). As mentioned above, the electron wavelength used for electron diffraction is much shorter than the wavelength of X-rays. This means that the radius of the Ewald sphere is much larger in electron diffraction experiments than in X-ray diffraction. As a consequence, the quite flat surface of the Ewald sphere can intersect a considerable number of reciprocal lattice rods, as shown in Figure 5.40. Diffraction occurs by reciprocal lattice vectors terminating on the surface of the Ewald sphere. Since the electrons have a very small wavelength, the diffraction angles are also much smaller than those of X-ray diffraction. Unlike X-ray diffraction, the electron diffraction arises from planes parallel to the incident beam. It allows a two-dimensional distribution of the reciprocal lattice to be revealed. The diffraction pattern is recorded on a fluorescent screen or photographic film (Fig. 5.41). If the electron beam is incident along one of the symmetry axes of the crystal, the recorded diffraction spots also exhibit the corresponding symmetry. While the diffracted beams have appreciable intensity at small diffraction angles, the intensity falls off rapidly as the diffraction angle increases. The reciprocal lattice rods are more unlikely to touch the Ewald sphere when they are further away from the reciprocal space origin. Thus, the spots in a diffraction pattern are from the planes with small h, k, and l indices. Since the indexing of diffraction spots requires some knowledge on crystallography, it is not discussed in detail here. The central spot is caused by a transmitted beam. Although a limited number of spots are recorded, electron diffraction directly visualizes the reciprocal lattice of a crystal.

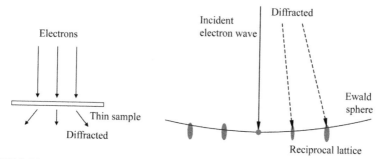

FIGURE 5.40   Principle of electron diffraction.

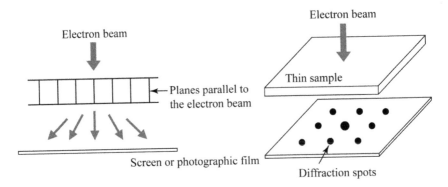

FIGURE 5.41   Formation of diffraction spots.

In TEM, a single crystal grain or particle may be selected for the diffraction experiments. If the sample is tilted with respect to the incident electron beam, we can obtain diffraction patterns from several crystal orientations. Then, the reciprocal lattice of the crystal can be mapped in three dimensions. Although electron diffraction is a very powerful technique for analyzing the crystalline quality and orientation of a material, it is subject to several limitations. First of all, the sample to be studied should be electron transparent and be made very thin. The required thickness is usually less than 100 nm. Therefore, a careful and time-consuming sample preparation procedure is necessary. Furthermore, many samples are vulnerable to radiation damage induced by the incident electrons. For nanomaterials such as nanoparticles and nanowires, the overall preparation procedure becomes much easier. Figures 5.42, 5.43 show TEM images of Ag nanoparticles and nanowire and their electron diffraction patterns. The study of magnetic materials by electron diffraction is very complicated because electrons in magnetic fields are deflected by the Lorentz force. Details on electron diffraction can be found in references [15–17].

PROBLEMS

5.1 A collimated laser beam at 633 nm normally illuminates a slit of 1 mm width. A lens with $f = 10$ cm placed just behind the slit focuses the diffracted light onto a screen located at the focal plane. What is the distance from the center of the diffraction pattern (central maximum) to the first minimum?

  5.2 The image of a point formed by a lens is a diffraction pattern. Suppose that a lens of $f = 10$ cm has a diameter of 2 cm. When this lens is fully illuminated by a plane wave at 532 nm, what is the diameter of the Airy disk at the focal plane?

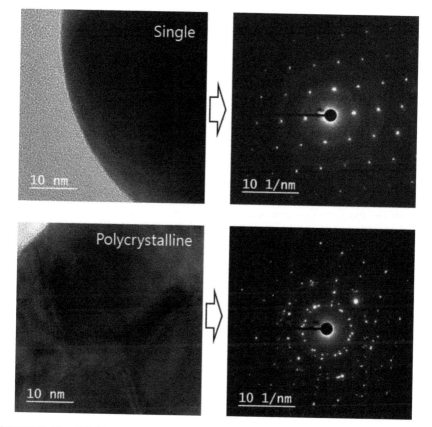

FIGURE 5.42　TEM images of Ag nanoparticles and their electron diffraction patterns.

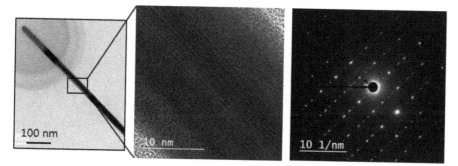

FIGURE 5.43　Electron diffraction pattern from an Ag nanowire.

5.3 Prove that in a transmission phase grating, eq 5.34 is independent of the refractive index of the grating medium.

5.4 Equation 5.34 shows that for normal incidence ($\theta_i = 0$), diffraction does not occur if $a < \lambda$. That is., the incident beam is directly transmitted or specular reflected. For oblique incidence, however, diffraction can occur even if $a < \lambda$. Prove it.

5.5 A transmission phase grating has periodicity of 1 μm. When light of $\lambda = 500$ nm is incident at an angle of 75° with respect to the grating normal, find the angles of diffracted beams.

5.6 The grating equation of eq 5.34 is valid only when both sides of the grating are in contact with air. Show that the more general equation is

$$a\left(n\sin\theta_m - n_i\sin\theta_i\right) = m\lambda. \tag{5.45}$$

Here n_i is the refractive index of the incident medium and n is the refractive index of the medium that contains diffracted beams (for reflection gratings, $n = n_i$). λ is the free-space wavelength.

5.7 Suppose that a grating structure is formed on the surface of a semi-infinite material of refractive index n. Show that the grating equation for this case is the same as eq 5.45.

5.8 Electron diffraction directly visualizes the two-dimensional reciprocal lattice of a crystal. When an electron beam is normally incident onto a thin BCC crystal whose surface is (001), draw the expected diffraction pattern.

5.9 In the zinc blende form of ZnS, a cubic unit cell contains four Zn atoms and four S atoms and their coordinates are Zn: (0,0,0), (1/2,1/2,0), (1/2,0,1/2), (0,1/2,1/2), and S: (1/4,1/4,1/4), (3/4,3/4,1/4), (3/4,1/4,3/4), (1/4,3/4,3/4). Derive simplified expressions for the structure factor. For which hkl reflections will the structure factor be zero?

5.10 As shown in Figure 5.44, $AuCu_3$ is an intermetallic compound that exhibits an order-disorder transition at 390°C. Below 390°C, the Au atoms occupy the corners of the cubic unit cell and the Cu atoms, the face centers. Above 390°C, both atoms are randomly positioned at the corners and face centers of the unit cell. Since there is no preferred position for Cu or Au, the probability that a particular atomic site is occupied by Au is 1/4, which is the atomic fraction of Au in the compound. Accordingly, the probability that the same site is occupied by Cu is 3/4. Derive the structure factor of $AuCu_3$ for the order and disorder states. For which hkl reflections, will the structure factor be identical in the two states?

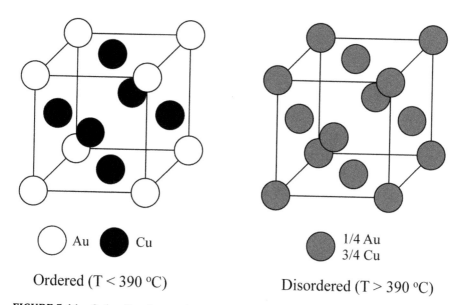

Ordered (T < 390 °C)

Disordered (T > 390 °C)

FIGURE 5.44 Order-disorder transition in the Au-Cu intermetallic compound system

REFERENCES

1. Hecht, E. *Optics,* 5th ed.; Pearson: Boston, 2016.
2. Pedrotti, F.; Pedrotti, L. M.; Pedrotti, L. S. *Introduction to Optics,* 3rd ed.; Addison-Wesley: Boston, 2006.
3. Fowles, G. *Introduction to Modern Optics,* 2nd ed.; Dover: New York, 1989.
4. Guenther, B. *Modern Optics,* 2nd ed.; Oxford University Press: New York, 2015.
5. Born, M.; Wolf, E. *Principles of Optics,* 7th ed.; Cambridge University Press: Cambridge, UK, 1999.
6. Saleh, B.; Teich, M. *Fundamentals of Photonics,* 2nd ed.; Wiley: Hoboken, NJ, 2007.
7. Brooker, G. *Modern Classical Optics*; Oxford University Press: New York, 2002.
8. O'Shea, D.; Suleski, T.; Kathman, A.; Prather, D. *Diffractive Optics*; SPIE Press: Washington, DC, 2004.
9. Kelly, A.; Knowles, K. *Crystallography and Crystal Defects,* 2nd ed.; Wiley: Chichester, UK, 2012.
10. McKie, D.; McKie, C. *Essentials of Crystallography*; Blackwell Scientific: Oxford, UK, 1986.
11. Hammond, C. *The Basics of Crystallography and Diffraction*; Oxford University Press: New York, 1997.
12. Lee, M. *X-Ray Diffraction for Materials Research*; Apple Academic Press: Oakville, Canada, 2016.
13. Cullity, B.; Stock, S. *Elements of X-ray Diffraction,* 3rd ed.; Prentice Hall: Upper Saddle River, N.J., 2001.
14. Warren, B. *X-Ray Diffraction*; Dover: New York, 1990.

15. Bozzola, J.; Russell, L. *Electron Microscopy,* 2nd ed.; Jones & Bartlett Publishers: Sudbury, Mass, 1998.
16. Goodhew, P.; Humphreys, J.; Beanland, R. *Electron Microscopy and Analysis,* 3rd ed.; Taylor & Francis: London, 2001.
17. Williams, D.; Carter, C. *Transmission Electron Microscopy,* 2nd ed.; Springer: New York, 2009.

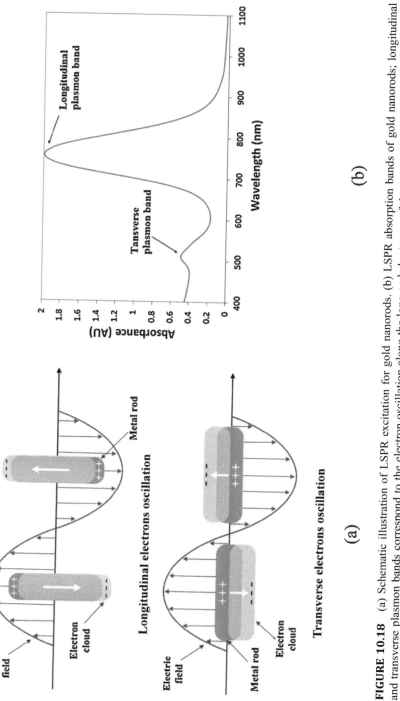

FIGURE 10.18 (a) Schematic illustration of LSPR excitation for gold nanorods. (b) LSPR absorption bands of gold nanorods; longitudinal and transverse plasmon bands correspond to the electron oscillation along the long and short axes of the nanorod, respectively. Reprinted with permission from Ref. [30]. Copyright© 2015 Elsevier.

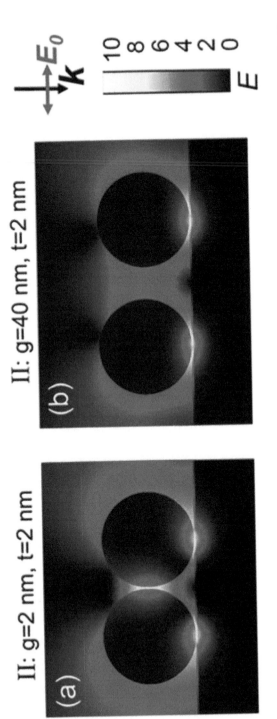

FIGURE 10.25 Local electric field distributions in a dimer of particles, when the gap width is (a) $g = 2$ nm and (b) $g = 40$ nm. Reprinted with permission from Ref. [43]. Copyright© 2016 Macmillan Publishers Limited.

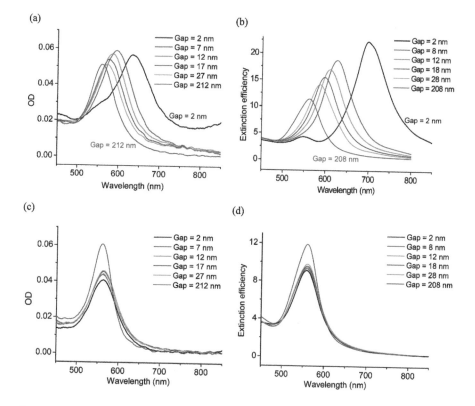

FIGURE 10.28 (a, c) Microabsorption and (b, d) DDA-simulated extinction efficiency spectra of Au nanodisc pairs for varying interparticle separation gap for incident light polarization direction (a, b) parallel and (c, d) perpendicular to the interparticle axis. Reprinted with permission from Ref. [46]. Copyright© 2007 American Chemical Society.

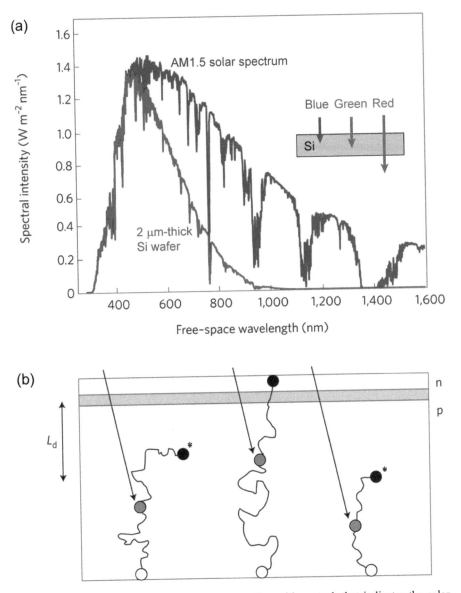

FIGURE 10.31 (a) AM1.5 solar spectrum, together with a graph that indicates the solar energy absorbed in a 2-μm-thick crystalline Si film (assuming single-pass absorption and no reflection). Clearly, a large fraction of the incident light in the spectral range 600–1100 nm is not absorbed in a thin crystalline Si solar cell. (b) Schematic indicating carrier diffusion from the region where photocarriers are generated to the p–n junction. Charge carriers generated far away (more than the diffusion length L_d) from the p–n junction are not effectively collected, owing to bulk recombination (indicated by the asterisk). Reprinted with permission from Ref. [64]. Copyright© 2010 Springer Nature.

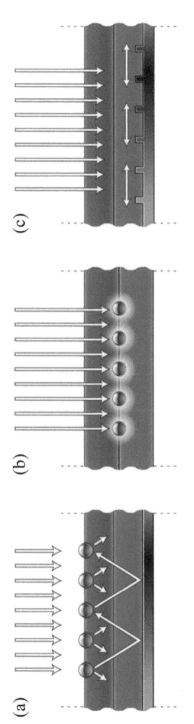

FIGURE 10.32 Plasmonic light-trapping schemes for thin-film solar cells. (a) Light trapping by scattering from metal nanoparticles at the surface of the solar cell. Light is preferentially scattered and trapped into the semiconductor thin film by multiple and high-angle scattering, causing an increase in the effective optical path length in the cell. (b) Light trapping by the excitation of localized surface plasmons in metal nanoparticles embedded in the semiconductor. The excited particles' near-field causes the creation of electron–hole pairs in the semiconductor. (c) Light trapping by the excitation of surface plasmon polaritons at the metal/semiconductor interface. A corrugated or textured metal film on the rear surface of a semiconductor absorption layer can couple light into SPP modes supported at the metal/semiconductor interface. Reprinted with permission from Ref. [64]. Copyright© 2010 Springer Nature.

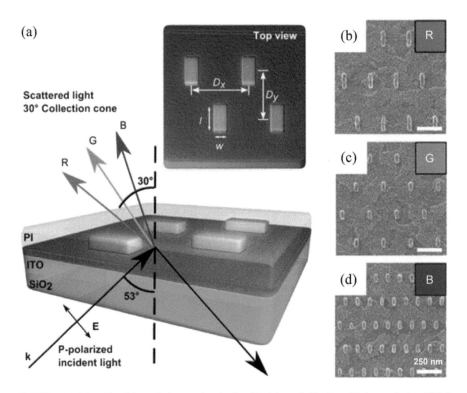

FIGURE 10.33 (a) Al-based plasmonic pixels. (b), (c), and (d) show high-resolution SEM images of portions of a red ($l = 153$ nm, $D_x = 256$ nm, $D_y = 410$ nm), green ($l = 88.5$ nm, $D_x = 272$ nm, $D_y = 272$ nm), and blue ($l = 81$ nm, $D_x = 118$ nm, $D_y = 227$ nm) pixel, respectively. The insets show the color created by a pixel of dimensions shown in each respective SEM image. Reprinted with permission from Ref. [80]. Copyright© 2016 American Chemical Society.

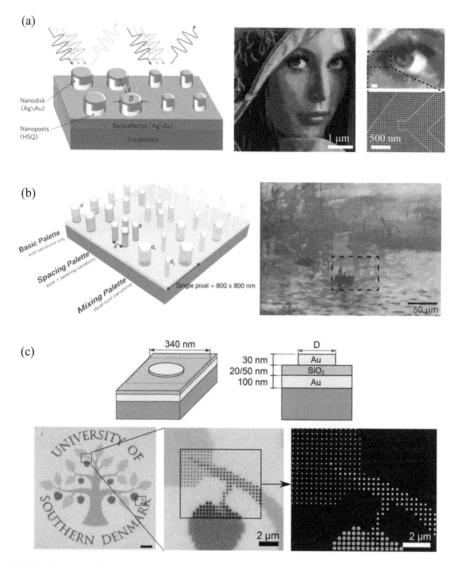

FIGURE 10.34 Designs and images of sub-wavelength plasmonic printing. (a) Each pixel consists of four nanodisks that support particle resonances. These disks are raised above equally sized holes on a back-reflector by hydrogen silsesquioxane (HSQ) nanoposts. Reprinted with permission from Ref. [86]. Copyright© 2012 Springer Nature. (b) An Al-based pixel design and the corresponding color image, where different colors are produced by spatially mixing and adjusting the space between discrete aluminum nanostructures. Reprinted with permission from Ref. [87]. Copyright© 2014 American Chemical Society. (c) A metal-insulator-metal structure with gold nanodisks coupled to a continuous gold film. Reprinted with permission from Ref. [88]. Copyright© 2014 American Chemical Society.

CHAPTER 6

Light Propagation in Anisotropic Media

6.1 INTRODUCTION

In isotropic media such as air and glass, light behaves the same way no matter which direction it is propagating in the medium. This means that the refractive index of an isotropic medium is independent of the propagation direction of light. Many crystals are naturally anisotropic, and in some media (such as liquid crystals) it is possible to induce anisotropy by applying an external electric field. The optical properties of anisotropic media depend on the propagation direction of light and its polarization state. In this chapter, we discuss the behavior of light in anisotropic media. We shall see that in such media, the electric vector $\mathbf{E}$ of a propagating electromagnetic wave is generally not parallel to its polarization direction, which is defined by the direction of its displacement vector $\mathbf{D}$. We will also show that the wave vector $\mathbf{k}$ is no longer perpendicular to $\mathbf{E}$, and the Poynting vector $\mathbf{S}$ is no longer parallel to $\mathbf{k}$. The latter means that the direction of energy flow may not be parallel to the direction of wave propagation in anisotropic media. However, Maxwell's equations require that $\mathbf{D}$ and $\mathbf{k}$ always be perpendicular to each other, and the definition of the Poynting vector also requires that $\mathbf{E}$ and $\mathbf{S}$ be perpendicular to one another. When plane waves propagate in a particular direction inside an anisotropic medium, two distinctly different polarization states (i.e., $\mathbf{D}$ directions) are allowed, and waves of these polarization states propagate with different velocities. The two $\mathbf{D}$ vectors correspond to the two normal modes of propagation with different indices of refraction. We will introduce an ellipsoidal surface called the index ellipsoid (or the indicatrix) and show how the allowed polarization states and their corresponding refractive indices can be determined for wave propagation in a given direction. Anisotropic media exhibit many peculiar optical phenomena, including double refraction and optical rotation. Many optical devices are made of anisotropic media utilizing these phenomena. Optically anisotropic media are generally crystalline. A crystal is a solid material whose constituent atoms are periodically arranged in three dimensions. The regularity of atomic arrangement can be described by symmetry elements, which ultimately determines the physical properties of a crystal. Crystals are classified into seven crystal systems according to their rotational symmetry elements: *triclinic, monoclinic, orthorhombic, trigonal, tetragonal, hexagonal*, and *cubic*. The optical properties of crystals are closely related to their crystallographic symmetry. A cubic crystal is optically isotropic. Important anisotropic media are uniaxial crystals. Trigonal, tetragonal, and hexagonal crystals are all uniaxial crystals. They possess a unique symmetry axis (e.g., threefold rotation axis, fourfold rotation axis). Therefore, the physical properties parallel and perpendicular to this symmetry axis are generally different. However, the properties of uniaxial crystals are identical for all directions perpendicular to their characteristic symmetry axis. A uniaxial crystal exhibits two refractive indices: "ordinary" index (n_o) and "extraordinary" index (n_e). Since many modern optical devices involve the use of uniaxial crystals, this chapter is mainly focused on the propagation of light in these media.

6.2   DIELECTRIC TENSOR OF AN ANISTROPIC MEDIUM

In a linear isotropic medium, the displacement vector **D** is always parallel to the applied electric field **E** and is related to it by a scalar quantity (the electrical permittivity ε), as shown in eq 1.59. In such a medium, the electrical permittivity is independent of the direction along which the **E** field is applied. However, this is no longer true for anisotropic media in which the induced polarization **P** depends, both in its magnitude and direction, on the direction of **E**. As a consequence, **D** may not be parallel to **E**. In a linear anisotropic crystal, each component of **D** is linearly related to the three components of **E**.

$$D_x = \varepsilon_{xx}E_x + \varepsilon_{xy}E_y + \varepsilon_{xz}E_z$$
$$D_y = \varepsilon_{yx}E_x + \varepsilon_{yy}E_y + \varepsilon_{yz}E_z \qquad (6.1)$$
$$D_z = \varepsilon_{zx}E_x + \varepsilon_{zy}E_y + \varepsilon_{zz}E_z$$

The 3×3 array of the coefficients ε_{ij} is called the *electric permittivity tensor* (or simply the dielectric tensor). The electric permittivity tensor is symmetric, that is, $\varepsilon_{ij} = \varepsilon_{ji}$, and therefore has only six independent coefficients. The energy density stored in an electric field[1,2] is

$$U_e = \frac{1}{2}\mathbf{E}\cdot\mathbf{D} = \frac{1}{2}(\varepsilon_{xx}E_x^2 + \varepsilon_{yy}E_y^2 + \varepsilon_{zz}E_z^2 + 2\varepsilon_{yz}E_yE_z + 2\varepsilon_{xz}E_xE_z + 2\varepsilon_{xy}E_xE_y) \cdot \quad (6.2)$$

Although the magnitudes of the coefficients ε_{ij} depend on how the x, y, and z directions are defined in the crystal structure, it is always possible to choose x, y, and z axes that make the off-diagonal coefficients vanish. These axes are called the principal dielectric axes of the crystal. For the remainder of this chapter, x, y, and z represent the principal axes of the crystal. In the principal dielectric coordinate system, the stored energy density U_e becomes

$$2U_e = \varepsilon_x E_x^2 + \varepsilon_y E_y^2 + \varepsilon_z E_z^2. \tag{6.3}$$

The corresponding permittivity tensor is diagonal and is given by

$$\begin{bmatrix} D_x \\ D_y \\ D_z \end{bmatrix} = \begin{bmatrix} \varepsilon_x & 0 & 0 \\ 0 & \varepsilon_y & 0 \\ 0 & 0 & \varepsilon_z \end{bmatrix} \begin{bmatrix} E_x \\ E_y \\ E_z \end{bmatrix}. \tag{6.4}$$

Here ε_x, ε_y, and ε_z are called the principal permittivities. An isotropic cubic crystal has $\varepsilon_x = \varepsilon_y = \varepsilon_z$, thus making **D** always parallel to **E**. It follows from eq 6.4 that **D** and **E** will have different directions in an anisotropic crystal, unless **E** coincides with one of the principal axes. Both cases are graphically illustrated in Figure 6.1.

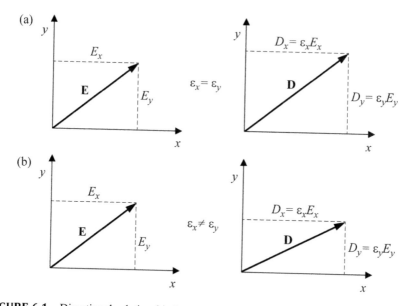

FIGURE 6.1 Directional relationship between **E** and **D** in the x–y plane, when (a) $\varepsilon_x = \varepsilon_y$ and (b) $\varepsilon_x \neq \varepsilon_y$.

We now consider the propagation of a plane wave in an anisotropic crystal. In this chapter, we assume that the medium is nonmagnetic ($\mu = \mu_o$) without any free charges or currents present. In an anisotropic crystal, the phase velocity of light depends not only on the direction of propagation but also on its polarization state. In addition, the polarization state of a plane wave may vary as it propagates through the crystal. Consider an electromagnetic plane wave propagating in the direction of wave vector $\mathbf{k}$. Substituting the plane-wave solution of eq 1.31 into Maxwell's equations, $\nabla \cdot \mathbf{D} = 0$ and $\nabla \cdot \mathbf{B} = 0$, gives

$$\mathbf{\nabla} \cdot \mathbf{D} = i\mathbf{k} \cdot \mathbf{D} = 0$$
$$\mathbf{\nabla} \cdot \mathbf{B} = \mu_0 i\mathbf{k} \cdot \mathbf{H} = 0 \qquad (6.5)$$

It follows from the relations that both $\mathbf{D}$ and $\mathbf{H}$ (also $\mathbf{B} = \mu_o\mathbf{H}$) are perpendicular to $\mathbf{k}$. By substituting the plane-wave solution into $\nabla \times \mathbf{H} = \partial \mathbf{D}/\partial t$ and $\nabla \times \mathbf{E} = -\partial \mathbf{B}/\partial t$, we obtain

$$\mathbf{k} \times \mathbf{H} = -\omega \mathbf{D}$$
$$\mathbf{k} \times \mathbf{E} = \mu_0 \omega \mathbf{H} \qquad (6.6)$$

These relations require that $\mathbf{H}$ be perpendicular to $\mathbf{D}$, and $\mathbf{H}$ must also be perpendicular to both $\mathbf{k}$ and $\mathbf{E}$. Equations 6.5 and 6.6 mention nothing about the directional relationship between $\mathbf{E}$ and $\mathbf{D}$, indicating that $\mathbf{D}$ is not necessarily parallel to $\mathbf{E}$. The Poynting vector $\mathbf{S}$, defined by $\mathbf{S} = \mathbf{E} \times \mathbf{H}$, is perpendicular to both $\mathbf{E}$ and $\mathbf{H}$. Since $\mathbf{H}$ (also $\mathbf{B}$) is at right angles to $\mathbf{D}$, $\mathbf{E}$, $\mathbf{k}$, and $\mathbf{S}$, these four vectors should be coplanar. That is, they should lie in one plane to which $\mathbf{H}$ is normal, as illustrated in Figure 6.2. In this plane, $\mathbf{D}$ is perpendicular to $\mathbf{k}$, and $\mathbf{E}$ is perpendicular to $\mathbf{S}$. However, $\mathbf{S}$ is not necessarily parallel to $\mathbf{k}$. Thus the direction of energy flow may be different from the direction of wave propagation in an anisotropic crystal. The angle between $\mathbf{S}$ and $\mathbf{k}$ is the same as that between $\mathbf{E}$ and $\mathbf{D}$.

6.3   INDEX ELLIPSOID

The surface of constant energy density given by eq 6.3 can be rewritten as

$$1 = \frac{D_x^2}{2U_e\varepsilon_x} + \frac{D_y^2}{2U_e\varepsilon_y} + \frac{D_z^2}{2U_e\varepsilon_z}. \qquad (6.7)$$

This represents the energy density equation in $\mathbf{D}$-space. If we use x, y, and z in place of $D_x/\sqrt{(2U_e\varepsilon_o)}$, $D_y/\sqrt{(2U_e\varepsilon_o)}$, $D_z/\sqrt{(2U_e\varepsilon_o)}$ and define the principal indices of refraction by $n_i^2 = \varepsilon_i/\varepsilon_o$ ($i = x, y, z$), eq 6.7 becomes

$$\frac{x^2}{n_x^2} + \frac{y^2}{n_y^2} + \frac{z^2}{n_z^2} = 1. \tag{6.8}$$

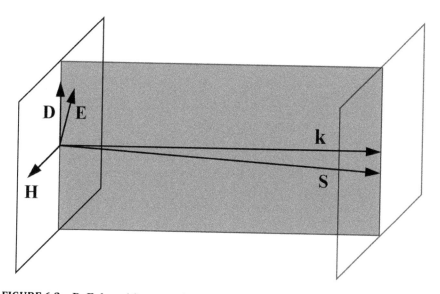

FIGURE 6.2 **D, E, k,** and **S** vectors should lie in one plane normal to **H**.

This is the equation of an ellipsoid whose semi-axes parallel to the x, y, and z directions are equal to the principal indices of refraction. The ellipsoid is called the *index ellipsoid* or the optical indicatrix. The index ellipsoid can be used to find the polarizations and refractive indices of a plane wave traveling along an arbitrary direction **k** in an anisotropic crystal. The procedure is explained with Figure 6.3. First we draw a plane passing through the origin of the ellipsoid, normal to **k**. The intersection of this plane with the ellipsoid is an ellipse. The intersection ellipse has two principal axes (long and short axes). These axes are parallel to the directions of two allowed **D** vectors, denoted as $\mathbf{D}_1$ and $\mathbf{D}_2$ in Figure 6.3. $\mathbf{D}_1$ and $\mathbf{D}_2$ are both perpendicular to **k**. The two polarizations $\mathbf{D}_1$ and $\mathbf{D}_2$ have refractive indices n_1 and n_2, respectively, which are the lengths of the principal semi-axes. The wave propagating along **k** can have its polarization (i.e., the direction of its **D** vector) only parallel to either of the principal axes of the intersection ellipse. No other **D** directions are allowed. This restriction vanishes in an isotropic crystal (such as a cubic crystal) where the curve of the intersection becomes a circle due to $n_x = n_y = n_z$.

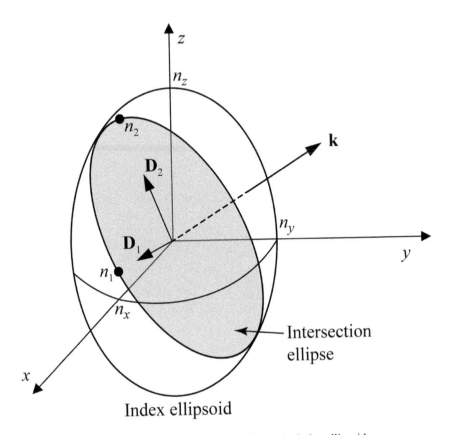

FIGURE 6.3   Two polarization modes determined from the index ellipsoid.

Consider a tetragonal crystal whose characteristic fourfold symmetry axis is along the z-direction. Since the x and y axes of the crystal are physically equivalent, we have $n_x = n_y \neq n_z$. Figure 6.4 depicts the behaviors of a plane wave when it propagates in different directions inside the crystal. When the wave propagates along the z axis of the crystal (Fig. 6.4a), the curve of the intersection is a circle whose radius equals $n_x = n_y = n_o$. Here n_o is called the *ordinary refractive index*. Since the intersection circle has no principal axes, the wave propagating along the z axis of the crystal can have any polarizations, that is, any **D** directions in the x–y plane. This wave has a refractive index of n_o, regardless of its polarization state. If the wave propagates along the y axis of the crystal (Fig. 6.4b), the curve of the intersection is an ellipse whose principal axes are parallel to the x and z axes of the crystal. These axes represent the directions of the allowed **D** vectors. An incident light wave polarized parallel to one of these directions will remain in the same polarization state as it propagates through the crystal. When the wave is polarized parallel to the x axis of the crystal, it has a refractive index of n_o. If the wave is polarized parallel to the z axis of the crystal, it has a refractive index of $n_z = n_e$. n_e is called the *extraordinary refractive index*. Even though the wave propagates in the same direction, the refractive index depends on its polarization state. When an unpolarized light wave is incident, it decomposes into two component waves whose polarizations are parallel to the x and z axes of the crystal, because **D** vectors are allowed only in those directions. Of course, the two component waves have different refractive indices and thus travel with different velocities. Figure 6.4c describes the behavior of a wave propagating along an arbitrary direction in the x–y plane. The intersection ellipse has the same shape and size as in Figure 6.4b. Thus, the wave has a refractive index of either n_o or n_e, depending on its polarization state. The configurations of Figure 6.4b,c are optically identical. In other words, a wave exhibits the same optical behavior for all propagation directions in a plane normal to the z axis of the crystal. Figure 6.4d depicts a more general case in which the wave vector **k** is along an arbitrary direction in the y–z plane. One of the allowed polarizations is directed parallel to the x axis of the crystal. The wave of this polarization (**D**$_o$) has a refractive index of n_o and is called the *ordinary wave*. The other polarization (**D**$_e$) has components parallel to both the y and z axes. The corresponding wave is called the *extraordinary wave*. While the refractive index of the ordinary wave is always equal to n_o, the extraordinary wave has a refractive index that depends on the propagation direction. When the angle between **k** and the z axis is θ, the refractive index $n(\theta)$ of the extraordinary wave is given by

$$\frac{1}{n^2(\theta)} = \frac{\cos^2\theta}{n_o^2} + \frac{\sin^2\theta}{n_e^2}. \tag{6.9}$$

Since the direction of **D**$_e$ depends on θ, $n(\theta)$ varies from n_o for $\theta = 0°$ to n_e for $\theta = 90°$. We need to distinguish the extraordinary wave from the extraordinary refractive index. When the wave contains a non-zero polarization component parallel to the z axis, it is referred to as the extraordinary wave. Therefore, the extraordinary wave can have not only n_e but also other values as its refractive index. However, n_o cannot be the refractive index of the extraordinary wave. When a wave propagates perpendicular to the z axis, it can be an ordinary or extraordinary wave depending on its polarization direction. However, there is no extraordinary wave propagating along the z axis. All waves propagating parallel to this axis are ordinary waves.

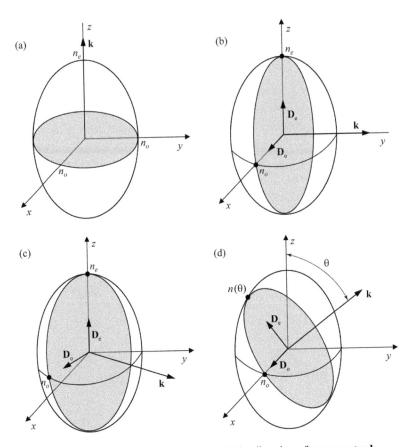

FIGURE 6.4   Polarization modes associated with the direction of wave vector **k**.

6.4   LIGHT PROPAGATION IN UNIAXIAL CRYSTALS

From eq 6.6 and the relation of $\mathbf{D} = \varepsilon\mathbf{E}$ (ε is the permittivity tensor), we obtain

$$\mathbf{k}\times(\mathbf{k}\times\mathbf{E})+\omega^2\mu_o\varepsilon\mathbf{E} = 0. \tag{6.10}$$

This equation can be rewritten in the following matrix form:

$$\begin{bmatrix} \omega^2\mu_o\varepsilon_x - k_y^2 - k_z^2 & k_xk_y & k_xk_z \\ k_yk_x & \omega^2\mu_o\varepsilon_y - k_x^2 - k_z^2 & k_yk_z \\ k_zk_x & k_zk_y & \omega^2\mu_o\varepsilon_z - k_x^2 - k_y^2 \end{bmatrix}\begin{bmatrix} E_x \\ E_y \\ E_z \end{bmatrix} = \begin{bmatrix} 0 \\ 0 \\ 0 \end{bmatrix}. \tag{6.11}$$

The equation represents a three-dimensional surface in **k**-space. The determinant of the matrix in eq 6.11 should be zero for nontrivial solutions to exist. Crystals are classified into seven crystal systems according to their symmetry elements. These are *triclinic, monoclinic, orthorhombic, trigonal, tetragonal, hexagonal,* and *cubic* systems. The optical properties of crystals are closely related to their crystallographic symmetry. A cubic crystal is optically isotropic. Since many modern optical devices involve the use of uniaxial crystals, they are important anisotropic media. A uniaxial crystal exhibits two refractive indices: "ordinary" index ($n_o = n_x = n_y$) and "extraordinary" index ($n_e = n_z$). Trigonal, tetragonal, and hexagonal crystals are optically uniaxial. When $n_e > n_o$, the crystal is said to be positive uniaxial. If $n_o > n_e$, it is negative uniaxial. The difference between n_o and n_e is called the birefringence. Uniaxial crystals have a single optic axis, which is parallel to their unique symmetry element (e.g., fourfold axis, sixfold axis). The direction of the optic axis is conventionally taken as the z direction.

For uniaxial crystals, the matrix determinant in eq 6.11 is simplified to

$$\left(\frac{k^2}{n_o^2} - \frac{\omega^2}{c^2} \right)\left(\frac{k_x^2 + k_y^2}{n_e^2} + \frac{k_z^2}{n_o^2} - \frac{\omega^2}{c^2} \right) = 0. \qquad (6.12)$$

Here k is the magnitude of the wave vector **k** and k_x, k_y, and k_z are its components along the x, y, and z directions, respectively. For a given direction of propagation, two solutions exist. One is a sphere and the other, an ellipsoid of revolution. The intersection of the k-surfaces with the y–z plane is drawn in Figure 6.5a. The two surfaces touch at two points on the z axis, which is the only optic axis. We can assume that the **k** vector lies in the y–z plane without loss of generality due to symmetry about the z axis (optic axis). As ω/c is a common factor, Figure 6.5a is redrawn without it in Figure 6.5b. This construction, now representing n-surfaces, is more convenient for finding the indices of refraction. For a given **k** direction, the refractive index is determined by finding the intersection of **k** with the n-surfaces. In Figure 6.5b, the wave vector **k** intersects the n-surfaces at two points. The distances from the origin to these points are the refractive indices of the wave. If the wave is ordinarily polarized, it has a refractive index of n_o. When the wave is extraordinarily polarized, the refractive index $n(\theta)$ depends on θ, as given in eq 6.9. Although the indices of refraction in a uniaxial crystal can also be found by the procedure described previously (see Fig. 6.4), the construction shown in Figure 6.5b provides a more convenient method. If the wave propagates along the y direction, it has a refractive index of either n_o or n_e, depending on its polarization direction. When the propagation direction is parallel to the optic axis, the wave has an index of n_o because its polarization directions are always perpendicular to the optic axis. In anisotropic media, the polarization direction of a wave is customarily defined by the direction of its displacement vector **D** because **D** is perpendicular to the wave vector **k**. Since **D** is always parallel to the electric field **E** in isotropic media, **E** is predominantly used to indicate the polarization state of light in air.

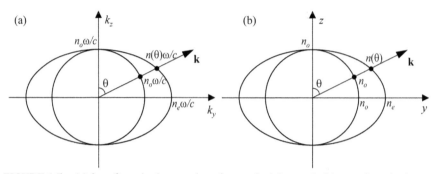

FIGURE 6.5   (a) k-surfaces in the y–z plane for a uniaxial crystal. (b) n-surfaces in the y–z plane.

When an electromagnetic wave propagates in an anisotropic crystal, its **E** and **D** vectors may not be parallel to each other. In order to see the physical origin of this effect, let us consider a hypothetical tetragonal crystal shown in Figure 6.6a. The characteristic fourfold symmetry axis of the crystal is taken as the z direction. We here assume that the crystal is positive uniaxial ($\varepsilon_z > \varepsilon_x = \varepsilon_y$, $n_e > n_o$). It follows from eq 6.4 that if an electric field **E** is applied along one of the principal axes (x, y, and z axes) of the crystal, the corresponding **D** vector is parallel to **E**, as illustrated in Figure 6.6b. When **E** has an arbitrary direction in the x–y plane (Fig. 6.6c), **D** is also parallel to **E** because of $\varepsilon_x = \varepsilon_y$. However, if **E** lies in the y–z plane, **D** will have a different direction in the same y–z plane because ε_y and ε_z are unequal. For a positive uniaxial crystal, ε_z is larger than ε_y. Therefore, **D** is more inclined toward the z axis, as shown in Figure 6.6d. The result will be opposite for a negative uniaxial crystal. For a given direction of **E**, the angle between **E** and **D** depends on the difference between ε_y and ε_z. Conversely, if ε_y and ε_z are known, the angular difference between **E** and **D** can be calculated. The ordinary and extraordinary indices of refraction are given by $n_o^2 = \varepsilon_y/\varepsilon_0$ and $n_e^2 = \varepsilon_z/\varepsilon_0$, respectively. This implies that when an electromagnetic wave propagates in a uniaxial crystal, its **E** and **D** vectors may have different directions. We now examine how the refractive index of a propagating wave and its directional relationship between **E** and **D** can be determined from the n-surfaces. When the wave propagates along the y axis, the wave vector **k** meets the n-surfaces at two points of n_o and n_e, as shown in Figure 6.7a,b. Therefore, two polarization modes can be supported: one parallel to the x axis and the other parallel to the z axis (see Fig. 6.4b). When the wave is ordinarily polarized with its **D** parallel to the x axis, it has a refractive index of n_o. The extraordinary wave, in which **D** is parallel to the z axis, has a refractive index of n_e. In both cases, **D** is parallel to **E**. When **k** is parallel to the z axis, the propagating wave is ordinary-polarized with a refractive index of n_o, as shown in Figure 6.7c. The polarization **D** of this ordinary wave can take any direction in the x–y plane. **E** is also parallel to **D** in this case, because the two vectors are always parallel to each other in the x–y plane due to $\varepsilon_x = \varepsilon_y$. When **k** makes an arbitrary angle θ with the optic axis, the propagating wave has a refractive index of either n_o or $n(\theta)$, depending on its polarization state. If the wave is ordinary-polarized with its **D** perpendicular to the optic axis, it has an index of n_o (Fig. 6.7d). **E** is parallel to **D** in this ordinary wave. In the other polarization state shown in Figure 6.7e, **D** has components both parallel and perpendicular to the optic axis. Therefore **E** may have a different direction than **D**, depending on the value of θ.

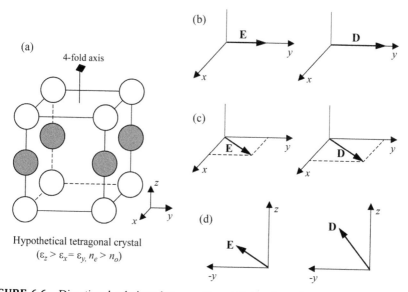

FIGURE 6.6 Directional relations between **E** and **D** in a hypothetical tetragonal crystal (a). When the electric field **E** is applied perpendicular to the characteristic fourfold axis, the displacement vector **D** is parallel to **E** (b,c). If **E** is applied in the y–z plane, **D** lies in the same plane but has a different direction than **E** (d).

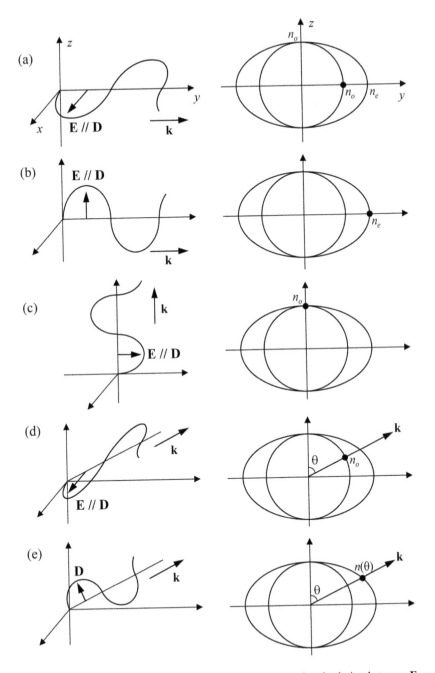

FIGURE 6.7   Determination of the refractive index and directional relation between **E** and **D** of a propagating wave from the *n*-surfaces.

As illustrated in Figure 6.2, **E**, **D**, **k**, and **S** should lie in one plane to which **H** is normal. In this plane, **D** is perpendicular to **k** and **E** is perpendicular to **S**. However, **S** is not necessarily parallel to **k**. Thus the direction of energy flow may be different from the direction of wave propagation in an anisotropic crystal. The directional relations between these four vectors are summarized in Figure 6.8. Before discussion, let us recall the definition of wavefront. It means an imaginary surface joining all points of equal phase in a wave. At each point in space, the wave vector **k** is normal to the wavefront that passes through the point. The wavefronts of a plane wave are parallel planes. The wave vector **k** of a plane wave thus has a single, straight direction, which is perpendicular to the planar wavefronts. The direction of **k** is defined as the direction of wave propagation. Figure 6.8 depicts the propagation of a plane wave whose wave vector **k** lies in the y–z plane and makes an angle θ with the optic axis. Since the optical properties of a uniaxial crystal are isotropic in a plane normal to its optic axis, the description that follows applies for any **k** direction as long as the angle between **k** and the optic axis is θ. When the wave vector **k** intersects the n-surface, the distance from the origin to the point of intersection is the refractive index of the wave. While **D** is always perpendicular to **k**, **E** has a direction tangential to the n-surface at the point of intersection. When the wave is ordinarily polarized, the n-surface is a sphere of radius n_o. Therefore, **E** is parallel to **D** and **S** is parallel to **k** in the ordinary wave. If the wave is extraordinarily polarized, the n-surface is an ellipsoid given by $x^2/n_e^2 + y^2/n_e^2 + z^2/n_o^2 = 1$. It is represented as an ellipse of $y^2/n_e^2 + z^2/n_o^2 = 1$ in Figure 6.8. When is $0°$, the wave is no longer extraordinary-polarized. It becomes an ordinary wave and has a refractive index of n_o. Since **E** has a direction tangential to the ellipse at the point of its intersection with **k**, the **E** vector of the extraordinary wave is not parallel to **D** except for $\theta = 90°$. As a consequence, the Poynting vector **S**, which is perpendicular to **E**, is not parallel to **k**. The angle between **k** and **S** is equal to the angle between **D** and **E**. This angular difference is known as the *walk-off angle*. The fact that **E** is in a direction tangential to the ellipse at the point of intersection can be easily proved from the tensor relation of eq 6.4. The proof is left as a problem (Problem 6.2). In summary, when a plane wave propagates along one of the principal axes of the uniaxial crystal, the wave has **E** // **D** and **k** // **S**, regardless of its polarization state. For a plane wave propagating in an arbitrary direction, the angle between **E** and **D** (i.e., the walk-off angle) depends on the polarization state and propagation direction of the wave. While the walk-off angle of the ordinary wave is always zero, independent of the propagation direction, the extraordinary wave has a walk-off angle that varies with θ.

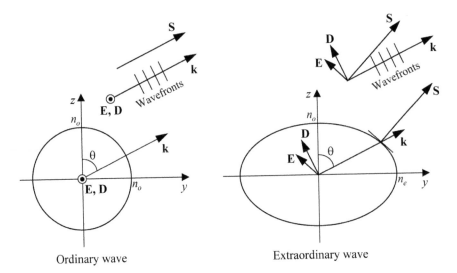

Ordinary wave   Extraordinary wave

FIGURE 6.8   In the ordinary wave, **E** // **D** and **k** // **S**. For the extraordinary wave, **E** is not parallel to **D** and **S** is not parallel to **k**, except for $\theta = 90°$. **E** is along a direction tangential to the index ellipse at the point of its intersection with **k**.

Suppose a plane-wave beam of finite size propagating inside a uniaxial crystal (Fig. 6.9). The Poynting vector **S**, by its definition, represents the direction of energy flow. Since the energy flows along the beam, **S** represents the direction of beam propagation. When the beam is ordinary-polarized, **k** is parallel to **S**. In other words, the direction of wave propagation coincides with the direction of energy propagation. In contrast, if the beam is extraordinary-polarized, **S** may have a different direction than **k**. In this case, the wavefronts of the beam are not perpendicular to **S**. It is important to note that the law of refraction at an interface is governed by the values of **k** vectors, not **S** vectors. Figure 6.10 depicts the distribution of **E**- and **D**-fields in an extraordinary beam whose **k** and **S** are not parallel. **D** is perpendicular to **k** and **E** is perpendicular to **S** at all times. The direction of **k** is defined as the direction normal to the wavefront. While **D** is parallel to the wavefront, **E** is inclined from it. In isotropic media, **D** is related to **E** by a scalar quantity. Therefore, they are always parallel to each other. In anisotropic crystals, the two vectors are related by a tensor and are not always parallel. Crystals belonging to the trigonal, tetragonal, and hexagonal systems have their atoms symmetrically arranged about a specific direction. These crystals have only one such direction, which is usually taken as the z direction. The optic axis corresponds to a unique direction about which the atoms are symmetrically arranged. Since trigonal, tetragonal, and hexagonal crystals have a single optic axis, they are called uniaxial crystals. Light propagating in an arbitrary direction that deviates from the unique symmetry direction (i.e., the direction of the optic axis) will encounter an asymmetric environment. That is why such crystals are optically anisotropic. Table 6.1 lists the refractive indices of some uniaxial crystals.

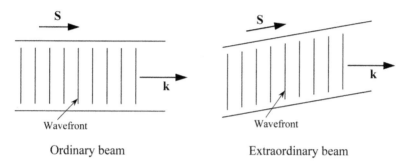

Ordinary beam   Extraordinary beam

FIGURE 6.9 Directions of **S** and **k** in ordinary and extraordinary beams.

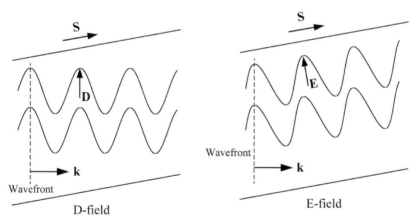

D-field   E-field

FIGURE 6.10 Distribution of **E**- and **D**-fields in an extraordinary beam whose **k** and **S** are not parallel.

TABLE 6.1 Refractive Indices of Some Uniaxial Crystals.

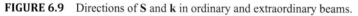

	n_o	n_e
KDP (KH_2PO_4)	1.513	1.471
Rutile (TiO_2)	2.616	2.903
$LiNbO_3$	2.300	2.208
$BaTiO_3$	2.416	2.364
Calcite	1.658	1.486

6.5 DOUBLE REFRACTION

We now examine the refraction of a plane wave at the surface of a uniaxial crystal, that is, at the boundary between air and a uniaxial crystal. The law of refraction requires that the wave vector components tangential to the boundary be conserved. This means that the projections of the wave vectors along the boundary must be equal for both the incident and refracted waves. Suppose that a plane wave of wave vector $\mathbf{k}_i$ is incident, at an angle θ_i, from air into a uniaxial crystal whose optic axis is normal to the boundary, as shown in Figure 6.11a. The magnitude of a wave vector $\mathbf{k}$ is $k = 2\pi n/\lambda_o$, where λ_o is the vacuum wavelength and n is the refractive index of the medium. Thus, the incident wave vector $\mathbf{k}_i$ has an actual magnitude of $k_i = 2\pi/\lambda_o$. If we let $\mathbf{k}_i$ has unit magnitude (i.e., $k_i = 1$), the wave vector inside the crystal has a magnitude equal to the refractive index of the crystal. The tangential component of the incident wave vector is $\sin\theta_i$. There are two possible wave vectors $\mathbf{k}_o$ and $\mathbf{k}_e$ inside the crystal whose tangential components to the boundary are equal to $\sin\theta_i$. These two wave vectors, supported by two different polarization modes, have different directions and magnitudes. $\mathbf{k}_o$ corresponds to an ordinary wave of refractive index n_o and $\mathbf{k}_e$, an extraordinary wave of index $n(\theta_e)$. Snell's law of refraction is satisfied for both refracted waves: $\sin\theta_i = n_o \sin\theta_o = n(\theta_e)\sin\theta_e$. Thus, refraction can occur in two different directions. The effect is called *double refraction*. When the incident wave is TE-polarized, it is refracted in the direction of $\mathbf{k}_o$ and propagates as an ordinary wave inside the crystal. If the incident wave is TM-polarized, it propagates in the direction of $\mathbf{k}_e$ as an extraordinary wave. An unpolarized light wave can be regarded as a mixture of TE and TM polarizations. Thus, when the incident light is unpolarized, it will be split into ordinary and extraordinary waves, traveling in different directions (Fig. 6.11b). Air is an isotropic medium with $n = 1$. If we set the magnitude of $\mathbf{k}_i$ to n, the construction shown in Figure 6.11a equally applies for the refraction of a plane wave at the interface between an isotropic medium of refractive index n and an anisotropic medium of indices n_o and n_e.

As described previously, the direction of energy propagation is not always parallel to the direction of wave propagation in an anisotropic medium. This is more obviously illustrated in the following case. Suppose that a plane-wave beam is normally incident into a uniaxial crystal whose optic axis is inclined from the crystal surface in the plane of incidence, as shown in Figure 6.12a. Since the incident wave vector $\mathbf{k}_i$ has no tangential component, the wave vector inside the crystal should also be normal to the boundary. There are two such wave vectors $\mathbf{k}_o$ and $\mathbf{k}_e$, and their magnitudes are $k_o = n_o$ and $k_e = n(\theta)$. Here θ is the angle between the wave vector and the optic axis. As before,

$\mathbf{k}_o$ corresponds to an ordinary wave and $\mathbf{k}_e$, an extraordinary wave. If the incident beam is TE-polarized, the beam propagates into the crystal with its $\mathbf{S}_o$ vector parallel to $\mathbf{k}_o$. When a TM-polarized beam is incident, however, it is deflected at the boundary by a walk-off angle, which is the angular difference between $\mathbf{k}_e$ and $\mathbf{S}_e$. It is to be noted that the wavefronts of this extraordinary beam are still parallel to the crystal surface because they are perpendicular to $\mathbf{k}_e$, as shown in Figure 6.12b. If an unpolarized beam, which contains both TE and TM polarizations, is incident, it can be spatially separated into two beams after passing through the crystal (Fig. 6.12c). The two beams have mutually orthogonal polarizations. One beam is TE-polarized with its electric field perpendicular to the plane of incidence and the other becomes TM-polarized with the electric field parallel to it. Then we can obtain linearly polarized beams from an unpolarized beam. In the configuration of Figure 6.12a, the walk-off angle of the extraordinary beam is functions of θ and the birefringence $\Delta n = (n_e - n_o)$ of the crystal. If the crystal is oriented so that θ is $0°$ or $90°$, the walk-off angle is zero regardless of the birefringence value. For a given crystal, there is an optimum value of θ that maximizes the walk-off angle. The maximum walk-off angles of most uniaxial crystals are less than several degrees. In order to make the transmitted beams completely separated without an overlapping region, the used crystal should be thicker in proportion to the size of the incident beam.

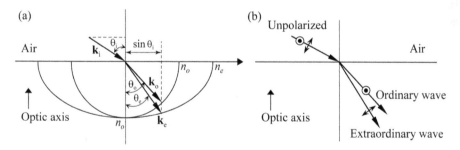

FIGURE 6.11 (a) Application of the law of refraction at the boundary between air and a uniaxial crystal. (b) Double refraction of unpolarized light.

PROBLEMS

6.1 An isotropic solid is a solid material in which physical properties do not depend on its orientation. Anisotropy refers to different properties in different directions, as opposed to isotropy, and many crystals are naturally anisotropic.

The current density (J) of a material is related to the applied electric field (E) by a proportionality constant called the electrical conductivity (σ). While J, E, and σ are all represented by scalar quantities in isotropic materials, their relationship in anisotropic media is given by the following matrix form:

$$
\begin{bmatrix} J_x \\ J_y \\ J_z \end{bmatrix} = \begin{bmatrix} \sigma_x & 0 & 0 \\ 0 & \sigma_y & 0 \\ 0 & 0 & \sigma_z \end{bmatrix} \begin{bmatrix} E_x \\ E_y \\ E_z \end{bmatrix}
$$

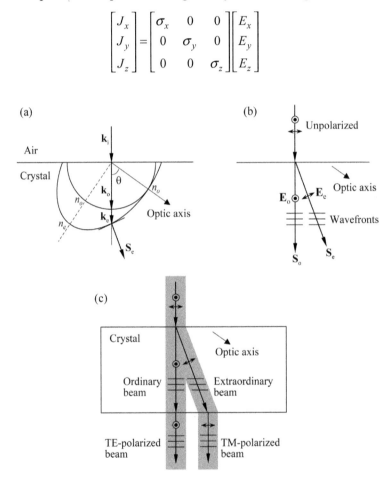

FIGURE 6.12   (a) Normal incidence of a plane wave into a uniaxial crystal whose optic axis is inclined from the crystal surface. (b) Split of an unpolarized beam into ordinary and extraordinary beams. (c) Spatial separation of an unpolarized beam into two linearly polarized beams.

There is a tetragonal crystal whose fourfold rotational symmetry axis is parallel to the z axis. Using the above relation, show that the electrical property of this crystal is isotropic, that is, direction-independent in the x–y plane.

6.2 When an extraordinary wave propagates in a uniaxial crystal, its **D** and **E** vectors are not parallel to each other, except for the case when the wave propagates perpendicular to the optic axis of the crystal. Figure 6.13 shows the index ellipse of a uniaxial crystal having refractive indices n_o and n_e. Suppose that the wave vector **k** of an arbitrary extraordinary wave intersects the ellipse at point (y, z), as illustrated. Maxwell's equations require that **D** is always perpendicular to **k**. Then prove that the **E** vector of the wave is tangential to the ellipse at the point of intersection.

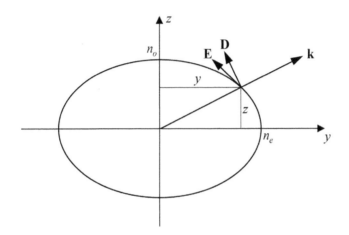

FIGURE 6.13 Index ellipse of a uniaxial crystal

6.3 Unpolarized light is incident on a uniaxial crystal at an incidence angle of 45°, as shown below. The incident light will reveal double refraction (Fig. 6.14).

(a) What is the angle of refraction for each wave?
(b) What is the walk-off angle of the extraordinary wave?
(c) If the incident light is monochromatic with $\lambda_o = 633$ nm, what will the wavelength of each refracted wave be?

6.4 Calculate the maximum walk-off angle available in a negative uniaxial crystal of $n_o = 2.2$ and $n_e = 2.0$ and state the conditions (propagation direction and polarization state) to take advantage of it.

6.5 A plane wave is incident from free space onto a uniaxial crystal ($n_o = 2.0$, $n_e = 2.2$) at an incidence angle of 45°. The optic axis of the crystal is in the plane of incidence and is perpendicular to the propagation direction of the incident wave. Determine the propagation directions of two refracted waves within the crystal.

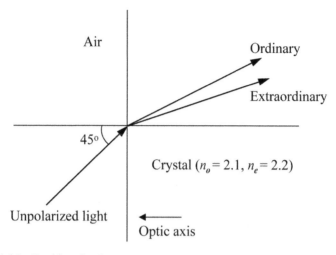

FIGURE 6.14   Double refraction

6.6 Lithium niobate ($LiNbO_3$) is an important material for optical wave-guides, mobile phones, piezoelectric sensors, optical modulators, and other various optical applications. It has a trigonal crystal system and is negative uniaxial ($n_o = 2.30$, $n_e = 2.21$).

Unpolarized light at $\lambda_o = 590$ nm is incident on a $LiNbO_3$ crystal at an angle of 60°, as shown in Figure 6.15. The optic axis lies in the plane of incidence and is parallel to the crystal surface. What are the refraction angles for ordinary and extraordinary waves?

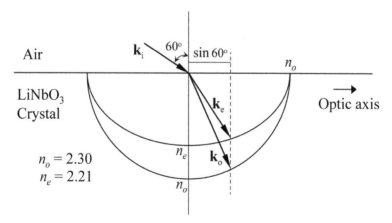

FIGURE 6.15   Double refraction in a $LiNbO_3$ crystal.

REFERENCES

1. Griffiths, D. *Introduction to Electrodynamics*, 4th ed.; Addison-Wesley: London, 2012.
2. Reitz, J.; Milford, F.; Christy, R. *Foundations of Electromagnetic Theory,* 4th ed.; Pearson/Addison-Wesley: Reading, Massachusetts, 2009.

CHAPTER 7

Polarization

7.1 POLARIZATION STATES OF LIGHT

Understanding and manipulating the polarization of light is crucial for many optical applications. Light is an electromagnetic wave. In free space, both the electric field and magnetic field are oscillating perpendicular to the direction of propagation. By convention, the polarization of light refers to the direction of the electric field. If the direction of the electric field of light is well defined, it is called *polarized*. The most common source of polarized light is a laser. Light is said to be *unpolarized* when the electric field has randomly fluctuating directions. Many light sources such as sun, lamps, and incandescent bulbs produce unpolarized light. Such ordinary light is produced by a number of independent atomic sources whose radiation is not synchronized. The resultant electric field does not maintain a constant direction of oscillation. As discussed in the previous chapter, an isotropic medium preserves the polarization of a wave because it does not differentiate between polarization states. However, an anisotropic medium (such as a birefringent crystal) can modify the polarization state of a wave. In addition, some crystals preferentially absorb light polarized in particular directions. Polarization is an important property of light for many optical systems. The polarization of light is classified into three types, depending on how the electric field is oriented: *linear polarization, circular polarization,* and *elliptical polarization.*

7.1.1 LINEAR POLARIZATION

In linearly polarized light, the electric field is confined to a single plane running parallel to the direction of wave propagation. Consider two waves given by

$$\mathbf{E}_x(z,t) = \mathbf{i}E_{ox}\cos(kz - \omega t) \qquad (7.1)$$

and

$$\mathbf{E}_y\left(z,t\right) = \mathbf{j}E_{oy}\cos\left(kz - \omega t + \delta\right). \tag{7.2}$$

One wave propagates in the z-direction with its electric field oscillating along the x-direction, and the other propagates in the same direction with the electric field oriented along the y-direction. Here $\mathbf{i}$ and $\mathbf{j}$ represent the unit vectors in the x- and y-directions, respectively, and δ is the relative phase difference between the waves. If δ is zero or an integral multiple of 2π, the two waves are in phase, as depicted in Figure 7.1. In this case, their vector sum becomes

$$\mathbf{E}\left(z,t\right) = \left(\mathbf{i}E_{ox} + \mathbf{j}E_{oy}\right)\cos\left(kz - \omega t\right). \tag{7.3}$$

The resultant wave has an amplitude vector of fixed magnitude and direction and thus is a linearly polarized wave. As the wave propagates, the electric field oscillates in a plane parallel to the propagation direction, as shown in Figure 7.1. The angle that the plane of oscillation makes with the y-axis depends on the relative magnitudes of E_{ox} and E_{oy}. The superposition of a number of linearly polarized waves, which are in phase with one another, results in a linearly polarized wave. This process can be reversely carried out. That is, any linearly polarized wave can be resolved into its linearly polarized component waves of equal phase. If δ is an odd integer multiple of π, we have

$$\mathbf{E}\left(z,t\right) = \left(\mathbf{i}E_{ox} - \mathbf{j}E_{oy}\right)\cos\left(kz - \omega t\right). \tag{7.4}$$

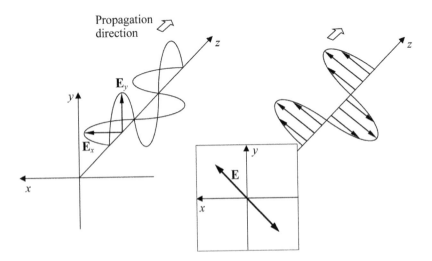

FIGURE 7.1   Linearly polarized light.

The resultant wave is also linearly polarized. But its plane of oscillation is rotated from that of the previous case by some angle, which is not necessarily 90°. If the plane of oscillation makes an angle of α with the y-axis in the case of $\delta = 0$, it makes an angle of $-\alpha$ with the y-axis when $\delta = \pi$. As the phase difference between the two waves changes from 0 to π, the plane of oscillation is rotated by 2α.

7.1.2   CIRCULAR POLARIZATION

Suppose that the two constituent waves have equal amplitudes, that is, $E_{ox} = E_{oy} = E_o$. Additionally, if their relative phase difference is $\pi/2$, $5\pi/2$, $9\pi/2$, and so on, as shown in Figure 7.2a, the resultant wave is

$$\mathbf{E}(z,t) = E_o \left\{ \mathbf{i} \cos(kz - \omega t) - \mathbf{j} \sin(kz - \omega t) \right\}. \tag{7.5}$$

This wave possesses two important features that distinguish it from the previously mentioned, linearly polarized wave. In the linearly polarized waves given by eqs 7.3 and 7.4, $\mathbf{E}$ has a fixed oscillation direction and its scalar magnitude sinusoidally varies with $(kz - \omega t)$. In the wave given by eq 7.5, however, the scalar magnitude of $\mathbf{E}$ is a constant and its direction is not confined to a single plane. As the wave propagates, the direction of $\mathbf{E}$ continuously varies. The $\mathbf{E}$ vector makes one complete rotation as the wave advances by one wavelength. The spatial distribution of $\mathbf{E}$ at some arbitrary time is plotted in Figure 7.2b. Note that the electric field at a fixed position z would appear to rotate counterclockwise when an observer looks at the approaching wave. For instance, at $z = 0$, the time-dependence of the electric field is given by $\mathbf{E}(t) = E_o(\mathbf{i} \cos \omega t + \mathbf{j} \sin \omega t)$. At $t = 0$, $\mathbf{E}$ is along the x-axis in Figure 7.2b. At a later time of $t = \pi/(2\omega)$, it lies along the y-axis. Thus if we look at the wave moving toward us, the electric field rotates counterclockwise with time, as depicted. Such a wave is then said to be *left-circularly polarized*. In contrast, if δ is $-\pi/2$, $3\pi/2$, $7\pi/2$, and so on, the composite wave is given by

$$\mathbf{E}(z,t) = E_o \left\{ \mathbf{i} \cos(kz - \omega t) + \mathbf{j} \sin(kz - \omega t) \right\}. \tag{7.6}$$

The scalar magnitude of $\mathbf{E}$ is the same as before, but $\mathbf{E}$ now rotates clockwise. The wave is then referred to as *right-circularly polarized*. Circularly polarized light consists of two perpendicular, linearly polarized waves of equal amplitude but differing in phase by 90°. As will be discussed later in this chapter, circularly polarized light can be produced by passing linearly polarized light through a quarter-wave plate. If we combine the left-circular wave of eq 7.5 with the right-circular wave of eq 7.6, a linearly polarized wave whose amplitude vector is $2E_o\mathbf{i}$ is obtained.

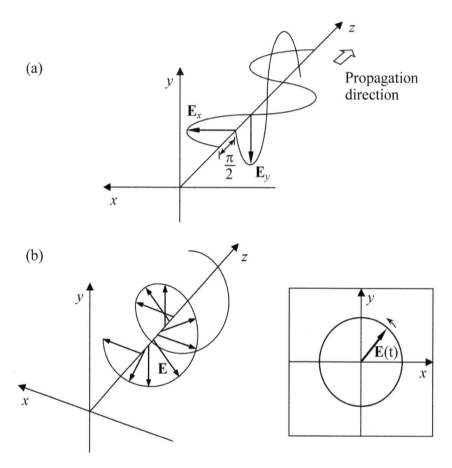

FIGURE 7.2   (a) Two constituent waves of equal amplitude and differing in phase by $\pi/2$. (b) Resulting left-circularly polarized light.

7.1.3   *ELLIPTICAL POLARIZATION*

We now describe the most general case in which the two constituent waves have different amplitudes and an arbitrary phase difference. This leads to an *elliptically polarized wave*. Linearly and circularly polarized light can be regarded as special cases of elliptically polarized light. In elliptically polarized light, the resultant electric field **E** rotates and changes in magnitude as well. The endpoint of **E** traces out an ellipse at a fixed point in space. At a fixed time t, the locus of the end of **E** follows a helical trajectory in space. The addition of eqs 7.1 and 7.2 gives

$$\mathbf{E}(z,t) = \mathbf{i}E_{ox}\cos(kz - \omega t) + \mathbf{j}E_{oy}\cos(kz - \omega t + \delta) = \mathbf{i}E_x + \mathbf{j}E_y . \qquad (7.7)$$

From the expression, we have $E_y/E_{oy} = \cos(kz - \omega t)\cos\delta - \sin(kz - \omega t)\sin\delta$, $\cos(kz - \omega t) = E_x/E_{ox}$, and $\sin(kz - \omega t) = [1 - (E_x/E_{ox})^2]^{1/2}$. Arranging these relations yields

$$\left(\frac{E_x}{E_{ox}}\right)^2 + \left(\frac{E_y}{E_{oy}}\right)^2 - 2\left(\frac{E_x}{E_{ox}}\right)\left(\frac{E_y}{E_{oy}}\right)\cos\delta = \sin^2\delta. \qquad (7.8)$$

This is the equation of an ellipse that makes an angle α with the coordinate system, as shown in Figure 7.3. The angle α is given by

$$\tan 2\alpha = \left(2E_{ox}E_{oy}\cos\delta\right)/\left(E_{ox}^2 - E_{oy}^2\right). \qquad (7.9)$$

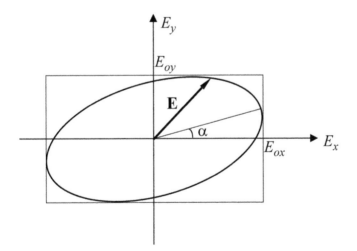

FIGURE 7.3   Elliptically polarized light.

The shape of the ellipse depends on two parameters: the ratio of E_{oy}/E_{ox} and the phase difference δ. If $\delta = \pm\pi/2$, eq 7.8 simplifies to

$$\left(\frac{E_x}{E_{ox}}\right)^2 + \left(\frac{E_y}{E_{oy}}\right)^2 = 1. \qquad (7.10)$$

Now, the principal axes of the ellipse are aligned with the coordinates axes x and y. Figure 7.4 shows the trajectory of the $\mathbf{E}$ vector when $\delta = \pi/2$ and $E_{oy} > E_{ox}$. Additionally, if $E_{ox} = E_{oy} = E_o$, eq 7.10 represents a circle and we have a circularly polarized wave. If δ is an even multiple of π, we have $E_y = (E_{oy}/E_{ox})E_x$. When δ is an odd multiple of π, $E_y = -(E_{oy}/E_{ox})E_x$. These two cases correspond to the linearly polarized waves of eqs 7.3 and 7.4, respectively. In this sense, circularly and linearly polarized light can be viewed as special cases of elliptically polarized light.

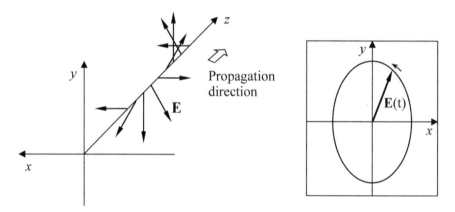

FIGURE 7.4   Trajectory of the $\mathbf{E}$ vector.

7.2   MATRIX TREATMENT OF POLARIZATION

The polarized state of light may change as it passes through an optical element. This can be mathematically described using a simple matrix technique developed by R. C. Jones.[1] In the mathematical description of polarization, polarized light is represented by a *Jones vector* and a linear optical element is represented by a *Jones matrix*. When a monochromatic light wave crosses an optical element, the polarization of the emerging light can be found by taking the product of the Jones matrix of the optical element and the Jones vector of the incident light. The same formalism applies for any complex system consisting of many optical elements. It is to be noted, however, that the Jones matrix technique is only applicable to monochromatic light that is already polarized. To appreciate how the technique is used, let us first examine Jones vectors that describe linear, circular, and elliptical polarizations. Suppose a polarized light beam propagating in the positive z-direction and let the $\mathbf{E}$-field of the beam represented by the vector shown in

Figure 7.5. Since the **E**-field varies continuously in magnitude and direction, Figure 7.5 depicts the magnitude and direction of **E** at a particular instant. As discussed in the previous section, the oscillation of the **E** vector can be described by a linear combination of the oscillations of two orthogonal components E_x and E_y.

$$\mathbf{E} = \mathbf{i}E_x + \mathbf{j}E_y = \mathbf{i}E_{ox}e^{i(kz-\omega t+\phi_x)} + \mathbf{j}E_{oy}e^{i(kz-\omega t+\phi_y)} \tag{7.11}$$

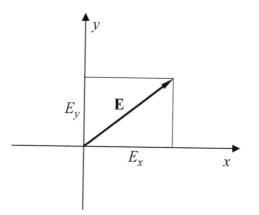

FIGURE 7.5 Instantaneous **E** vector of light propagating in the z direction.

Here E_{ox} and E_{oy} represent the amplitudes of the component waves E_x and E_y, respectively. ϕ_x and ϕ_y are their phases. Equation 7.11 may be rewritten as

$$\mathbf{E} = \left[\mathbf{i}E_{ox}e^{i\phi_x} + \mathbf{j}E_{oy}e^{i\phi_y} \right] e^{i(kz-\omega t)}. \tag{7.12}$$

The bracketed quantity is the complex amplitude vector $\mathbf{E}_a$ of the polarized light. The quantity $\mathbf{E}_a$ has both x- and y-components. Since the polarization state of the light is completely determined by the relative amplitudes and phases of the two component waves, we have only to concentrate on the complex amplitude $\mathbf{E}_a$. The *Jones vector* of the light is written as a 2×1 matrix:

$$\mathbf{E}_a = \begin{bmatrix} E_{ox}e^{i\phi_x} \\ E_{oy}e^{i\phi_y} \end{bmatrix}. \tag{7.13}$$

Thus, the Jones vector represents the amplitude and phase of the electric field in the x- and y-directions. Note that the physical electric field is the real part of this vector. When the light is linearly polarized with its **E** vector oscillating along the x-axis, as shown in Figure 7.6a, we have

$$\mathbf{E}_a = E_{ox} \begin{bmatrix} 1 \\ 0 \end{bmatrix}. \tag{7.14}$$

In the case of $E_{oy} = 0$, the phase ϕ_x can be set to zero for convenience. For the light linearly polarized along the y-axis (Fig. 7.6b), the corresponding Jones vector is

$$\mathbf{E}_a = E_{oy} \begin{bmatrix} 0 \\ 1 \end{bmatrix}. \tag{7.15}$$

In Figure 7.6c, the electric vector $\mathbf{E}$ oscillates along a line inclined at an angle θ to the x-axis. To produce the resultant polarization, the two component waves should be in phase with one another and their amplitudes E_{ox} and E_{oy} are both nonzero. Since it corresponds to a phase difference of $2\pi m$ (m is an integer), we can simply set $\phi_x = \phi_y = 0$. Let E_o be the amplitude of the propagating light. The Jones vector is then given by

$$\mathbf{E}_a = \begin{bmatrix} E_{ox} \\ E_{oy} \end{bmatrix} = E_o \begin{bmatrix} \cos\theta \\ \sin\theta \end{bmatrix}. \tag{7.16}$$

In Figure 7.6c, the $\mathbf{E}$ vector lies in the first and third quadrants. This is because the two orthogonal component waves are in phase. If they are 180° out of phase, saying $\phi_x = 0$ and $\phi_y = \pi$, $\mathbf{E}$ oscillates between the second and fourth quadrants. This means that when the x-component in eq 7.13 increases from the origin along the positive x-direction, the y-component increases along its negative direction.

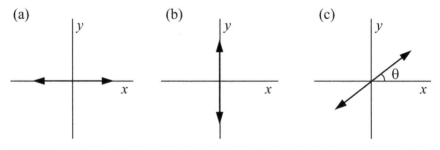

FIGURE 7.6   Three different $\mathbf{E}$ vector directions of linearly polarized light propagating in the z direction.

Left-circularly polarized light has $E_{ox} = E_{oy} = E_o$, $\phi_x = 0$ and $\phi_y = \pi/2$. Then we have

$$\mathbf{E}_a = E_o \begin{bmatrix} 1 \\ i \end{bmatrix}. \tag{7.17}$$

Right-circular polarized light is obtained when $E_{ox} = E_{oy} = E_o$, $\phi_x = 0$ and $\phi_y = -\pi/2$. The corresponding Jones vector is

$$\mathbf{E}_a = E_o \begin{bmatrix} 1 \\ -i \end{bmatrix}. \tag{7.18}$$

Circular polarization is a special case of elliptical polarization. Suppose that the two component waves have unequal amplitudes and a phase difference of $(\phi_y - \phi_x) = \pi/2$. Then, the electric vector $\mathbf{E}$ traces an ellipse whose principal axes are aligned with the x and y axes. In this case, $\mathbf{E}$ sweeps the ellipse counterclockwise, as shown in Figure 7.4. The corresponding Jones vector becomes

$$\mathbf{E}_a = \begin{bmatrix} E_{ox} \\ iE_{oy} \end{bmatrix}. \tag{7.19}$$

If the phase difference is $-\pi/2$, $\mathbf{E}$ traces the ellipse clockwise and the Jones vector is given by

$$\mathbf{E}_a = \begin{bmatrix} E_{ox} \\ -iE_{oy} \end{bmatrix}. \tag{7.20}$$

When the phase difference is any value other than $m\pi$ or $(m + 1/2)\pi$, the principal axes of the ellipse are inclined to the x- and y-axes. Note that if it is $m\pi$, linearly polarized light is obtained. Let the phase difference be ϕ (i.e., $\phi_x = 0$, $\phi_y = \delta$). Then the Jones vector is

$$\mathbf{E}_a = \begin{bmatrix} E_{ox} \\ E_{oy}e^{i\delta} \end{bmatrix} = \begin{bmatrix} E_{ox} \\ E_{oy}\cos\delta + iE_{oy}\sin\delta \end{bmatrix}. \tag{7.21}$$

As shown in eq 7.21, the most general form of the Jones vector is

$$\mathbf{E}_a = \begin{bmatrix} A \\ B + iC \end{bmatrix}. \tag{7.22}$$

Here A, B, and C are parameters that depend on the amplitudes of the component waves and their phase difference. The sum of the squared magnitudes of the two component elements of Jones vectors is proportional to the intensity of light. It is customary to normalize the intensity to unity by dividing both elements in the vector by $(A^2 + B^2 + C^2)^{1/2}$. The normalized Jones vector is

$$\mathbf{E}_a = \frac{1}{\sqrt{A^2 + B^2 + C^2}} \begin{bmatrix} A \\ B + iC \end{bmatrix}. \tag{7.23}$$

When light is linearly polarized along the x-direction, as shown in Figure 7.6a, $A = E_{ox}$ and $B = C = 0$. Then the Jones vector reduces to

$$\mathbf{E}_a = \begin{bmatrix} 1 \\ 0 \end{bmatrix}. \tag{7.24}$$

Equations 7.14 and 7.24 are basically of the same form. Similarly, the Jones vector for light polarized along the y-direction is

$$\mathbf{E}_a = \begin{bmatrix} 0 \\ 1 \end{bmatrix}. \tag{7.25}$$

When light is linearly polarized at 45° from the x-axis,

$$\mathbf{E}_a = \frac{1}{\sqrt{2}} \begin{bmatrix} 1 \\ 1 \end{bmatrix}. \tag{7.26}$$

If light is linearly polarized at 60° from the x-axis ($\theta = 60°$ in Fig. 7.6c),

$$\mathbf{E}_a = \frac{1}{2} \begin{bmatrix} 1 \\ \sqrt{3} \end{bmatrix}. \tag{7.27}$$

For left-circularly polarized light,

$$\mathbf{E}_a = \frac{1}{\sqrt{2}} \begin{bmatrix} 1 \\ i \end{bmatrix} \tag{7.28}$$

and for right-circularly polarized light,

$$\mathbf{E}_a = \frac{1}{\sqrt{2}} \begin{bmatrix} 1 \\ -i \end{bmatrix}. \tag{7.29}$$

The prefactor $1/\sqrt{2}$ or $1/2$ is often neglected in many calculations because it does not affect the polarization state of light. Note that for circularly polarized light, one of the two elements in the Jones vector is pure imaginary. It is common to assign the first element to a real number. As shown above, the polarization state of light is determined by the relative ratio of the two component elements of its Jones vector. Suppose that left-circularly polarized light is combined with right-circularly polarized light of the same amplitude. The addition of their Jones vectors gives

$$\begin{bmatrix} 1 \\ i \end{bmatrix} + \begin{bmatrix} 1 \\ -i \end{bmatrix} = 2 \begin{bmatrix} 1 \\ 0 \end{bmatrix}.$$

The resultant is linearly polarized light having twice the amplitude. This is a simple example that demonstrates the usefulness of the Jones vectors.

The polarization state of light can be modified by various optical elements. While polarized light is represented by a Jones vector, linear optical elements are represented by Jones matrices. When light passes through an optical element, the polarization of the emerging light can be found by taking the product of the Jones matrix of the optical element and the Jones vector of the incident light. The basic principle of this operation is here described with half- and quarter-wave plates. The underlying mechanisms of these polarization devices will be discussed in Section 7.4. When a polarized incident beam represented by its Jones vector $\mathbf{E}_i$ is transformed into a different polarization state $\mathbf{E}_t$ by the operation of an optical element, the process is mathematically expressed as

$$\mathbf{E}_t = M\mathbf{E}_i. \tag{7.30}$$

The transformation matrix M, known as the Jones matrix, is given by

$$M = \begin{bmatrix} a & b \\ c & d \end{bmatrix}. \tag{7.31}$$

Wave plates are typically made of a birefringent material (such as a uniaxial crystal). Since a birefringent crystal has two different indices of refraction, it is possible to induce a controlled phase shift between the ordinary and extraordinary polarization components of a light wave. Suppose that we have a polarized light beam propagating in the positive z-direction with its electric field oscillating in the x–y plane. The Jones vector for this light can be written as

$$\mathbf{E}_i = \begin{bmatrix} E_{ox}e^{i\phi_x} \\ E_{oy}e^{i\phi_y} \end{bmatrix}. \tag{7.32}$$

Here ϕ_x and ϕ_y are the initial phases of the x- and y-component waves. Assume that a wave plate shifts the phases of these two component waves by δ_x and δ_y, respectively. After passing through this wave plate, the light of eq 7.32 will have a Jones vector given by

$$\mathbf{E}_t = \begin{bmatrix} E_{ox}e^{i(\phi_x+\delta_x)} \\ E_{oy}e^{i(\phi_y+\delta_y)} \end{bmatrix}. \tag{7.33}$$

Equation 7.33 is rewritten as

$$\mathbf{E}_t = \begin{bmatrix} E_{ox}e^{i(\phi_x+\delta_x)} \\ E_{oy}e^{i(\phi_y+\delta_y)} \end{bmatrix} = \begin{bmatrix} e^{i\delta_x} & 0 \\ 0 & e^{i\delta_y} \end{bmatrix} \begin{bmatrix} E_{ox}e^{i\phi_x} \\ E_{oy}e^{i\phi_y} \end{bmatrix} = \begin{bmatrix} e^{i\delta_x} & 0 \\ 0 & e^{i\delta_y} \end{bmatrix} \mathbf{E}_i. \qquad (7.34)$$

The Jones matrix of the wave plate has the following form.

$$M = \begin{bmatrix} e^{i\delta_x} & 0 \\ 0 & e^{i\delta_y} \end{bmatrix}. \qquad (7.35)$$

A half-wave plate (often referred to as "$\lambda/2$-plate") rotates the polarization direction of linearly polarized light. Suppose that a light beam linearly polarized at θ from the x-axis is incident into a half-wave plate whose optic axis is along the y-axis, as shown in Figure 7.7. After passing through the half-wave plate, the beam has a polarization direction of $\theta' = -\theta$. The polarization directions of the incident and transmitted beams are symmetric with respect to the optic axis of the plate. Thus if the incident beam has either $\theta = 0°$ or $90°$, it is transmitted without any change in polarization direction. A half-wave plate induces a relative phase shift of π between the two components waves, that is, $|\delta_y - \delta_x| = \pi$. There are an infinite number of choices for δ_x and δ_y, including $\delta_x = 0$ and $\delta_y = \pi$, $\delta_x = -\pi/2$ and $\delta_y = \pi/2$, and so on. No matter what choice we make, the results are physically identical. When the optic axis of the half-wave plate is along the y-axis, the Jones matrix is

$$M = \begin{bmatrix} 1 & 0 \\ 0 & -1 \end{bmatrix}. \qquad (7.36)$$

For the case of $\theta = 45°$ and $\theta' = -45°$, the transformation is written as

$$\begin{bmatrix} 1 \\ -1 \end{bmatrix} = \begin{bmatrix} 1 & 0 \\ 0 & -1 \end{bmatrix}\begin{bmatrix} 1 \\ 1 \end{bmatrix}. \qquad (7.37)$$

If the incident light is linearly polarized at $60°$ from the x-axis, the transmitted light will have the following Jones vector.

$$\begin{bmatrix} 1 \\ -\sqrt{3} \end{bmatrix} = \begin{bmatrix} 1 & 0 \\ 0 & -1 \end{bmatrix}\begin{bmatrix} 1 \\ \sqrt{3} \end{bmatrix} \qquad (7.38)$$

On passing through the wave plate, the polarization direction is rotated by $60°$. When the incident light is polarized along the y-direction (i.e., vertical polarization), we have

$$\begin{bmatrix} 0 \\ -1 \end{bmatrix} = \begin{bmatrix} 1 & 0 \\ 0 & -1 \end{bmatrix} \begin{bmatrix} 0 \\ 1 \end{bmatrix}. \tag{7.39}$$

Despite the minus sign, the polarization direction remains unaltered because the Jones vector of the transmitted light has no x-component.

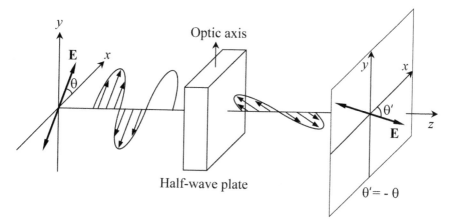

FIGURE 7.7 Rotation of polarization direction by the operation of a half-wave plate.

Circularly polarized light can be produced by introducing a phase difference of $\pi/2$ between two orthogonal components of linearly polarized light. A quarter-wave plate (also called "$\lambda/4$-plate") is a device that induces a relative phase shift of $\pi/2$ between the two components waves. Thus it can convert linearly polarized light into circularly polarized light. In Figure 7.8, a light beam linearly polarized at 45° from the x-axis is incident into a quarter-wave plate with its optic axis oriented along the y-axis. The light entering the quarter-wave plate is resolved into two orthogonal linearly polarized components of equal amplitude. On emerging from the plate, these two components become out of phase by $\pi/2$. Therefore the emerging light is circularly polarized. The sense of circular polarization depends on the optic axis orientation of the plate. Assume that the quarter-wave plate is made of a positive uniaxial crystal ($n_e > n_o$). If the optic axis of the plate is oriented parallel to the y-axis, as illustrated in Figure 7.8, the relative phase shift between the two components waves is $(\delta_y - \delta_x) = \pi/2$. Then, the transformation matrix becomes

$$M = \begin{bmatrix} 1 & 0 \\ 0 & i \end{bmatrix}. \tag{7.40}$$

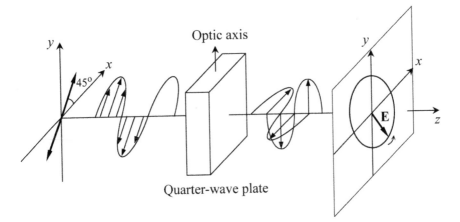

FIGURE 7.8 Conversion of linearly polarized light into circularly polarized light by the operation of a quarter-wave plate.

The Jones vector of the emerging light is given by

$$\begin{bmatrix} 1 \\ i \end{bmatrix} = \begin{bmatrix} 1 & 0 \\ 0 & i \end{bmatrix}\begin{bmatrix} 1 \\ 1 \end{bmatrix}. \tag{7.41}$$

Hence left-circularly polarized light is obtained. Suppose that the optic axis of the quarter-wave plate has been rotated by 90° and thus made parallel to the x-axis. This leads to $(\delta_y - \delta_x) = -\pi/2$. In this case, the transformation matrix is

$$M = \begin{bmatrix} 1 & 0 \\ 0 & -i \end{bmatrix}. \tag{7.42}$$

As a consequence, the same incident light will be right-circularly polarized. When two quarter-wave plates whose optic axes are along the y-axis are placed in the beam path of linearly polarized light, the corresponding operation is

$$\begin{bmatrix} 1 & 0 \\ 0 & i \end{bmatrix}\begin{bmatrix} 1 & 0 \\ 0 & i \end{bmatrix}\begin{bmatrix} 1 \\ 1 \end{bmatrix} = \begin{bmatrix} 1 \\ -1 \end{bmatrix}. \tag{7.43}$$

Two quarter-wave plates of the same orientation plays a role of a single half-wave plate, thus rotating the polarization direction of linearly polarized light. This is obvious from the fact that the thickness of a quarter-wave plate is half the thickness of a half-wave plate. If the first quarter-wave plate is replaced with a half-wave plate, we have

$$\begin{bmatrix} 1 & 0 \\ 0 & i \end{bmatrix}\begin{bmatrix} 1 & 0 \\ 0 & -1 \end{bmatrix}\begin{bmatrix} 1 \\ 1 \end{bmatrix} = \begin{bmatrix} 1 & 0 \\ 0 & -i \end{bmatrix}\begin{bmatrix} 1 \\ 1 \end{bmatrix} = \begin{bmatrix} 1 \\ -i \end{bmatrix}. \tag{7.44}$$

A linearly polarized beam is converted into a right-circularly polarized beam by this combination. By combining various optical elements, we can control not only the polarization state of a light beam but also the transmitted intensity. This will be described in Section 7.4. In the configuration shown in Figure 7.8, the incident light has a polarization angle of $\theta = 45°$. If θ is any value other than ±45°, the polarization of the emerging light will be elliptical rather than circular because two orthogonal polarization components have unequal amplitudes. Of course, when θ is either 0° or 90°, there is no change in polarization state since only one polarization component exists within the quarter-wave plate. Figure 7.9 illustrates that the polarization state of the emerging light depends on the angle between the polarization direction of the incident light and the optic axis of the quarter-wave plate.

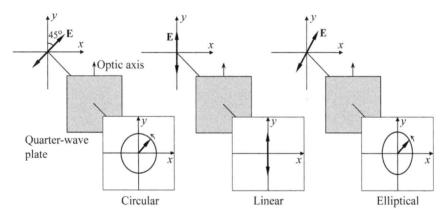

FIGURE 7.9 Dependence of the polarization state of emerging light on the polarization direction of incident light.

7.3 PRODUCTION OF POLARIZED LIGHT

Most common sources of visible light, including sun, lamps, and incandescent bulbs, produce incoherent light. Radiation is produced by a large number of independent atoms or molecules whose emissions are uncorrelated. Consequently, the electric field of light has random directions of oscillation. This state is said to be *unpolarized* (or *randomly polarized*). The current section describes several methods of producing polarized light from such ordinary unpolarized light. The basic principles of most of the methods have already been discussed earlier.

7.3.1 POLARIZATION BY REFLECTION

When light is incident into an interface between two media of different refractive index, it is partially reflected and partially transmitted. The reflection amplitude coefficients are different for waves polarized parallel and perpendicular to the plane of incidence. The Fresnel equations show that TM-polarized wave (with its **E**-field parallel to the plane of incidence) is not reflected if the wave is incident at the Brewster angle θ_B (Fig. 7.10a). The physical mechanism may be qualitatively understood from the electron-oscillator model. When a light wave is incident on the interface, it is refracted and enters the transmitting medium at some angle θ_t. The electric field of the incoming wave oscillates the bound electrons of that medium, which in turn re-radiate. The oscillating dipoles generate a transmitted (refracted) wave. In general, a portion of the radiated energy appears in the form of a reflected wave. The polarization of light is always perpendicular to the direction in which the light is travelling. The dipoles that produce the transmitted (refracted) wave oscillate in the polarization direction of that wave. The reflected wave is also generated by the same dipoles. However, dipoles radiate no energy along their vibration axis. Thus, when a TM-polarized wave is incident at the Brewster angle, there is no reflected wave because its expected propagation direction is parallel to the dipole vibration axis. In the case of TE polarization, the dipole vibration is perpendicular to the propagation direction of the reflected wave. Therefore, such a situation of *zero reflectance* does not occur. Unpolarized light can be viewed as an equal mixture of TE and TM polarization components. Thus, when an unpolarized light beam is incident at the Brewster angle, the reflected beam is completely TE-polarized, as illustrated in Figure 7.10b. For this reason, the Brewster angle is sometimes called the polarization angle. A glass plate or a stack of plates placed at the Brewster angle can be used as a polarizer. When the incident medium is air ($n_i = 1.0$) and the transmitting medium is glass ($n_t = 1.5$), the polarization angle is about 56°. Therefore, if an unpolarized beam enters a glass at an angle of 56°, the reflected beam will be completely polarized with its **E**-field perpendicular to the plane of incidence, that is, parallel to the glass surface. However, a single plate is not effective as a polarizer. It is to be noted that although fully TE-polarized, the reflected beam is weak because the reflection amplitude coefficient at the Brewster angle is not

unity. The stronger transmitted beam still contains both TM and TE components. A more useful polarizer can be obtained by placing a stack of glass
plates at the Brewster angle, as shown in Figure 2.12. Then, a fraction of the
TE-polarized light is reflected from each surface of each plate (air–glass and
glass–air interfaces). For a stack of plates, each reflection depletes the TE
component present in the beam, leaving a greater fraction of TM-polarized
light in the transmitted beam. In this stacked structure, the reflected beam is
spread out and may not be so useful. However, the transmitted beam is fully
TM-polarized and has a single propagation direction.

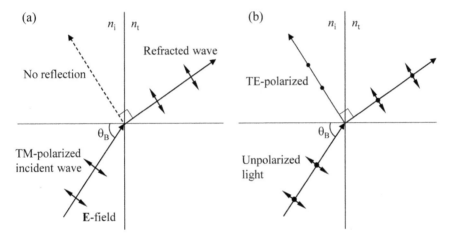

FIGURE 7.10   (a) Incidence of a TM-polarized wave at the Brewster angle θ_B. (b) Reflected
beam completely TE-polarized.

7.3.2   *POLARIZATION BY SELECTIVE ABSORPTION*

The simplest linear polarizer of this type is the wire-grid polarizer, which
consists of many fine parallel metal wires placed on a sheet of transparent
material. A wire-grid polarizer converts an unpolarized beam of light into
a linearly polarized beam. Suppose an unpolarized electromagnetic wave
impinging on the grid (Fig. 7.11). The electric field of the wave can be
resolved into two orthogonal components, one parallel to the wires and
the other perpendicular to them. The component parallel to the wires will
induce the oscillatory motion of conduction electrons along the length of
the wires. As the electrons are accelerated, they collide with lattice atoms
and lose some energy, thereby heating the wires (joule heat). The oscillating
electrons also radiate electromagnetic wave in all directions, except the

direction of the oscillation itself. The wave radiated in the forward direction turns out to be 180° out of phase with the incident wave, resulting in little transmission of the *E*-field component parallel to the wires. The wave radiated in the backward direction simply appears as a reflected wave. For the electric field perpendicular to the wires, the electrons are not free to move far across the width of the wires and their oscillatory motion is inhibited. In this case, the corresponding field is essentially unaltered and passes through the grid. The grid behaves likes a dielectric material when the electric field is perpendicular to the wires, whereas it exhibits a metallic behavior for the field parallel to the wires. The *transmission axis* (TA) of the grid is perpendicular to the wires. We may simply guess that the parallel component of the *E*-field easily slips through the spacing between the wires. What actually happens is opposite to it. For practical purposes, the wire-to-wire separation must be less than the wavelength of the incident light. In addition, the width of each wire should be small compared to the inter-wire separation. Therefore, the wire-grid polarizer is particularly useful for infrared and microwave radiations. Nowadays, advanced lithographic techniques enable the fabrication of very fine metallic grids suitable for visible light. In wire-grid polarizers, the degree of polarization is little dependent on wavelength and angle of incidence.

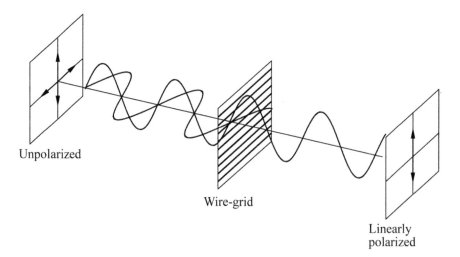

Unpolarized

Wire-grid

Linearly
polarized

FIGURE 7.11   Linear polarization by a wire-grid polarizer.

Some crystals such as tourmaline and herapathite exhibit *dichroism*, preferential absorption of light that is polarized in particular directions.

While the original meaning of dichroism is the split of a light beam into two beams of different wavelength (i.e., color), polarization-dependent absorption is also referred to as dichroism. Despite the preferential absorption, these dichroic crystals are seldom used as a linear polarizer. Since the dichroic effect is wavelength-dependent, the crystals appear colored. Another practical limitation is the difficulty of growing them into large sizes. The most widely used linear polarizer is Polaroid *H*-sheet, which was invented by Land in 1938. When a sheet of polyvinyl alcohol (PVA) is heated and stretched in a certain direction, its long, hydrocarbon molecules are aligned in the direction of stretching and form an array of linear molecular chains. If the sheet is then dipped into a solution of iodine, the iodine attaches to the PVA molecules and makes them conducting along the length of the chains. Valence electrons from the iodine dopant are able to move along the polymer chains, but not transverse to them. Just as in a wire-grid polarizer, light polarized perpendicular to the conducting chains is transmitted while light polarized parallel to them is absorbed. The TA of the polarizer is thus perpendicular to the direction in which the sheet is stretched. The Polaroid *H*-sheet polarizer, much cheaper than other types of linear polarizer, is widely used for sunglasses, photographic filters, and display devices.

7.3.3 POLARIZATION BY BIREFRINGENCE

Other linear polarizers make use of the birefringent properties of uniaxial crystals. When a beam of unpolarized light is normally incident onto a uniaxial crystal whose optic axis is inclined from the crystal surface, it is split into two beams after passing through the crystal, as shown in Figure 6.12. The two beams have mutually orthogonal polarizations. One beam is TE-polarized with its electric field perpendicular to the plane of incidence and the other becomes TM-polarized with the electric field parallel to it. In order to physically separate the two beams, the birefringence of the crystal must be large and its thickness should also increase in proportion to the size of the incident beam. Things become easier if the emerging beams are made to propagate in different directions. This can be accomplished by various beam-splitting polarizers. One of the most commonly used beam splitters is shown in Figure 7.12, where two prisms are combined with an air gap between them. The prisms are made of a birefringent material and their optic axes are aligned perpendicular to the plane of incidence. An unpolarized light beam has two orthogonal polarization components in the prism, one perpendicular to the optic axis (ordinary wave) and the other parallel to it (extraordinary wave). The two components have the same angle of incidence at the prism–air interface, but their refractive indices are different. As a result, these two component waves have different critical angles for total internal reflection. The angle of incidence is equal to the apex angle of the prisms. Thus if the prisms are made to have an apex angle intermediate between the two critical angles, one of the waves is totally reflected from the first prism while the other transmits into the second prism. Which polarization component is internally reflected depends on whether the prisms are made of a positive uniaxial or negative uniaxial crystal. The air gap may be filled with a thin transparent adhesive material. In this case, the apex angle of the prisms should be modified. In Figure 7.12, the optic axes of the prisms are parallel to one another. Several other designs can also be constructed for polarizing beam splitters. For instance, a Wollaston beam splitter consists of two prisms with mutually perpendicular optic axes. It separates unpolarized light into two orthogonal linearly polarized beams that propagate in different directions (see Problem 7.1).

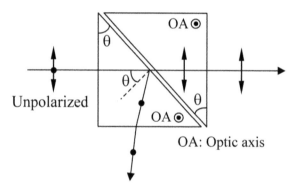

FIGURE 7.12   A polarizing beam splitter.

7.4   POLARIZATION DEVICES

7.4.1   *LINEAR POLARIZER*

As shown in the previous section, a wire-gird or sheet polarizer can linearly polarize unpolarized light. The linear polarizer selectively absorbs the E-field of light oscillating along a specific direction, while transmitting the field perpendicular to that direction. This preferential direction of transmission is called the TA of the polarizer. In the ideal polarizer, the transmitted light will be completely linearly polarized in the same direction as the TA of the polarizer. In reality, however, the selectivity may not be 100%. The completeness of polarization can be easily tested by introducing a second identical polarizer, which is called an analyzer. In Figure 7.13, the TA of the analyzer is rotated by θ from the vertical TA of the polarizer. Suppose that both the polarizer and analyzer are ideal linear polarizers. If the amplitude of the E-field transmitted by the polarizer is E_0, its component parallel to the TA of the analyzer, $E_0 \cos \theta$, will pass through the analyzer. Then the intensity measured by a detector placed behind the analyzer is given by

$$I(\theta) = I(0)\cos^2 \theta. \tag{7.45}$$

This is known as Malus's law. $I(0)$ is the intensity when θ is zero. It corresponds to the maximum intensity obtained when the TA of the analyzer is coincident with the TA of the polarizer. As the analyzer is rotated from $\theta = 0$, the measured intensity will gradually decrease, reaching a minimum at $\theta = 90°$. The minimum intensity will be zero only when the polarizer and analyzer are ideal linear polarizers. When the transmission axes of two linear polarizers are arranged perpendicular to one another, the arrangement is said to be *crossed*. In the ideal case, no light can pass through two crossed linear polarizers.

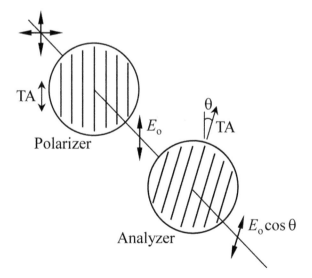

FIGURE 7.13   Two linear polarizers with nonparallel transmission axes.

7.4.2 PHASE RETARDER

We now describe a different type of polarization devices known as phase retarders. The phase retarder does not involve the absorption of a specific polarization component but induces a phase difference between two orthogonal polarization components. Phase retarders are mostly made of uniaxial crystals. A typical phase retarder is simply a uniaxial crystal plate with a carefully chosen orientation and thickness. The crystal is cut into a plate so that the optic axis of the crystal is parallel to the surface of the plate. Unlike linear polarizers, phase retarders are designed to work at specific wavelengths because the induced phase difference depends on the wavelength of light. They are used for monochromatic light sources such as laser beams. When a light wave of wavelength λ_o is normally incident onto the plate, it is decomposed into ordinary and extraordinary waves. The ordinary wave whose polarization is perpendicular to the optic axis travels with a speed of $v_o = c/n_o$, while the extraordinary wave polarized parallel to the optic axis travels at $v_e = c/n_e$. In addition, the ordinary and extraordinary waves have different wavelengths inside the plate, which are given by λ_o/n_o and λ_o/n_e, respectively. This leads to a phase difference between the two component waves when they exit the crystal. As a consequence, the emerging light may have a different polarization state than the incident light. The phase difference ($\Delta\delta$) depends on the thickness (d) of the plate and is given by

$$\Delta\delta = \frac{2\pi\Delta n d}{\lambda_o}. \tag{7.46}$$

Here $\Delta n = |n_e - n_o|$ is the birefringence of the crystal and λ_o is the vacuum wavelength of the light. A retarding plate that induces a phase difference of π is called a half-wave plate. The thickness of the half-wave plate is $d = \lambda_o/(2\,\Delta n)$. Figure 7.14 shows how the polarization direction of linearly polarized light can be rotated by a half-wave plate, where the light is propagating in the z-direction and the optic axis of the plate is along the y-direction. In anisotropic crystals, the polarization direction is defined by the direction of the displacement vector $\mathbf{D}$, rather than the electric field $\mathbf{E}$. However, when the wave propagates in a direction perpendicular to the optic axis of the crystal, $\mathbf{E}$ is always parallel to $\mathbf{D}$. The direction of the electric field is here used for convenience. Suppose that the $\mathbf{E}$-field of the incident light makes an angle θ with the y-axis. The electric field is resolved into two orthogonal components $\mathbf{E}_e$ and $\mathbf{E}_o$ in the plate; $\mathbf{E}_e$ is parallel to the optic axis (extraordinary wave) and E_o, perpendicular to it (ordinary wave). The ordinary and extraordinary waves are in phase with one another on the entrance surface of the plate. As the two waves propagate through the plate, their phase difference gradually increases. On the exit surface of the plate, they will have a phase difference of π. Since a relative phase shift of π has been introduced into the two waves, the emerging light has a polarization direction that is symmetric to the original polarization direction with respect to the optic axis. Obviously, there is no polarization rotation when θ is 0 or 90°. The mathematical representation for the polarization rotation by a half-wave plate was already described in section 7.2.

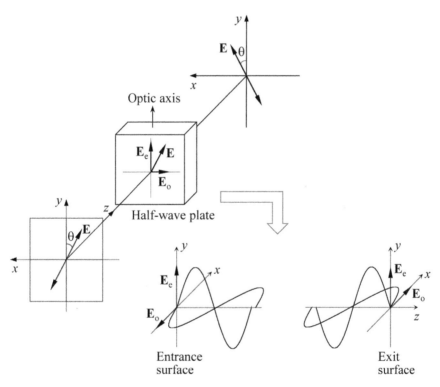

FIGURE 7.14 Rotation of the polarization direction of linearly polarized light by a half-wave plate.

A quarter-wave plate introduces a phase difference of $\pi/2$ between the ordinary and extraordinary waves. Thus it can convert linearly polarized light into circularly polarized light and vice versa. The quarter-wave has half a thickness of the half-wave plate, when they are made of the same crystal. In Figure 7.15, a light beam linearly polarized at 45° from the *y*-axis is incident into a quarter-wave plate with its optic axis oriented along the *y*-axis. As in a half-wave plate, the incident light is resolved into two orthogonal linearly polarized components. Since the polarization angle is 45°, these two components are of equal amplitude. On emerging from the plate, the two components become out of phase by $\pi/2$. This makes the emerging light circularly polarized. The sense of circular polarization depends on the orientation of the optic axis of the plate, as discussed in Section 7.2. If left-circularly polarized light was obtained when the optic axis was oriented parallel to the *y*-axis, right-circularly polarized light can be produced by rotating the plate by 90° about the *z*-axis. Note that unless the polarization angle θ is ±45°, elliptically polarized light will emerge since the two orthogonal polarization components have unequal amplitudes.

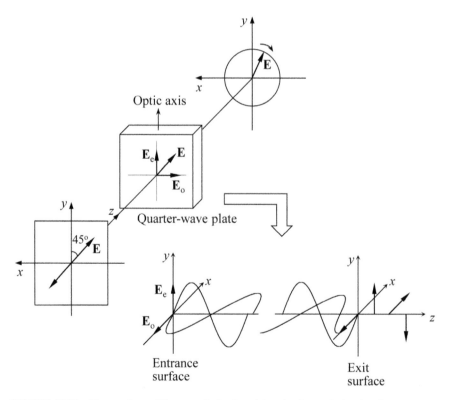

FIGURE 7.15   Conversion of linear polarization into circular polarization by a quarter-wave plate.

7.4.3 COMBINATION OF POLARIZATION DEVICES

Multiple polarization devices can be combined to control the transmitted intensity as well as the polarization state of a light beam. Figure 7.16 shows an optical system consisting of a phase retarder placed between two crossed linear polarizers. In the absence of a phase retarder, no light is transmitted since the transmission axes of the polarizers are orthogonal. Suppose that the inserted phase retarder is a half-wave plate made of a positive uniaxial crystal ($n_e > n_o$) and its optic axis makes an angle θ with the y-axis, that is, the TA of the first polarizer. After passing through the half-wave plate, the incident light will be linearly polarized at 2θ from the y-axis. The polarization component parallel to the TA of the second polarizer is cos $(90° - 2\theta)$. The transmitted intensity is then

$$I(\theta) = I_o \sin^2 2\theta. \tag{7.47}$$

The transmitted intensity of the light can be controlled by rotating the half-wave plate. Here I_o is the maximum transmission intensity of the system, which is obtained when $\theta = \pm 45°$. When the optic axis of the half-wave plate is vertical, that is, $\theta = 0$, its Jones matrix is given in eq 7.36. For an arbitrary θ value, the matrix is

$$M = \begin{bmatrix} \cos 2\theta & \sin 2\theta \\ \sin 2\theta & -\cos 2\theta \end{bmatrix} \tag{7.48}$$

The Jones matrices of the linear polarizers transmitting light with electric fields parallel to the x and y axes are

$$M_x = \begin{bmatrix} 1 & 0 \\ 0 & 0 \end{bmatrix} \text{ and } M_y = \begin{bmatrix} 0 & 0 \\ 0 & 1 \end{bmatrix}. \tag{7.49}$$

Let the incident beam be unpolarized. After passing through the first polarizer, its electric field can be represented by the following Jones vector:

$$\frac{1}{\sqrt{2}} \begin{bmatrix} 0 \\ 1 \end{bmatrix}$$

We here assume that the intensity of the incident beam is unity and half of the initial intensity passes through the polarizer. The Jones vector of the transmitted beam is

$$E_t = \begin{bmatrix} 1 & 0 \\ 0 & 0 \end{bmatrix} \begin{bmatrix} \cos 2\theta & \sin 2\theta \\ \sin 2\theta & -\cos 2\theta \end{bmatrix} \frac{1}{\sqrt{2}} \begin{bmatrix} 0 \\ 1 \end{bmatrix} = \frac{1}{\sqrt{2}} \begin{bmatrix} \sin 2\theta \\ 0 \end{bmatrix}. \tag{7.50}$$

The transmitted beam has no E-component parallel to the y-axis, thus being horizontally polarized. The intensity is the same as eq 7.47, where I_o is 1/2 since we assumed that the intensity of the incident beam is unity and half of the incident intensity passes through the first polarizer.

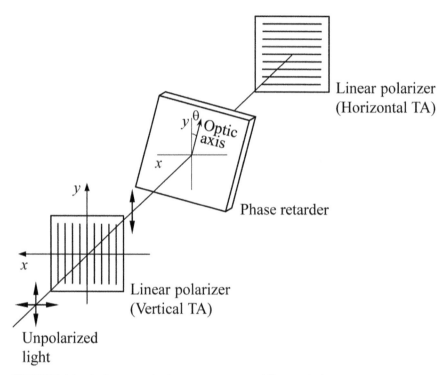

FIGURE 7.16 A phase retarder between two crossed linear polarizers.

Suppose that the retarder in Figure 7.16 is a quarter-wave plate with its optic axis at 45° with respect to the y-axis ($\theta = 45°$). In this case, the Jones matrix of the quarter-wave plate is represented as

$$M = \begin{bmatrix} 1/\sqrt{2} & -i/\sqrt{2} \\ -i/\sqrt{2} & 1/\sqrt{2} \end{bmatrix}. \tag{7.51}$$

The derivation of eq 7.51 remains as an exercise (see Problem 7.2). The Jones vector of the transmitted beam is then

$$\mathbf{E}_t = \begin{bmatrix} 1 & 0 \\ 0 & 0 \end{bmatrix} \begin{bmatrix} 1/\sqrt{2} & -i/\sqrt{2} \\ -i/\sqrt{2} & 1/\sqrt{2} \end{bmatrix} \frac{1}{\sqrt{2}} \begin{bmatrix} 0 \\ 1 \end{bmatrix} = \frac{-i}{2} \begin{bmatrix} 1 \\ 0 \end{bmatrix}. \tag{7.52}$$

While the transmitted light is also horizontally polarized, its intensity is a quarter of the intensity of the incident unpolarized light. After passing through the quarter-wave plate, the light becomes circularly polarized. Thus its E-field component parallel to the x-axis can pass through the second polarizer. An optical isolator is a device that allows the transmission of light in only one direction. It is useful for preventing reflected light from returning back to the source. A simple isolator system can be constructed by a quarter-wave plate combined with a polarizing beam splitter. Figure 7.17 illustrates a simplified diagram of the isolator system used for optical storage devices (CD, DVD). Linearly polarized light from a laser source passes through a polarizing beam splitter and then becomes circularly polarized by a quarter-wave plate. The optic axis of the quarter-wave plate is at 45° with respect to the polarization direction of the incident light. After reflection from the storage medium, light crosses the quarter-wave plate again. This converts circularly polarized light back into linearly polarized light. Since the returning light has passed through the quarter-wave plate twice, it is linearly polarized but has a polarization direction orthogonal to that of the incident light. The polarizing beam splitter blocks the reflected light and deflects it toward a detector, as shown. In this system, the quarter-wave plate rotates the polarization direction of the incoming light by 90° and thus acts as a half-wave plate.

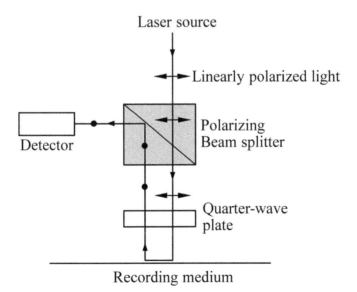

FIGURE 7.17   Isolator system used for optical storage devices.

PROBLEMS

7.1 The Wollaston polarizing beam splitter consists of two triangular prisms cemented together to orient their optic axes perpendicular to one another. It separates an unpolarized light beam into two orthogonal linearly polarized outgoing beams. Figure 7.18 illustrates the trajectory of light in the case when the prisms are made of a positive uniaxial crystal ($n_e > n_o$).

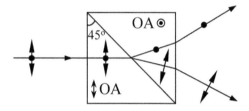

FIGURE 7.18   Wollaston polarizing beam splitter.

The operation principle can be better understood with the schematics shown in Figure 7.19. When unpolarized light is incident on the beam splitter, the first prism decomposes it into ordinary and extraordinary waves. The ordinary wave is refracted at the inter-prism interface and enters the second prism as an extraordinary wave. Note that while the wave is ordinarily polarized in the first prism, it is extraordinarily polarized in the second prism. Since the refractive index (n_e) for the extraordinary wave is larger than the refractive index (n_o) for the ordinary wave, the wave incident onto the interface is refracted upward. The result is opposite in the case when the decomposed extraordinary wave is incident onto the interface. In contrast, the wave extraordinarily polarized by the first prism is refracted into the second prism as an ordinary wave.

Then, calculate the angle between the two outgoing beams if the prisms are made of a uniaxial crystal with indices $n_o = 2.0$ and $n_e = 2.2$.

7.2 Derive the Jones matrices of eqs. 7.48 and 7.51.

7.3 Let us consider we are making half-wave plates with a uniaxial crystal of $n_e = 1.659$ and $n_o = 1.654$. What are the plate thicknesses required for He–Ne laser ($\lambda_o = 632.8$ nm) and Nd:YAG laser ($\lambda_o = 1064$ nm), respectively?

7.4 A beam of linearly polarized light at $\lambda_o = 600$ nm is changed into circularly polarized light after passing through a 0.03-mm thick slice of a uniaxial crystal. Calculate the birefringence of the crystal, assuming that 0.03 mm is the minimum thickness showing the polarization conversion effect.

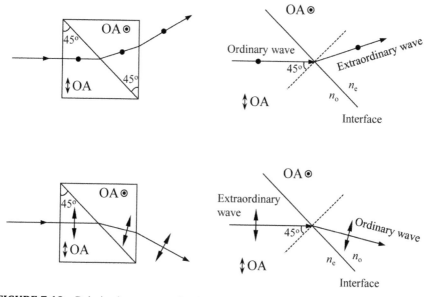

FIGURE 7.19 Polarization states and trajectories of two outgoing beams in the Wollaston beam splitter.

7.5 The optic axis of a half-wave plate makes an angle θ with the y-axis. The Jones matrix of the plate is given by eq 7.48. Show that this half-wave plate will convert left-circularly polarized light into right-circularly polarized light, regardless of the value of θ.

7.6 State what happens to unpolarized light incident on the following beam splitter. Assume that the prisms are made of a positive uniaxial crystal (Fig. 7.20).

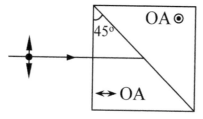

FIGURE 7.20 Polarizing beam splitter

REFERENCE

1. Jones, R. A New Calculus for the Treatment of Optical Systems. *J. Opt. Soc. Am.* **1941,** *31*, 488.

CHAPTER 8

Thin-Film Optics and Reflectance Control

Light reflection and transmission at the planar surface of a dielectric material are determined by its refractive index. However, there are many circumstances under which the surface reflectance of light should be suppressed or enhanced. In this chapter, we describe how light can be controlled at the surfaces and interfaces of materials. Multilayer thin films are widely used to control the reflectance and transmittance of light. These films are usually deposited on glass, silicon or metal substrates by evaporation. It is possible to design multilayer films with special characteristics. Useful applications of such films include antireflection (AR) coatings, band-pass filters, and highly reflecting dielectric mirrors. Micro/nano-scale structures formed into the surface of a substrate can also play a role to suppress the surface reflectance. Unlike multilayered AR coatings, this structure is inherently broadband and works at a wide range of incidence angles. An alternative approach to reduce light reflection is to produce a porous layer on the surface or coat it with a thin film of graded refractive index.

8.1 MULTILAYER THIN FILMS

8.1.1 THEORY

The theory of interference in a single-layer dielectric film has been treated in Section 4.5 for the case of normal incidence. Figure 4.16 shows a thin dielectric layer of refractive index n_1 and thickness d placed between two semi-infinite media of indices n_0 and n_s. This single-layer system has two boundaries denoted as I and II. As shown in eq 4.43, the net electric and magnetic fields at one boundary are related to those at the other as follows.

$$\begin{bmatrix} E_{\mathrm{I}} \\ H_{\mathrm{I}} \end{bmatrix} = \begin{bmatrix} \cos\delta & i\sin\delta/Y_1 \\ iY_1\sin\delta & \cos\delta \end{bmatrix} \begin{bmatrix} E_{\mathrm{II}} \\ H_{\mathrm{II}} \end{bmatrix} = M_1 \begin{bmatrix} E_{\mathrm{II}} \\ H_{\mathrm{II}} \end{bmatrix} \tag{8.1}$$

The characteristic matrix M_1 relates the fields at the two boundaries. Many useful and interesting effects make use of multilayer stacks. Figure 8.1 shows a multilayer film consisting of N layers, where each layer has a particular refractive index and thickness. The number of boundaries at this system is $(N + 1)$. Equation 8.1 is still valid for the fields at boundaries I and II. The fields E_{II} and H_{II} at the boundary II are now related to the fields E_{III} and H_{III} at the boundary III by a second characteristic matrix M_2.

$$\begin{bmatrix} E_{II} \\ H_{II} \end{bmatrix} = M_2 \begin{bmatrix} E_{III} \\ H_{III} \end{bmatrix}$$

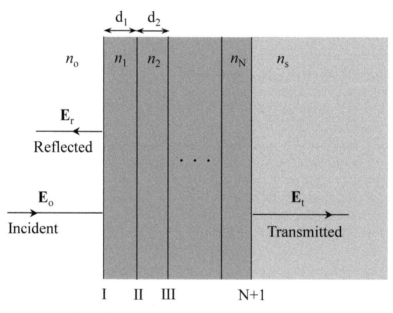

FIGURE 8.1   Multilayer film.

In this way, the first boundary is related to the last boundary by

$$\begin{bmatrix} E_I \\ H_I \end{bmatrix} = M_1 M_2 \cdots M_N \begin{bmatrix} E_{(N+1)} \\ H_{(N+1)} \end{bmatrix}. \tag{8.2}$$

The characteristic matrix of the entire system is then

$$M = M_1 M_2 \cdots M_N = \begin{bmatrix} m_{11} & m_{12} \\ m_{21} & m_{22} \end{bmatrix}. \tag{8.3}$$

Equation 8.2 is rewritten as

$$\begin{bmatrix} E_o + E_r \\ Y_o(E_o - E_r) \end{bmatrix} = \begin{bmatrix} m_{11} & m_{12} \\ m_{21} & m_{22} \end{bmatrix} \begin{bmatrix} E_t \\ Y_s E_t \end{bmatrix}.$$ (8.4)

By expanding the matrices, we can obtain the reflection and transmission amplitude coefficients as follows.

$$r = \frac{E_r}{E_o} = \frac{Y_o m_{11} + Y_o Y_s m_{12} - m_{21} - Y_s m_{22}}{Y_o m_{11} + Y_o Y_s m_{12} + m_{21} + Y_s m_{22}}$$ (8.5)

$$t = \frac{E_t}{E_o} = \frac{2Y_o}{Y_o m_{11} + Y_o Y_s m_{12} + m_{21} + Y_s m_{22}}$$ (8.6)

These equations make it possible to find the reflection and transmission amplitude coefficients for any configuration of multilayer films. To obtain the four matrix elements of m_{11}, m_{12}, m_{21}, and m_{22}, we need to compute the characteristic matrices for individual layers and multiply them.

8.1.2 AR FILMS

AR coatings are used in a wide variety of applications where light passes through an optical surface and low reflection is desired. The well-known examples include AR coatings for camera lenses and other optical instruments, and AR coatings for solar cells and light-emitting devices. AR coatings on glasses are of practical importance, since they are used as the substrates for many optical and optoelectronic devices. The simplest AR coating consists of a single layer of transparent material whose refractive index equals the square root of the substrate's refractive index. In air, such an AR coating theoretically gives zero reflectance at a light wavelength (in the coating) equal to four times the layer thickness, as shown in eq 4.49. A layer is called a "quarter-wave layer" or "$\lambda/4$ layer," when its thickness is equal to a quarter of the wavelength of light within the layer. Glasses have an index of refraction of about 1.52. Thus, an optimum single layer for zero reflectance should have a refractive index of 1.23. However, there are no solid materials with such a low refractive index. MgF_2, which has a refractive index of 1.38, is the closest material with good physical properties. As shown in Figure 4.18, a single layer of MgF_2 can reduce the reflectance of glass from about 4% to less than 1.5%, over the visible spectrum.

It is possible to obtain essentially zero reflectance at a wavelength if two layers are coated using available materials. We here consider the case of normal incidence. For a two-layer film, the resulting characteristic matrix is

$$
M = M_1 M_2 = \begin{bmatrix} \cos\delta_1 & i\sin\delta_1/Y_1 \\ iY_1\sin\delta_1 & \cos\delta_1 \end{bmatrix} \begin{bmatrix} \cos\delta_2 & i\sin\delta_2/Y_2 \\ iY_2\sin\delta_2 & \cos\delta_2 \end{bmatrix}.
$$

When both layers are a quarter-wave layer ($d_1 = \lambda_0/4n_1$, $d_2 = \lambda_0/4n_2$), δ_1 and δ_2 are both $\pi/2$. The matrix product then becomes

$$
M = \begin{bmatrix} 0 & i/Y_1 \\ iY_1 & 0 \end{bmatrix} \begin{bmatrix} 0 & i/Y_2 \\ iY_2 & 0 \end{bmatrix} = \begin{bmatrix} -n_2/n_1 & 0 \\ 0 & -n_1/n_2 \end{bmatrix}.
$$

Substituting the matrix elements into eq 8.5 leads to

$$
R = \left[\frac{n_0 n_2{}^2 - n_s n_1{}^2}{n_0 n_2{}^2 + n_s n_1{}^2} \right]^2. \tag{8.7}
$$

In air ($n_0 = 1$), $R = 0$ is obtained when

$$
n_2/n_1 = \sqrt{n_s}. \tag{8.8}
$$

The first layer (from the air side) should have a lower refractive index than the second layer. On a glass substrate ($n_s = 1.52$), the ideal ratio is $n_2/n_1 = 1.23$. This requirement can be well satisfied by combining materials that have different indices of refraction. TiO_2 ($n = 2.40$), ZrO_2 ($n = 2.10$), and ZnS ($n = 2.35$) are commonly used for high-index layers (*H*-layers), while CeF_3 ($n = 1.65$), SiO_2 ($n = 1.46$), and MgF_2 ($n = 1.38$) serve as low-index materials (*L*-layers). Some coating materials such as Y_2O_3 ($n = 1.82$) have an intermediate refractive index (*M*-layers). The notations *H*, *M*, and *L* are relatively used; when a SiO_2 layer is employed as an *L*-layer, a CeF_3 layer can be regarded as an *M*-layer. Figure 8.2a illustrates a two-layer coating system where a quarter-wave MgF_2 layer ($n_1 = 1.38$) is employed as the first layer and a quarter-wave CeF_3 layer ($n_2 = 1.65$), as the second layer. In the given system, *R* is as low as 0.094% because the index ratio, $n_2/n_1 \approx 1.20$, is close to the ideal ratio. Of course, this essentially zero reflectance is achieved at a wavelength of $\lambda_0 = 4n_1 d_1 = 4n_2 d_2$. For other wavelengths, higher *R* values are obtained.

The characteristic matrix for a three-layer film can be solved in a similar way. When all three layers are a quarter-wave layer, we have

$$M = \begin{bmatrix} 0 & i/Y_1 \\ iY_1 & 0 \end{bmatrix} \begin{bmatrix} 0 & i/Y_2 \\ iY_2 & 0 \end{bmatrix} \begin{bmatrix} 0 & i/Y_3 \\ iY_3 & 0 \end{bmatrix} = \begin{bmatrix} 0 & -in_2/(n_1 n_3) \\ -i(n_1 n_3)/n_2 & 0 \end{bmatrix}.$$

The reflectance is then

$$R = \left[\frac{n_0 n_s n_2^2 - (n_1 n_3)^2}{n_0 n_s n_2^2 + (n_1 n_3)^2} \right]^2. \tag{8.9}$$

No reflection occurs if the indices satisfy

$$(n_1 n_3)/n_2 = \sqrt{n_0 n_s}. \tag{8.10}$$

In the three-layer AR system shown in Figure 8.2b, R is 0.092% at λ_o. Compared to two-layer AR coatings, three-layer coatings can maintain low reflectance values over a wider range of wavelengths.

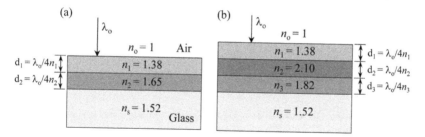

FIGURE 8.2 (a) A two-layer AR coating system. (b) A three-layer AR coating.

8.1.3 HIGH-REFLECTANCE FILMS

High-reflectance coatings can also be designed. A simple approach is depicted in Figure 8.3, where quarter-wave H- and L-layers are alternately stacked on a substrate. The matrix product of two adjacent layers is

$$M_{LH} = \begin{bmatrix} 0 & i/Y_L \\ iY_L & 0 \end{bmatrix} \begin{bmatrix} 0 & i/Y_H \\ iY_H & 0 \end{bmatrix} = \begin{bmatrix} -n_H/n_L & 0 \\ 0 & -n_L/n_H \end{bmatrix}.$$

If the periodic stack consists of $2N$ layers (N H-layers and N L-layers), the characteristic matrix of the entire system is then

$$M = \begin{bmatrix} (-n_H/n_L)^N & 0 \\ 0 & (-n_L/n_H)^N \end{bmatrix}. \tag{8.11}$$

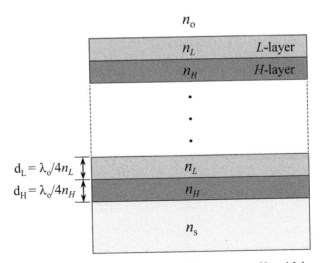

FIGURE 8.3 High-reflectance coating consisting of quarter-wave H- and L-layers alternately stacked on a substrate.

By substituting the matrix elements into eq 8.5, we have

$$R = \left[\frac{(n_0/n_S)(n_H/n_L)^{2N} - 1}{(n_0/n_S)(n_H/n_L)^{2N} + 1} \right]^2. \tag{8.12}$$

The reflectance R approaches unity for large N. For instance, a twelve-layer stack ($N = 6$) of TiO_2 ($n_H = 2.40$) and SiO_2 ($n_L = 1.46$) deposited on a glass substrate gives $R = 98.4\%$, while an eight-layer stack ($N = 4$) results in $R = 89.0\%$. When N is further increased to 8, R becomes 99.8% and the resulting stack behaves like a perfect mirror. If the multilayer stack is free standing in air, we can set $n_0 = n_s = 1$. When the stacking sequence is reversed, the term (n_H/n_L) in eq 8.12 is replaced with (n_L/n_H). But the reflectance still approaches unity as N increases. The high reflectance is a consequence of the constructive interference between light reflecting off the layer interfaces, as illustrated in Figure 8.4. Consider light rays reflected from two adjacent interfaces within the stack. The two rays have a phase difference of π resulting from their optical path difference. However, one of the rays undergoes a phase shift of π upon reflection. Note that when one ray is externally reflected, the other is internally reflected. Since the two rays have a net phase difference that equals an integer multiple of 2π, they constructively interfere. As the number of layers increases, interference occurs more strongly, increasing the reflectance. Of course, eq 8.12 represents the maximum reflectance observed at the wavelength λ_0, for which the layers have optical thicknesses of $\lambda_o/4$ (actual thicknesses are $d_H = \lambda_o/4n_H$ and $d_L = \lambda_o/4n_L$). For other wavelengths, the reflectance must be calculated using the general form of the characteristic matrix, which contains the wavelength-dependent phase differences. Figure 8.5 shows a typical reflectance curve obtained when H- and L-layers are alternately coated on a glass substrate. The width of the high-reflectance range increases as the ratio n_H/n_L increases.

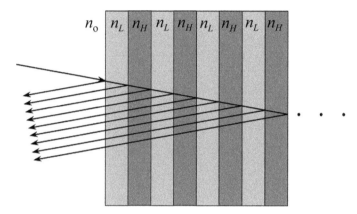

FIGURE 8.4   Constructive interference between light reflecting off the layer interfaces.

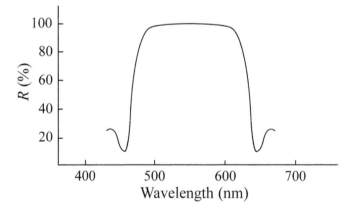

FIGURE 8.5   Reflectance curve for multilayer high-reflectance coating.

8.1.4   DESIGN OF OPTICAL COATINGS

The concept of *Admittance* is very useful for designing optical coatings.[1,2] Let us go back to Figure 4.16, in which a thin dielectric film of refractive index n_1 is deposited on a substrate of index n_s. Equation 4.43 relates the net electric and magnetic fields at the interface I to those at the interface II. We can normalize eq 4.43 by dividing both sides by E_{II} to obtain

$$\begin{bmatrix} E_I/E_{II} \\ H_I/E_{II} \end{bmatrix} = \begin{bmatrix} B \\ C \end{bmatrix} = \begin{bmatrix} \cos\delta & i\sin\delta/Y_1 \\ iY_1\sin\delta & \cos\delta \end{bmatrix} \begin{bmatrix} 1 \\ Y_s \end{bmatrix}. \tag{8.13}$$

Here B and C are the normalized electric and magnetic fields at the front interface I. Y_1, Y_s, and Y_0 are defined in Section 4.5. The common factor can be neglected so that $Y_1 = n_1$, $Y_s = n_s$, and $Y_0 = n_0$. From the quantities B and C, we can extract the reflection and transmission characteristics of the coating system. The admittance Y is defined as the ratio of the normalized magnetic field to the normalized electric field. The admittance at the film-substrate interface is $Y_s = n_s$. The admittance at the front interface is generally complex, given by

$$Y = H_I/E_I = C/B = [Y_s\cos\delta + iY_1\sin\delta]/[\cos\delta + i(Y_s/Y_1)\sin\delta] = x + iy. \tag{8.14}$$

Thus, the admittance Y depends on n_1, n_s, and δ. From eq 4.44, we have $Y = Y_0(E_0 - E_r)/(E_0 + E_r)$. As the reflection amplitude coefficient is given by $r = E_r/E_0 = (Y_0 - Y)/(Y_0 + Y)$, the reflectance R is

$$R = |r|^2 = \left| \frac{Y_0 - Y}{Y_0 + Y} \right|^2. \tag{8.15}$$

For a multilayer film, the matrix in eq 8.13 is just replaced by the characteristic matrix of the entire system. Thus eq 8.15 holds for any configuration. When the substrate is uncoated, $Y = Y_s$. Therefore, the reflectance of a bare substrate is given by $R = (n_0 - n_s)^2/(n_0 + n_s)^2$, which is equivalent to eq 2.10. If the substrate is an absorbing medium whose refractive index has both real and imaginary parts ($n_s = n_R + in_I$), then the reflectance is $R = [(n_0 - n_R)^2 + n_I^2]/[(n_0 + n_R)^2 + n_I^2]$. This is equivalent to eq 2.64. As a layer grows on the substrate and its thickness increases, the admittance Y gradually varies from its starting point $Y_s = n_s$. This consequently alters the reflectance.

We now examine how the reflectance changes by combinations of layers and their thicknesses. It is assumed that the incident medium is air (i.e., $Y_0 = n_0 = 1$). The substrate material (i.e., the transmitting medium) is glass unless stated otherwise. When transparent dielectric layers are coated onto the substrate, the transmittance (T) of the coating system is given by the relation of $T = 1 - R$. Since the admittance is generally a complex number, its variation can be represented on the complex plane. This is called the *admittance diagram.* Consider a simple case when a thin layer of index n_1 is coated onto the glass substrate. From eq 8.14, we have the following relations.

$$(x - n_s)\cos\delta = (n_s/n_1)y\sin\delta$$
$$y\cos\delta = (n_1 - xn_s/n_1)\sin\delta$$

Eliminating δ yields

$$[x - (n_s^2 + n_1^2)/2n_s]^2 + y^2 = [(n_s^2 - n_1^2)/2n_s]^2. \qquad (8.16)$$

Equation 8.16 represents a circle centered at $((n_s^2 + n_1^2)/2n_s, 0)$. The radius of the circle is $|(n_s^2 - n_1^2)/2n_s|$. The circle passes through $(n_s, 0)$ and $(n_1^2/n_s, 0)$. The locus of the admittance (x, y) starts from the point $(n_s, 0)$ and is traced out clockwise, which can be shown by setting $\delta = 0$. That is., the admittance Y starts from the substrate admittance $(n_s, 0)$ and rotates clockwise on the complex plane as the layer thickness increases. It arrives on $(n_1^2/n_s, 0)$ when the layer has a quarter-wave thickness $(d = \lambda_0/4n_1, \delta = \pi/2)$. When an additional condition of $n_1 = n_s^{1/2}$ is satisfied, the reflectance becomes zero. If the thickness is further increased, the admittance continues to rotate clockwise on the circle and returns to the starting point when the layer has a half-wave thickness $(d = \lambda_0/2n_1, \delta = \pi)$. As shown in eq 8.15, the reflectance decreases as Y gets closer to the admittance of air, $Y_o = (1, 0)$, and increases in the opposite case. Figure 8.6 compares the loci of Y for two different cases of $n_1 < n_s$ and $n_1 > n_s$. The reflectance is reduced (when $n_1 < n_s$) or enhanced (when $n_1 > n_s$) from that of bare substrate, depending on the refractive index of the layer. That is why AR coating based on a single layer requires a material whose index is lower than the substrate index. It is to be noted that when the layer has a half-wave thickness (in fact, when δ is a multiple of π), the coating system has the same reflectance as the bare substrate, regardless of whether $n_1 < n_s$ or $n_1 > n_s$.

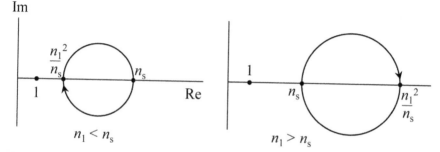

FIGURE 8.6 Admittance loci for $n_1 < n_s$ and $n_1 > n_s$.

Equation 8.15 is still valid for multilayer structures if the matrix in eq 8.13 is replaced by the product of characteristic matrices for each layer. The multilayer structure shown in Figure 8.3 is obtained by alternately coating H- and L-layers onto the substrate. After the first H-layer is coated, eq 8.13 becomes

$$\begin{bmatrix} B \\ C \end{bmatrix} = \begin{bmatrix} 0 & i/n_H \\ in_H & 0 \end{bmatrix} \begin{bmatrix} 1 \\ n_s \end{bmatrix} = \begin{bmatrix} in_s/n_H \\ in_H \end{bmatrix}.$$

Now the admittance is $Y_H = C/B = n_H^2/n_s$. Y_H is larger than $Y_s = n_s$ and thus the reflectance increases. When the second L-layer is over-coated, we have

$$\begin{bmatrix} B \\ C \end{bmatrix} = \begin{bmatrix} 0 & i/n_L \\ in_L & 0 \end{bmatrix} \begin{bmatrix} 0 & i/n_H \\ in_H & 0 \end{bmatrix} \begin{bmatrix} 1 \\ n_s \end{bmatrix} = \begin{bmatrix} -n_H/n_L \\ -(n_s n_L)/n_H \end{bmatrix} = \begin{bmatrix} 0 & i/n_L \\ in_L & 0 \end{bmatrix} \begin{bmatrix} 1 \\ n_H^2/n_s \end{bmatrix}. \quad (8.17)$$

With the over-coated L-layer, the admittance changes to $Y_{LH} = n_s(n_L/n_H)^2$. Since n_H is larger than n_s, Y_{LH} is smaller than Y_H. This will lower the reflectance. It follows from eq 8.17 that this double-layer system is equivalent to a single L-layer coated onto a new substrate whose refractive index is n_H^2/n_s. The new substrate is the original substrate coated with a single H-layer. Similarly, the third H-layer can be regarded as an H-layer coated onto another new substrate that has a refractive index of $n_s(n_L/n_H)^2$. The addition of the third layer gives rise to $Y_{HLH} = n_H^4/(n_s n_L^2)$ and increases the reflectance again. The fourth L-layer yields $Y_{LHLH} = n_s(n_L/n_H)^4$. When each H-layer is coated, the admittance increases, while the L-layer decreases it. This is graphically depicted in Figure 8.7a. As the H- and L-layers are alternately coated, R repeats to increase and then decrease. However, the increment in R with the H-layer is larger than the decrement caused by the L-layer. Therefore, R approaches unity for a large number of layers, as illustrated in Figure 8.7b. For instance, when the multilayer stack consists of TiO_2 ($n_H = 2.40$) and SiO_2 ($n_L = 1.46$), we have $Y_H = 3.79$, $Y_{LH} = 0.56$, $Y_{HLH} = 10.24$, and $Y_{LHLH} = 0.21$. The corresponding R values are $R_H = 0.34$, $R_{LH} = 0.08$, $R_{HLH} = 0.68$, and $R_{LHLH} = 0.43$. Note that although Y_{LHLH} is smaller than Y_{LH}, R_{LHLH} is larger than R_{LH}. This is because Y_{LHLH} is farther away from the admittance of air, $Y_0 = 1$. As shown in eq 8.15, the reflectance R increases to unity when the admittance Y approaches either infinity or zero. The admittance of the multilayer stack shown in Figure 8.3 is $Y_{LH...LH} = n_s(n_L/n_H)^{2N}$, where $2N$ is the number of layers. In this [air|$(LH)^N$|glass] system, the position of the admittance is very close to zero on the real axis of the complex plane. If another H-layer is coated on top of the film so that both ends of the stack consist of the same-type layer, the corresponding [air|$H(LH)^N$|glass] system has $Y = (n_H/n_L)^{2N}(n_H^2/n_s)$. The admittance is very far away from Y_0. In either case, R becomes essentially 100% if N is sufficiently large. The reflectance given here represents the maximum reflectance observed at some target wavelength λ_0, for which the coated layers have an optical thickness of $\lambda_0/4$. For other wavelengths, the layers are no longer a quarter-wave layer. Then, the semi-circles shown in Figure 8.7a become incomplete semi-circles. They will be shortened or lengthened depending on the wavelength of incident light, giving rise to complex admittances. However, when the index ratio n_H/n_L is large enough and the wavelength is not much deviating from λ_0, the admittance ends up on a position still far away from $Y_0 = (1, 0)$ or close to $(0, 0)$. This makes the high-reflectance region fairly wide, as shown in Figure 8.5.

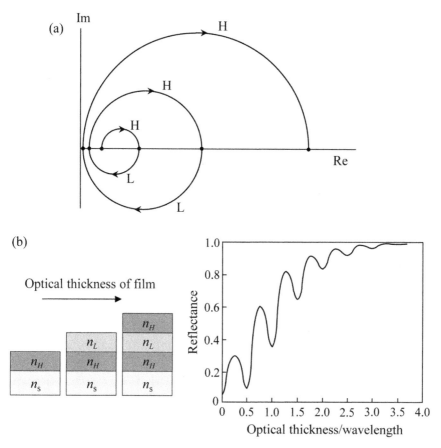

FIGURE 8.7 (a) The locus of admittance when H- and L-layers are alternately coated. (b) Variation of the reflectance R with increasing optical thickness. R approaches unity for a large number of layers.

The admittance diagram is also useful for designing AR coatings and predicting their performance. Suppose that a single MgF_2 layer ($n_1 = 1.38$) is coated on the glass substrate ($n_s = 1.52$) and this layer has a quarter-wave thickness at λ_0, as shown in Figure 8.8a. When the wavelength (λ) of incident light equals λ_0, the admittance is located on (1.25, 0), as depicted in Figure 8.8b. For λ greater than λ_0, the quarter-wave thickness, $\lambda/4n_1$, is larger than the actual layer thickness of $\lambda_0/4n_1$. The resulting phase difference is thus less than $\pi/2$. The admittance at λ is located in the fourth quadrant of the complex plane, as shown in Figure 8.8c. For the opposite case when λ is shorter than λ_0, the admittance arrives on a position in the first quadrant (Fig. 8.8d). The distances from these admittances to (1, 0) are not much different from the distance from (1.25, 0) to (1, 0). Therefore, low R values can be maintained over quite a wide range of wavelengths, as shown in Figure 4.18. As already discussed, $R = 0$ is not achieved with a single layer due to the absence of low-index materials that satisfy the condition of $n_1 = n_s^{1/2}$. To obtain $R = 0$, the admittance of the coating should arrive on (1, 0), which is the admittance of air. Consider the double-layer coating shown in Figure 8.2a. The admittance diagram for this [air|LM|glass] system is plotted in Figure 8.9a. When the M-layer of index n_2 is coated on the substrate, the admittance arrives on (1.79, 0). After the L-layer of index n_1 is over-coated, it arrives on (1.06, 0). Since the final admittance is very close to (1, 0), the obtained reflectance, $R = 0.09\%$, is nearly zero. This reflectance is achieved at the target wavelength of $\lambda_0 = 4n_1d_1 = 4n_2d_2$. For other wavelengths, the layers are no longer a quarter-wave layer. The admittances thus arrive on some positions (marked "A" and "B" in Fig. 8.9b,c) that are away from (1.06, 0). As the wavelength more deviates from λ_0, the distances from "A" and "B" to (1, 0) more increase, thereby increasing the reflectance. As a consequence, the reflectance vs. wavelength spectrum exhibits a "V-shape" curve. This is called a *V-coating*.

Two-layer AR coatings exhibit a single minimum in R, which can be theoretically zero at $\lambda = \lambda_0$. On either side of the minimum, the reflectance rises more abruptly than for the single-layer coating. Thus far, our discussion has been restricted to quarter-wave thicknesses, that is, $\lambda/4$ layers. The introduction of layers with different optical thicknesses can make the two-layer coatings work over a wide range of wavelengths. Before going into this matter in detail, let us examine how the admittance locus of a layer depends on the underlying structure. In the configuration of Figure 8.10a where a $\lambda/4$ L-layer is coated on the glass substrate, the admittance locus of the layer starts from (1.52, 0) and ends on (1.25, 0). Then the semi-circle has a radius of 0.135. As shown in Figure 8.10b, when the same L layer is located on top of an M-layer of $\lambda/4$ thickness, its admittance arrives on

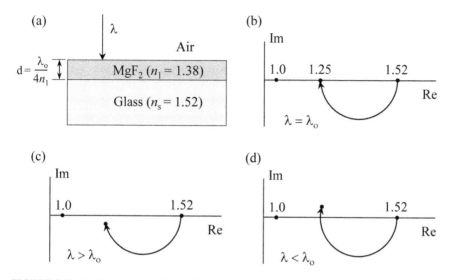

FIGURE 8.8 (a) Single-layer AR coating. (b) At the target wavelength of λ_o, the admittance is located on (1.25, 0). It deviates from this position for other wavelengths (c and d).

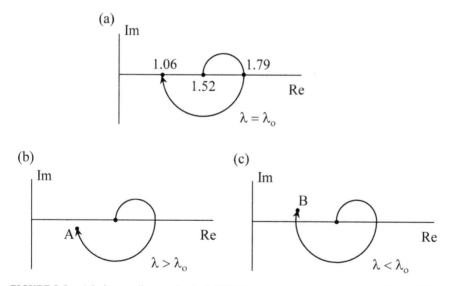

FIGURE 8.9 Admittance diagram for the [air|LM|glass] coating system shown in Figure 8.2(a).

(1.06, 0), starting from (1.79, 0). The semi-circle radius is now 0.365. If the M-layer has a $\lambda/2$ thickness (Fig. 8.10c), the admittance locus of the L-layer becomes identical to the case of Figure 8.10a. This indicates that the radius of the admittance locus of a layer depends on its starting point.

In other words, if the admittance starts from different positions, the locus has different radii. The $\lambda/2$ M-layer has no effect on the reflectance at λ_0, because its admittance returns to the substrate admittance. Therefore, the double-layer system of Figure 8.10c behaves like a single $\lambda/4$ L-layer at this wavelength. However, the $\lambda/2$ M-layer helps maintain AR characteristics over a wider range of wavelengths. The way in which the $\lambda/2$ layer improves the AR performance can be readily understood from the admittance plots shown in Figure 8.11. At $\lambda = \lambda_0$, the admittance locus arrives on the point "C" along the A–B–A–C route. At some wavelength of $\lambda > \lambda_0$, the locus of the $\lambda/2$ layer ends on the point D. The locus of the $\lambda/4$ L-layer starts from this point and arrives on the point E. Since the locus DE has a larger radius than the semi-circle AC, the point E may be closer to $(1, 0)$ than the point C at a specific wavelength. The case of $\lambda < \lambda_0$ would give a similar effect with the points D and E above the real axis and the loci slightly longer than full circle and semi-circle. Two reflectance minima then appear on both sides of the wavelength λ_0. This "$\lambda/4$–$\lambda/2$"configuration is thus called a W-coating. Although the reflectance at λ_0 is higher compared to the $\lambda/4$–$\lambda/4$ coating, it remains at low values over a wide range of wavelengths. The structure given in Figure 8.10c is denoted as [air|LMM|glass], where "MM" means two $\lambda/4$ M-layers, that is, a single $\lambda/2$ layer. The insertion of a $\lambda/2$ H-layer into the [air|LM|glass] system can also extend the AR range. The admittance of the corresponding [air|$LHHM$|glass] system is shown in Figure 8.12. This "$\lambda/4$–$\lambda/2$–$\lambda/4$"configuration can give extremely low R values over a fairly wide range around λ_0. Typical reflectance spectra obtained with different configurations are compared in Figure 8.13. In each configuration, the curve shape somewhat varies with the refractive indices of the layers.

8.2 SURFACE STRUCTURES FOR ANTIREFLECTION

In the previous Section 8.1, we have shown that the reflectance and transmittance of light can be controlled by use of multilayer dielectric films. Multilayer-based AR coatings consist of alternating layers of contrasting refractive index. Layer thicknesses are chosen to produce destructive interference in the reflected beams and constructive interference in the transmitting beams. This makes the AR performance of the coating change with incident angle, so that color effects often appear at oblique angles. A wavelength range must be specified when designing such coatings. Micro/nano-scale structures formed into the surface of a substrate can also suppress light reflection significantly. Unlike multilayer AR coatings, these structures are inherently broadband and work at a wide range of incident angles. This section introduces various surface structures utilized for AR.

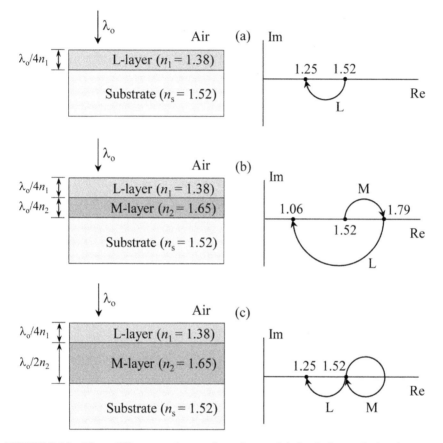

FIGURE 8.10   Three different coating configurations and their admittance loci at the target wavelength λ_o.

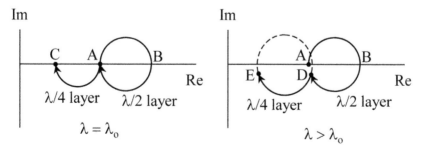

FIGURE 8.11   Admittance loci for the [air|LMM|glass] system.

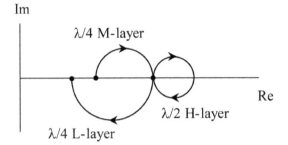

FIGURE 8.12   Admittance diagram for the [air|*LHHM*|glass] coating system.

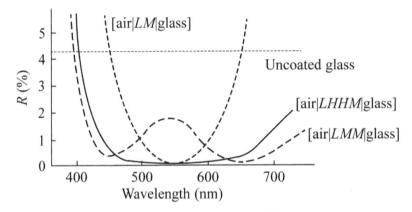

FIGURE 8.13   Typical reflectance spectra obtained with different configurations.

8.2.1 SURFACE TEXTURE

Surface texturing, either in combination with an AR coating or by itself, can be used to minimize reflection. Any "roughening" of the surface reduces reflection by increasing the chance of reflected light bouncing back onto the surface, as illustrated in Figure 8.14. When incoming light strikes a flat surface, it is reflected from the surface at an angle identical to the angle of incidence, where the reflectance R is determined by the Fresnel equations. In a textured surface, the reflected light strikes the surface again, thus reducing the reflectance to R^2. Si has a high surface reflectance (>35%) due to its large refractive index of around 4.0. For photovoltaic applications, the surface of Si is textured to reduce reflection loss and thus enhance light trapping. Surface texturing can be accomplished in a number of ways. It is well known that (100)-oriented single-crystalline Si (s-Si) wafer can be textured into randomly distributed pyramids by alkali etchants such as KOH or NaOH.[3–9] Figure 8.15 shows random-pyramid textures formed on the surface of s-Si wafer. The pyramid structure, revealed by etching anisotropy, reflects incident light twice before it escapes from the surface, thus reducing the total reflectance to less than 20%. Texturing the surface by anisotropic etching is commonly used as a standard technique in the manufacturing of s-Si solar cells. The textured surface makes a normally incident light beam obliquely enter the absorption layer of the cell. Thus it can enhance the probability of light trapping by increasing the optical thickness of the absorption layer. Another type of surface texturing is known as "inverted-pyramid texturing." Compared to the random-pyramid structure where pyramids have a wide distribution in size and shape, the inverted-pyramid structure is a more efficient light-trapping geometry that can improve the energy conversion efficiencies of s-Si solar cells to ~25%.[10–13] Nevertheless, the standard texturing process for s-Si solar cells is alkali etching, because the inverted-pyramid structure requires lithographic processes that significantly increase the overall fabrication cost. Since polycrystalline Si (p-Si) wafer has multiple orientations, only a small fraction of the surface has the required orientation of (100). Consequently, it is impossible to uniformly control its surface structure by anisotropic alkali etching. Acid etching is generally used to reduce the surface reflectance of p-Si solar cells. An acid solution (for instance, HF/HNO$_3$ solution), which tends to etch the wafer isotropically, can result in surface features due to the inhomogeneity of etching speed. Although acid etching helps suppress the surface reflectance to some degree, the obtained irregular structures are less effective than the pyramid textures. p-Si wafers can be regularly textured by using a lithographic technique or mechanically sculpturing the front surface with saws or lasers. In particular, honeycomb textures are known to give very high efficiencies when applied to p-Si solar cells. The honeycomb structures can be fabricated by a number of different methods, including photolithography, interference lithography, and laser interference. Figure 8.16 shows a honeycomb texture fabricated by laser interference, along with its effect on the surface reflectance.

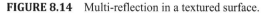

Textured surface

FIGURE 8.14   Multi-reflection in a textured surface.

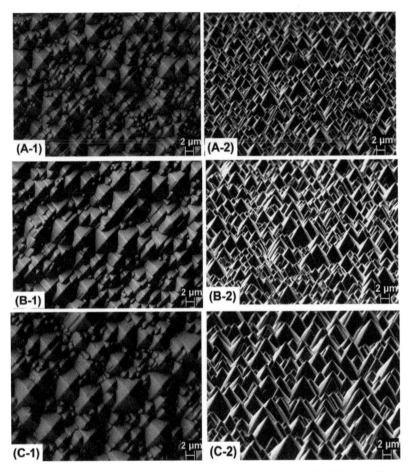

FIGURE 8.15   Random-pyramid textures formed on the surface of s-Si wafer. Reprinted with permission from Ref. [9]. Copyright © 2015 Elsevier.

8.2.2   MOTH'S EYE STRUCTURE

Moth's eye AR structures are arrays of protuberances, with dimensions smaller than the wavelength of the light incident upon them. These structures form a region of graded refractive index at the interface between two media, significantly reducing light reflection from the interface. This "moth's eye structure" has its origin in nature. Moths, well-known nocturnal insects, are able to see at night because their eyes absorb a high portion of light instead of reflecting it. The eyes of moths have two characteristic optical structures that advantageously function to increase their light sensitivity: the tapetal mirror and the corneal nipple array. The tapetal mirror situated behind the eye's photoreceptors reflects unabsorbed light back into the eye structure, thus providing a second change to absorb incoming photons. The corneal nipple array, which is known as the "moth's eye" structure, covers the micron-sized facets of the eye and acts as an AR coating. As a consequence, the moth is able to see in low-light conditions. Natural moth's eye structures consist of protuberances arranged in a hexagonal array, with heights up to 250 nm and spacings of 200–250 nm (Fig. 8.17). These unique structures help moths evade detection by predators in moonlight and maximize light capture to see in the dark. AR structures have also been found on the transparent wings of moths and on the corneal surfaces of butterflies.

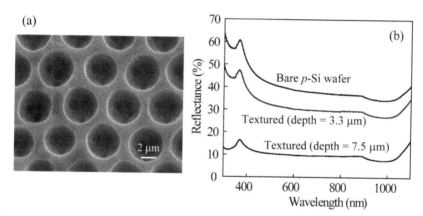

FIGURE 8.16 (a) Honeycomb texture fabricated on a p-Si wafer by laser interference. (b) Variation of the surface reflectance with the texture depth. Adapted with permission from Ref. [14]. Copyright © 2013 Elsevier.

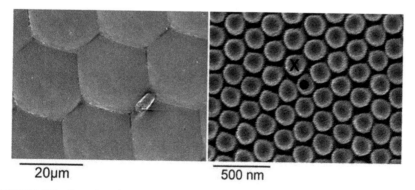

FIGURE 8.17 Scanning electron microscopy (SEM) images of natural moth's eye structures. Reprinted with permission from Ref. [15]. Copyright © 2016 Macmillan Publishers Limited.

Biomimetic, artificial AR structures have been fabricated on a variety of materials such as glasses, polymers, and silicon using different methods (Figs. 8.18 and 8.19). The reduction of reflection plays an important role in applications ranging from photovoltaic and display technologies to individual optical components. From a technological point of view, moth's eye AR structures refer to any artificial structures fabricated to increase the amount of light coupled through the surfaces of optical devices. This "moth's eye structure" was first reproduced by crossing three gratings at 120° using lithographic techniques, and has been used as an AR surface on glass windows[18]. Today the moth's eye structure can be made extremely accurately using electron-beam etching, and is used commercially on solid plastic and other lenses. The mechanism through which moth's eye structures reduce reflection is the removal of a stepwise index change at the air-material interface. When surface structures have feature sizes on the order of micrometers, the reduction of reflection can be explained with the help of geometric optics approximation. As shown in Figure 8.14, rays should be reflected at least twice before being reverted back. However, if the surface structure has feature sizes much smaller than the wavelength of incident light, the incident light cannot resolve the structure. Instead, the light behaves as if it were encountering a medium whose optical property is a weighted average of the optical properties of two involved media. In other words, the structure forms a so-called effective medium. This is known as *effective medium theory* (EMT) or *effective medium approximation*. Figure 8.20 depicts three different surface structures regularly arranged on a substrate material. A surface textured with ridges smaller than the wavelength of light will interact with the light as if it had a single effective layer of refractive index governed by the volume ratio between the ridges and channels (Fig. 8.20a). That is., this effective layer will have a refractive index value intermediate between the refractive index (n_0 =1) of air and the refractive index (n_s) of the substrate material. By the same token, a stepped surface profile will act as a multilayer film where different layers have different effective refractive indices (Fig. 8.20b). Similarly, a tapered profile will behave like an infinite stack of infinitesimally thin layers, exhibiting a gradual index change as light propagates through the profile (Fig. 8.20c). A moth's eye structure is thus regarded as an effective medium of graded index, which plays a role to smooth the transition between one medium and another. Proper design of such structures can result in extremely low reflectances due to the absence of an abrupt change in the refractive index. This can be understood in such a way that the effective medium is a stack of many thin layers, and so the destructive interference between reflections from each layer cancels out all reflected light, maximizing the proportion of light transmitted. Moth's eye structures maintain very low reflectances over a broad range of wavelengths and angles. The broadband and omnidirectional AR characteristics also benefits from the gradual index change.

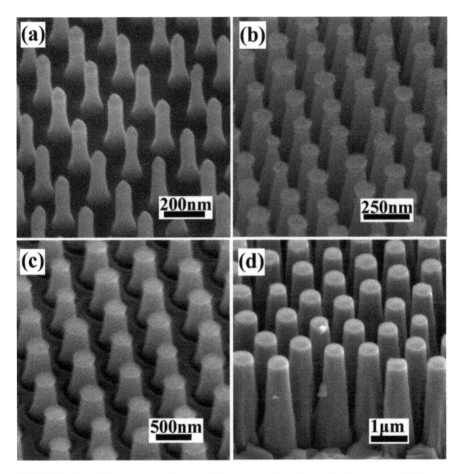

FIGURE 8.18 SEM images of Si nanopillar arrays with uniform tip diameters of (a) 60 nm, (b) 125 nm, (c) 300 nm, and (d) 600 nm. Reprinted with permission from Ref. [16]. Copyright © 2008 American Institute of Physics.

The reflection characteristics of AR surfaces with continuous profiles can be analyzed using two methods: film-stack method and rigorous coupled-wave analysis (RCWA). The film-stack method uses EMT. For a specific surface profile, the effective index of refraction is derived as a function of the depth *z* into the profile. The resulting refractive index gradient is approximated by a stack composed of hundreds of layers. The reflection characteristics of this layered structure are then determined by using the multilayer theory discussed in Section 8.1. RCWA is a differential-based modeling method that is typically applied to solve scattering from periodic dielectric structures. It is the most commonly employed method for analyzing the reflection characteristics of moth's eye structures. RCWA, which does not involve the use of EMT, represents any periodic feature profile as a Fourier series, from which a set of differential equations are developed. RCWA is widely used to model artificial subwavelength AR structures with good agreement between theory and experiment.[19,20] RCWA simulation results have been experimentally verified. This makes RCWA a very effective technique for the design of artificial moth's eye structures. In pillar arrays, the period, shape, and height of the pillars are all crucial factors determining the AR performance. Therefore, these factors must be optimized when designing artificial AR structures for specific purposes. For instance, there will be an optimal period and height for Si solar cells. Figure 8.21a is a contour plot showing the dependence of reflectance at $\lambda = 1$ μm on the period *d* and height *h* in Si pillar arrays. For small periods ($d \ll \lambda$), the structure behaves like an effective medium that creates a refractive index gradient. That is., the array of tapered pillars exhibits properties similar to those expected for an interface of graded index. At small pillar heights, the reflectance is still high because the interface appears abrupt to incoming light. As the height is increased, the refractive index varies more gradually. As a consequence, the reflectance sharply drops. For periods approaching the wavelength λ, rather a complicated behavior is observed due to resonance effects. Between these two regions, there is a low-reflectance band at relatively modest pillar heights, representing a "sweet spot" for the design of artificial moth's eye structures. In general, increasing the pillar height decreases the reflectance. However, the fabrications of taller features are more difficult and costly. For solar cell applications, surface features of high aspect ratio may have an adverse effect of increasing the probability of charge recombination. The existence of a sweet spot thus has important implications for the design of AR surfaces. Pillar heights above approximately half of the wavelength have no practical merits. Natural moths have pillar heights less than 250 nm. Although AR with moth's eye structures is broadband, there are local maximum and minimum in reflectance. Their positions can be tuned by varying the period of the pillars (Fig. 8.21b). The shape of the pillars also has a dramatic effect on the reflectance characteristics. Sharp pillars in the form of a thumbtack lead to a sharp minimum in the reflectance spectrum, making them unsuitable for broadband applications. At the other extreme when pillars have near-vertical side walls and flat tops, the array acts as a single effective layer, as shown in Figure 8.20a. RCWA simulations have shown that smoothly tapered pillars are most optimal for broadband applications[21]. Moth's eye arrays designed for visible wavelengths have periods on the scale of a few hundreds of nm. The scale is too small for standard photolithography to be applicable for the fabrication of such arrays. Several nanofabrication techniques have been employed to create moth's eye structures in a range of materials. In electron-beam lithography (e-beam lithography), a focused beam of electrons is scanned over a photoresist layer coated on the surface of a substrate. A development process leaves behind the exposed resist in the form of an array of dots. Etching is then carried out to transfer a designed pattern into the underlying substrate. Although the period and feature size of the pattern can be precisely controlled by e-beam lithography, it is a time-consuming and highly expensive technique, making this technique unsuitable for commercial and scalable fabrication. As will be discussed in Chapter 9, the interference between two or more laser beams can generate a periodic series of fringes with a period less than the wavelength of the used laser. The interference pattern can be used to expose a photoresist layer coated on the surface of a material. The exposed resist is then developed to leave behind a mask that will be used for a subsequent etching process. This interference lithography is a fast and cheap method of fabricating subwavelength surface structures over large areas and can be effectively utilized to produce artificial moth's eye arrays. Another simple method is nanoimprinting, sometimes called nanoimprint lithography. A stamp, fabricated by conventional lithographic techniques, is pressed into a polymer film coated on a substrate. Removing the stamp leaves an imprint pattern in the polymer film, and the imprinted film is used as an etch mask to transfer the pattern into the substrate. Nanosphere lithography is a bottom-up, self-assembly approach that involves the formation of an etch mask with polymer or silica spheres deposited from a colloidal solution (Fig. 8.19). In nanosphere lithography, the size of mask spheres determines the period of pillar arrays.

Applications for moth's eye structures exist wherever light passes through an optical surface and low reflection is desired. Application areas include solar cells, displays, smart phones, touch panels, and optical elements. Moth's eye AR structures can also be applied to improve the light extraction efficiency of light-emitting diodes by decreasing the number of photons trapped in the device due to total internal reflection.[22] The development of scalable techniques such as nanoimprinting and interference lithography makes it possible to fabricate moth's eye arrays over large areas. However, the susceptibility of moth's eye structures to damage by abrasion remains an issue. In this respect, moth's eye AR arrays may be more suited on inner surfaces or at the interface between materials rather than on exposed surfaces.

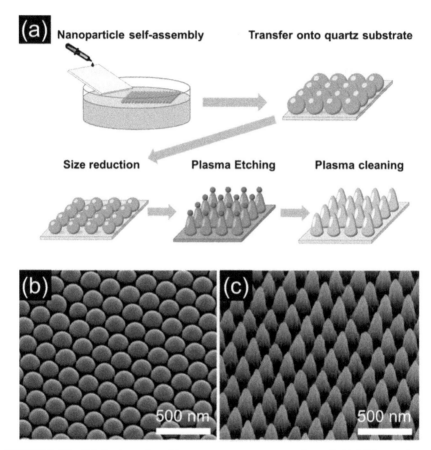

FIGURE 8.19   Fabrication of moth's eye structures on quartz by colloidal lithography. (a) Schematic of the fabrication process. (b) SEM image of polystyrene (PS) nanoparticle array with high magnification. (c) SEM image of moth eyes mimicking nanopillar arrays on quartz substrate with high magnification. Reprinted with permission from Ref. [17]. Copyright © 2013 American Chemical Society.

8.2.3   THIN FILMS OF GRADED INDEX

The broadband AR properties of moth's eye structures originate from the fact that the effective index of refraction is gradually varying in these structures.

We can anticipate similar effects from thin films of graded refractive index. Let us consider a single layer on the surface of a glass substrate and assume that the refractive index of this layer is 1.26, which is an intermediate value between the index of air and the index (n_s = 1.52) of the glass substrate. The reflectance versus wavelength spectra in this configuration can be easily derived using the theory of interference discussed in Section 4.5. The derived spectra are shown in Figure 8.22a. As illustrated, a single reflectance minimum appears from the configuration and its position redshifts as the layer thickness increases. Suppose that the layer is divided into two layers of refractive indices n_1 = 1.17 and n_2 = 1.35 so that the refractive index change becomes less abrupt to incoming light. The corresponding spectra can also be calculated using the multilayer theory discussed in Section 8.1, although the calculation is a bit lengthy compared to the case of a single layer. As shown in Figure 8.22b, the calculated spectra have two reflectance minima, exhibiting AR over a broad range of wavelengths. More subdivided layers will result in lower reflectances over a wider range, once the layers have refractive indices gradually increasing from 1 to n_s. In the calculated spectra of Figure 8.22, the refractive indices were assumed constant, independent of the wavelength.

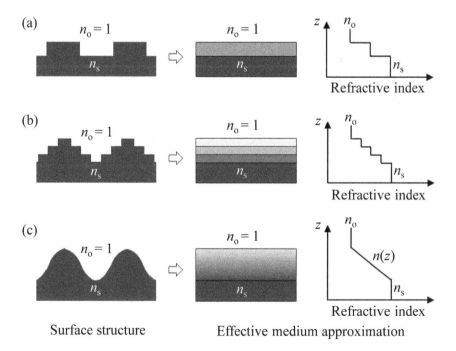

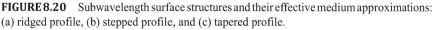

FIGURE 8.20 Subwavelength surface structures and their effective medium approximations: (a) ridged profile, (b) stepped profile, and (c) tapered profile.

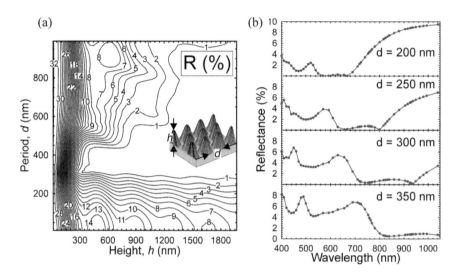

FIGURE 8.21 (a) Contour plot for the dependence of reflectance at $\lambda = 1$ μm on the period d and height h in Si pillar arrays. (b) Simulated reflectance spectra for Si moth-eye arrays with a pillar height of 400 nm and various periods. Reprinted with permission from Ref. [21]. Copyright © 2008 American Institute of Physics.

Single-layer AR coatings work only at a wavelength and at normal incidence. On the contrary, graded-index coatings exhibit broadband, omnidirectional AR characteristics. This definitely requires materials with low refractive indices. However, dense solids with refractive indices less than 1.35 are hardly available. A strategy to overcome this limitation is to make use of porous nanostructures. For instance, bulk silica (SiO_2) has a refractive index of 1.46 but its nanostructures can have a refractive index less than 1.1.[23,24] As the density varies, so does the effective refractive index of the material. Oblique-angle deposition is a viable vapor-deposition technique to grow porous thin films, which is enabled by self-shadowing effects and surface diffusion. Random growth fluctuations on the substrate generate a shadow region that incident vapor flux cannot reach, and a non-shadow region where incident flux deposits preferentially. This creates a porous structure consisting of tilted nanorods. The grown film has a very low refractive index as a result of its nanoporous nature. It has been shown that the refractive index of a TiO_2 nanorod layer can be controllably varied from 2.7 to 1.3 by changing the vapor incident angle.[25] The refractive index of a SiO_2 nanorod layer can be varied from 1.46 to 1.05. Figure 8.23a shows a graded-index coating consisting of TiO_2 nanorod layers and SiO_2 nanorod layers grown by oblique-angle deposition on an AlN substrate ($n_s = 2.05$).

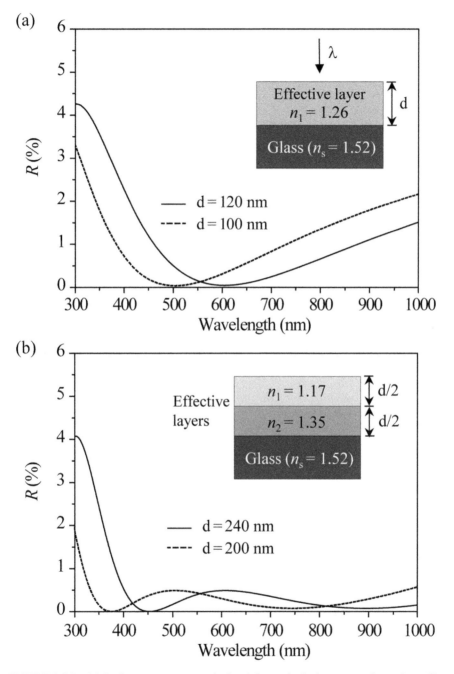

FIGURE 8.22 (a) Reflectance spectra calculated for a single layer coated on glass. (b) Spectra calculated for a two-layer coating system.

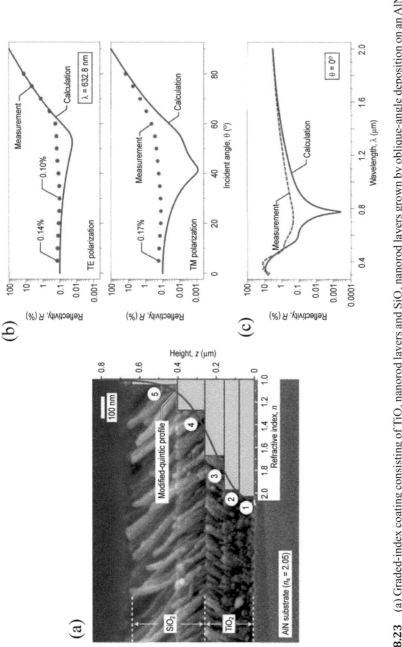

FIGURE 8.23 (a) Graded-index coating consisting of TiO$_2$ nanorod layers and SiO$_2$ nanorod layers grown by oblique-angle deposition on an AlN substrate ($n_s = 2.05$). (b) Incident angle-dependent reflectivity measured at $\lambda = 632.8$ nm for TE and TM polarizations. (c) Wavelength-dependent reflectivity, measured at normal incidence. Reprinted with permission from Ref. [25]. Copyright © 2007 Springer Nature.

The coating consists of five layers with varying refractive indices; the lower three layers are made of TiO_2 nanorods and the upper two layers are made of SiO_2 nanorods. All layers have well-defined interfaces, and the feature size (<50 nm) of individual nanorods is much smaller than the wavelength of visible light. The bottom TiO_2 layer has a refractive index of 2.03, which closely matches the refractive index of AlN. The top SiO_2 layer has a refractive index of 1.05, very close to the index of air. Therefore, this graded-index structure yields excellent AR characteristics. Figure 8.23b shows the incident angle-dependent reflectivity measured at λ = 632.8 nm for TE and TM polarizations. The graded-index coating has very low reflectivity for all incident angles except for angles close to 90°. The reflectivity remains below 0.3% at incident angles between 0° and 55° for both polarizations. The wavelength-dependent reflectivity, measured at normal incidence, is shown in Figure 8.23c. Experimental results agree well with theoretical calculations. The angular dependence and spectral dependence of the reflectivity confirms the broadband, omnidirectional AR characteristics of the graded-index coating.

8.2.4 POROUS SURFACE LAYER

Porous AR coatings were first observed in the 19th century by Fraunhofer upon etching a glass surface with an acid solution[26]. Etching created a mixture of media (air and glass), and the produced porous layer served to reduce reflection at the glass surface. For a porous material with porosity *f*, the refractive index n_p is given by

$$n_p^2 = n^2(1-f)+f. \tag{8.18}$$

Here, *n* is the refractive index when the material is dense. Equation 8.18 refers to a weighted average of the electrical permittivities of the material and air. Since the permittivity of a material is related to its density, the refractive index can be lowered by introducing porosity. In order for the relation of eq 8.18 to be valid, however, the pore size should be substantially smaller than the wavelength of light, and the pore distribution must also be homogeneous. To make a porous layer with n_p = 1.23 on glass (*n* = 1.52), a porosity of 0.6 is needed. When this layer is etched to a quarter-wave thickness, it acts as a single-layer AR coating. Similar results can be obtained by depositing porous oxide films using sol–gel processes. The sol–gel process makes it possible to obtain coatings with high homogeneity and tailored inhomogeneity. Single-layer AR coatings are only effective at one wavelength. Increasing the number of layers increases the bandwidth over which low reflection losses can be obtained. Multi-layer AR coatings deposited by sol–gel methods have been demonstrated on many substrates including glass. An alternative method for achieving broadband AR characteristics is to use a single layer with graded porosity, which results in a gradual index change through the thickness of the layer. This is the foundation of the moth's eye AR structures, where small projections create a graded effective medium serving to reduce index mismatch at the interface and eliminate reflection. As described above, oblique-angle deposition is an effective technique to grow porous thin films. It is well known that thin films deposited at oblique angles of incidence grow with densities less than that of the bulk material. At sufficiently large angles, self-shadowing effect becomes the dominant growth mechanism resulting in extremely porous films.[27,28] In this technique, the angle of incidence is utilized as a means of controlling the porosity and hence refractive index of the coating. The most ideal case is to produce a graded-index AR coating by depositing the same material as the substrate so that the refractive indices at the substrate interface would be naturally matched.

Polymer materials have also been used for the AR coating layers because they are more easily coated over larger areas compared with inorganic materials. Since most polymers have refractive indices similar to 1.5, they are good AR coating materials for both glass and plastic substrates. Porous polymer structures can be accomplished in various ways. A method based on the phase separation of a macromolecular liquid to generate nanoporous polymer films is demonstrated.[29] This approach makes use of the demixing of a binary polymer blend consisting of polystyrene (PS) and polymethylmethacrylate (PMMA) and the subsequent removal of the PS phase with a selective solvent. Since the effective refractive index is a function of the pore volume fraction in the film, a variation of the volume fraction of PS in the blend varies the refractive index of the produced nanoporous film. The underlying principle can work for a large number of polymers. Porous polymer films can be employed for broadband and omnidirectional AR coatings, when these films form a gradient-index structure. A method of producing such films involves the spin-coating of a PS-block-PMMA/PMMA blend solution into a surface-modified glass substrate, which results in a gradient distribution of PMMA domains through the thickness of the coated layer.[30] The removal of the PMMA domains produced a PS structure of graded porosity.

8.3 PROBLEMS

8.1 The characteristic matrix for a two-layer coating has the following general form:

$$M = \begin{bmatrix} \cos\delta_1 & i\sin\delta_1/n_1 \\ in_1\sin\delta_1 & \cos\delta_1 \end{bmatrix} \begin{bmatrix} \cos\delta_2 & i\sin\delta_2/n_2 \\ in_2\sin\delta_2 & \cos\delta_2 \end{bmatrix}.$$

The phase terms are given by $\delta_1 = 2\pi n_1 d_1/\lambda$ and $\delta_2 = 2\pi n_2 d_2/\lambda$, where λ is the wavelength of incident light. Suppose that two layers of $n_1 = 1.17$, $d_1 = 120$ nm and $n_2 = 1.35$, $d_2 = 120$ nm are deposited on a glass substrate ($n_s = 1.52$). Then determine the reflectance R as a function of the wavelength in a range of $\lambda = 400$–800 nm.

8.2 A single layer that has an optical thickness (nd) of 125 nm is deposited on a glass substrate. When the layer has a refractive index of (a) 2.40, (b) 1.82, and (c) 1.46, determine the reflectance for normal incidence as a function of the wavelength.

8.3 A thin layer of MgF_2 ($n = 1.38$) is deposited on glass ($n = 1.52$) so that it is antireflecting at 550 nm under normal incidence. What wavelength will be minimally reflected when the light is incident at 30°? Assume that the refractive indices are independent of the wavelength.

8.4 Find the refractive index and thickness of a layer deposited on a substrate ($n_s = 2.1$) such that no normally incident light of 750 nm wavelength is reflected.

8.5. Calculate the peak reflectance of a high-reflectance multilayer stack consisting of four H-layers and four L-layers with $n_H = 2.35$ and $n_L = 1.46$. Assume that the substrate is glass.

8.6 (a) Derive the admittance Y when a single dielectric layer of n_1 is deposited on an absorbing substrate whose refractive index is given by $n_s = n_R + in_I$. (b) Draw the locus of Y in the complex plane. (c) Calculate the reflectance R when the layer has a quarter-wave thickness.

REFERENCES

1. Macleod, H. *Thin-Film Optical Filters*, 4th ed.; CRC Press: Boca Raton, USA, 2010.
2. Willey, R. *Practical Design and Production of Optical Thin Films,* 2nd ed.; CRC Press: New York, 2002.
3. Arndt, R.; Allison, J.; Haynos, J.; Meulenberg, A. *Optical Properties of the Comsat Non-reflective Cell*, Proceedings of the 11th IEEE International Photovoltaic Specialists Conference, New York, 1975; pp 40–43.

4. Vazsonyi, E.; De Clercq, K.; Einhaus, R.; Van Kerschayer, F.; Said, K.; Poortmans, J. Improved Anisotropic Etching Process for Industrial Texturing of Silicon Solar Cells. *Sol. Energy Mater. Sol. Cells* **1999**, *57*, 179–188.
5. Zubel, I.; Kramkowska, M. The Effect of Isopropyl Alcohol on Etching Rate and Roughness of (100) Si Surface Etched in KOH and TMAH Solutions. *Sens. Actuators A* **2001**, *93*, 138–147.
6. Xi, Z.; Yang, D.; Que, D. Texturization of Monocrystalline Silicon with Tribasic Sodium Phosphate. *Sol. Energy Mater. Sol. Cells* **2003**, *77*, 255–263.
7. Chu, A.; Wang, J.; Tsai, Z.; Lee, C. A Simple and Cost-effective Approach for Fabricating Pyramids on Crystalline Silicon Wafers. *Sol. Energy Mater. Sol. Cells* **2009**, *93*, 1276–1280.
8. Khanna, A., et al. Influence of Random Pyramid Surface Texture on Silver Screen-printed Contact Formation for Monocrystalline Silicon Wafer Solar Cells. *Sol. Energy Mater. Sol. Cells* **2015**, *132*, 589–596.
9. Babu, P.; Khanna, A.; Hameiri, Z. The Effect of Front Pyramid Heights on the Efficiency of Homogeneously Textured Monocrystalline Silicon Wafer Solar Cells. *Renew. Energy* **2015**, *78*, 590–598.
10. Campbell, P.; Green, M. Light Trapping Properties of Pyramidally Textured Surfaces. *J. Appl. Phys.* **1987**, *62*, 243–249.
11. Smith, A.; Rohatgi, A. Ray Tracing Analysis of the Inverted Pyramid Texturing Geometry for High Efficiency Silicon Solar Cells. *Sol. Energy Mater. Sol. Cells* **1993**, *29*, 37–49.
12. Green, M.; Zhao, J.; Wang, A.; Wenham, S. Very High Efficiency Silicon Solar Cells-science and Technology. *IEEE Trans. Electron. Dev.* **1999**, *46*, 1940–1947.
13. Zhao, J.; Wang, A.; Green, M. 24.5% Efficiency Silicon PERT Cells on MCZ Substrates and 24.7% Efficiency PERL Cells on FZ Substrates. *Prog. Photovolt. Res. Appl.* **1999**, *7*, 471–474.
14. Yang, B; Lee, M. Fabrication of Honeycomb Texture on Poly-Si by Laser Interference and Chemical Etching. *Appl. Surf. Sci.* **2013**, *284*, 565–568.
15. Lee, K.; Yu, Q.; Erb, U. Mesostructure of Ordered Corneal Nanonipple Arrays: The Role of 5-7 Coordination Defects. *Sci. Reports* **2016**, *6*, 28342.
16. Hsu, C.; Connor, S.; Tang, M.; Cui, Y. Wafer-scale Silicon Nanopillars and Nanocones by Langmuir-Blodgett Assembly and Etching. *Appl. Phys. Lett.* **2008**, *93*, 133109.
17. Ji, S.; Song, K.; Nguyen, T.; Kim, N.; Lim, H. Optimal Moth Eye Nanostructure Array on Transparent Glass Towards Broadband Antireflection. *ACS Appl. Mater. Interfaces* **2013**, *5*, 10731.
18. Gale, M. Diffraction, Beauty and Commerce. *Phys. World* **1989**, *2*, 24–28.
19. Toyota, H., Takahara, K.; Okano, M.; Yotsuya, T.; Kikuta, H. Fabrication of Microcone Array for Antireflection Structured Surface Using Metal Dotted Pattern. *Jpn. J. Appl. Phys.* **2001**, *40*, 747–749.
20. Boden, S.; Bagnall, D. Moth-eye Antireflective Structures. *Encyclopedia of Nanotechnology*, 2nd ed.; Springer: Dordrecht, Netherlands, 2016.
21. Boden, S.; Bagnall, D. Tunable Reflection Minima of Nanostructured Antireflective Surfaces. *Appl. Phys. Lett.* **2008**, *93*, 133108.
22. Brunner, R.; Sandfuchs, O.; Pacholski, C.; Morhard, C.; Spatz, J. Lessons from Nature: Biomimetic Subwavelength Structures for High-performance Optics. *Laser Photon. Rev.* **2012**, *6*, 641–659.
23. Xi, J.; Kim, J.; Schubert, E. Silica Nanorod-array Films with very Low Refractive Indices. *Nano Lett.* **2005**, *5*, 1385–1387.

24. Xi, J., et al. Very Low-refractive-index Optical Thin Films Consisting of an Array of SiO₂ Nanorods. *Opt. Lett.* **2006**, *31*, 601–603.

25. Xi, J., et al. Optical Thin-film Materials with Low Refractive Index for Broadband Elimination of Fresnel Reflection. *Nat. Photon.* **2007**, *1*, 176–179.

26. Fraunhofer, J. *Gesammelte Schriften*; F. Hommel: Munich, 1887.

27. Robbie, K.; Brett, M. Sculptured Thin Films and Glancing Angle Deposition: Growth Mechanisms and Applications. *J. Vac. Sci. Technol. A* **1997**, *15*, 1460–1465.

28. Kennedy, S.; Brett, M. Porous Broadband Antireflection Coating by Glancing Angle Deposition. *Appl. Optics* **2003**, *42*, 4573–4579.

29. Walheim, S.; Schaffer, E.; Mlynek, J.; Steiner, U. Nanophase-separated Polymer Films as High-performance Antireflection Coatings. *Science* **1999**, *283*, 520–522.

30. Li, X.; Gao, J.; Xue, L.; Han, Y. Porous Polymer Films with Gradient-refractive-index Structure for Broadband and Omnidirectional Antireflection Coatings. *Adv. Func. Mater.* **2010**, *20*, 259–265.

CHAPTER 9

Interference Lithography

9.1 INTRODUCTION

Interference lithography is a powerful technique for fabricating periodic structures. It allows regular arrays of fine features to be patterned without the use of complex optical systems or photomasks. With the advent of short-wavelength and high-power lasers, interference lithography became a very popular manufacturing tool. Photolithography, literally meaning *light-stone-writing* in Greek, refers to any process that uses light to transfer a geometric pattern from a photomask to a photosensitive material such as photoresist. After development by a series of treatments, the photoresist is used either as an etch mask for the underlying substrate or a template for the deposition of a new material. Although conventional photolithography is well established and widely used in microelectronic industry to create integrated circuits, the required optical system and photomasks are very expensive. Interference lithography, which is based on the interference of several coherent laser beams, can produce periodic patterns over large areas in a much simpler way. The basic principle of interference lithography is the same as in holography. Thus it is also called *holographic lithography*. An interference pattern between two or more coherent light waves is recorded in a photoresist. This interference pattern represents a periodic series of fringes consisting of intensity maxima and minima. A photoresist pattern corresponding to the periodic intensity pattern emerges upon post-exposure processing. A one-dimensional (1D) pattern is generated by two-beam interference, while three-beam interference produces a 2D pattern. By using four or more beams, 3D structures can be generated. Interference lithography requires that a coherent light source be employed. This is readily achieved with a laser but broadband light sources would require a filter. A laser beam is often used directly without any collimation step. It is first split into two or more beams and then recombined in the region where interference occurs. Beam splitting and recombination are typically carried out by mirrors, prisms, and diffraction gratings. Interfering high-power pulsed laser beams can induce periodic structures on the surface of materials (including metals, ceramics, and polymers) based on photothermal and/or photochemical mechanisms. Using this approach, the material can be directly structured just in a few seconds. Such surface patterns can be effectively utilized for a range of applications including tribology (wear and friction reduction), photovoltaics, and biotechnology. The applicability of interference lithography is limited to patterning arrayed features or uniformly distributed aperiodic structures only. Nevertheless, it has many advantages as a fabrication method for periodic structures. It is a mask-free, low-cost, and scalable process. In addition, it enables the shape and period of the structures to be controlled by adjusting optical parameters such as the number, polarization directions, and incident angles of interfering beams. In this chapter, we describe the basic concepts of interference lithography, together with how 1D, 2D, and 3D periodic structures can be created using this technique.

9.2 THEORY OF INTERFERENCE LITHOGRAPHY

9.2.1 TWO-BEAM INTERFERENCE

We start from the simplest case in which two nonparallel plane waves of the same frequency propagate simultaneously in space. The two waves are given by

$$\mathbf{E}_1(\mathbf{r},t) = \mathrm{Re}\{\mathbf{E}_{01}e^{i(\mathbf{k}_1 \cdot \mathbf{r}-\omega t)}\}$$
$$\mathbf{E}_2(\mathbf{r},t) = \mathrm{Re}\{\mathbf{E}_{02}e^{i(\mathbf{k}_2 \cdot \mathbf{r}-\omega t)}\},$$

where $\mathbf{E}_{01}$ and $\mathbf{E}_{02}$ are the complex amplitude vectors of the waves, and $\mathbf{k}_1$ and $\mathbf{k}_2$ are the corresponding wave vectors. The position vector $\mathbf{r}$ can be represented as $\mathbf{r} = x\mathbf{i} + y\mathbf{j} + z\mathbf{k}$, where $\mathbf{i}$, $\mathbf{j}$, and $\mathbf{k}$ are the unit vectors along the x, y, and z directions, respectively. The intensities of the individual waves are given by $I_1 = |\mathbf{E}_{01}|^2$ and $I_2 = |\mathbf{E}_{02}|^2$. Since I_1 and I_2 are constant, the intensity (i.e., energy) of each wave is homogeneously distributed in space if the waves propagate separately. When the two waves overlap in space, the principle of superposition is applied. Then, the total electric field is the sum of the electric fields of the individual waves.

$$\mathbf{E}_t(\mathbf{r},t) = \mathbf{E}_1(\mathbf{r},t) + \mathbf{E}_2(\mathbf{r},t) = \mathrm{Re}\{(\mathbf{E}_{01}e^{i\mathbf{k}_1 \cdot \mathbf{r}} + \mathbf{E}_{02}e^{i\mathbf{k}_2 \cdot \mathbf{r}})e^{-i\omega t}\} \quad (9.1)$$

The resultant wave is not a single plane wave but a more general type of wave. The amplitude $\mathbf{A}(\mathbf{r})$ of the resultant wave is position-dependent, giving rise to a nonuniform intensity distribution. The intensity of the resultant wave is given by

$$I(\mathbf{r}) = |\mathbf{A}(\mathbf{r})|^2 = \mathbf{A}(\mathbf{r}) \cdot \mathbf{A}^*(\mathbf{r}) = (\mathbf{E}_{01}e^{i\mathbf{k}_1 \cdot \mathbf{r}} + \mathbf{E}_{02}e^{i\mathbf{k}_2 \cdot \mathbf{r}}) \cdot (\mathbf{E}_{01}^* e^{-i\mathbf{k}_1 \cdot \mathbf{r}} + \mathbf{E}_{02}^* e^{-i\mathbf{k}_2 \cdot \mathbf{r}}). \quad (9.2)$$

By expanding eq 9.2, we have

$$I(\mathbf{r}) = I_1 + I_2 + (\mathbf{E}_{01} \cdot \mathbf{E}_{02}^*)e^{i(\mathbf{k}_1 - \mathbf{k}_2) \cdot \mathbf{r}} + (\mathbf{E}_{02} \cdot \mathbf{E}_{01}^*)e^{i(\mathbf{k}_2 - \mathbf{k}_1) \cdot \mathbf{r}}. \quad (9.3)$$

The intensity I of the resultant wave is not equal to the sum $(I_1 + I_2)$ of the intensities of the individual waves but contains two additional terms called the interference terms. These terms make the intensity position-dependent and create a nonuniform distribution of energy in space. The resultant intensity I may be either greater or less than $I_1 + I_2$, depending on the position. However, the spatial average of the intensity is equal to $(I_1 + I_2)$ and the energy is thus conserved. The wave vectors $\mathbf{k}_1$ and $\mathbf{k}_2$ have the same magnitude because the two waves propagate in the same medium. If the wave vectors are collinear, $\mathbf{k}_1 - \mathbf{k}_2 = 0$ and then the intensity is uniformly distributed. This means that the propagation directions of the waves should be nonparallel in order to obtain a spatially-varying intensity profile. When the amplitude vectors of the waves are perpendicular to one another, there is no interference effect because the dot products $\mathbf{E}_{01} \cdot \mathbf{E}^*_{02}$ and $\mathbf{E}_{02} \cdot \mathbf{E}^*_{01}$ become zero. Therefore, interference does not occur when the two plane waves involved have orthogonal polarizations.

To understand the basic concepts of two-beam interference, consider the configuration of Figure 9.1, where two linearly polarized plane waves with the same wavelength intersect at an angle of 2θ. As graphically illustrated, the interference of the two plane waves produces a 1D fringe pattern whose period "d" is directly proportional to the wavelength λ and inversely proportional to the sine of the incident angle θ. The interference pattern is observed wherever the two waves spatially overlap. The formation of the interference pattern can be more easily understood in reciprocal space. Since the interference fringes have the form of a 1D grating, the orientation and period of the grating can be characterized by a grating vector $\mathbf{g}$. The grating vector is equal to the difference of the incident wave vectors, that is, $\mathbf{g} = \mathbf{k}_1 - \mathbf{k}_2$. When two beams interfere, the wave vectors of the beams and the induced grating vector form a triangle where the grating vector connects the tips of the wave vectors. The grating vector is perpendicular to the induced fringe lines and has a magnitude of $g = 2\pi/d$. In other words, the lines of the induced grating bisect the wave vectors. Since $\mathbf{k}_1$ and $\mathbf{k}_2$ have the same magnitude of $k = 2\pi/\lambda$, the period of the interference fringes is

$$d = \frac{\lambda}{2\sin\theta}.$$ (9.4)

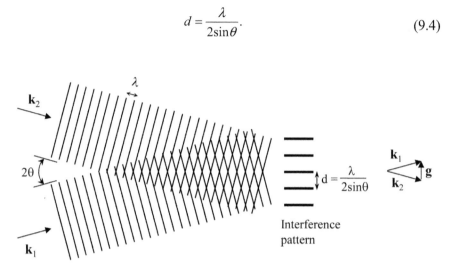

$$d = \frac{\lambda}{2\sin\theta}$$

Interference pattern

FIGURE 9.1   Interference between two plane waves.

A smaller angle between the wave vectors produces a longer-period interference pattern. Equation 9.4 can also be derived from eq 9.3. Suppose that the two planes waves of Figure 9.1 propagate in the x–y plane with their linear polarizations parallel to the z-axis, as depicted in Figure 9.2a. Here, the x-axis is chosen such that both $\mathbf{k}_1$ and $\mathbf{k}_2$ make an equal angle θ with it. For simplicity, we assume that the two waves have the same amplitude E_o. The amplitude vectors in eq 9.1 through eq 9.3 become $\mathbf{E}_{01} = \mathbf{E}_{02} = \mathbf{E}^*_{01} = \mathbf{E}^*_{02} = E_o\mathbf{k}$, where $\mathbf{k}$ is the unit vector along the z-axis. In the configuration of Figure 9.2a, the wave vector difference, $\mathbf{k}_1 - \mathbf{k}_2 = \mathbf{k}_{12}$, is directed along the y-axis and has a magnitude of $k_{12} = (4\pi \sin\theta)/\lambda$. The dot product $(\mathbf{k}_1 - \mathbf{k}_2) \cdot \mathbf{r}$ becomes $(y\,4\pi \sin\theta)/\lambda$. Then eq 9.3 reduces to

$$I(\mathbf{r}) = E_o^2 + E_o^2 + 2E_o^2\cos(y4\pi\sin\theta/\lambda) = 4I_o\cos^2(y2\pi\sin\theta/\lambda).$$ (9.5)

Here $I_o = E_o{}^2$ is the intensity of each wave in the absence of interference. The resulting intensity is a sinusoidal function that depends only on the y value (i.e., a 1D grating), as plotted in Figure 9.2b. Equation 9.5 shows that the period of the grating is $d = \lambda/(2\sin\theta)$, which is equal to $2\pi/k_{12}$. The lines of the induced grating are perpendicular to the y-axis. Suppose a vector that has a magnitude of $2\pi/d$ and is directed perpendicular to the grating lines. If this vector is defined as the grating vector $\mathbf{g}$, we have the relation of $\mathbf{k}_{12} = \mathbf{k}_1 - \mathbf{k}_2 = \mathbf{g}$, as illustrated in Figure 9.2c.

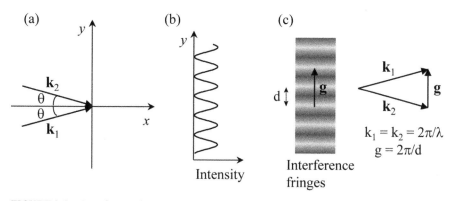

FIGURE 9.2   Interference between two plane waves. (a) Two wave vectors in the x–y planes. (b) The resulting sinusoidal intensity distribution. (c) Relation between the wave vectors $\mathbf{k}_1$, $\mathbf{k}_2$ and the grating vector $\mathbf{g}$.

Equation 9.5 can be derived in an alternative way. When the two plane waves have the same amplitude and polarization direction, the total electric field is simply written by

$$E_t = E_0 e^{i(\mathbf{k}_1 \cdot \mathbf{r} - \omega t)} + E_0 e^{i(\mathbf{k}_2 \cdot \mathbf{r} - \omega t)}. \tag{9.6}$$

From the geometry of Figure 9.2a, $\mathbf{k}_1 = k \cos\theta\,\mathbf{i} + k \sin\theta\,\mathbf{j}$ and $\mathbf{k}_2 = k \cos\theta\,\mathbf{i} - k \sin\theta\,\mathbf{j}$. Since the position vector is given by $\mathbf{r} = x\mathbf{i} + y\mathbf{j} + z\mathbf{k}$, eq 9.6 reduces to

$$E_t = 2E_0 \cos(k \sin\theta\, y) e^{i(k \cos\theta x - \omega t)}. \tag{9.7}$$

Since the intensity is the squared amplitude, we have

$$I = 4I_0 \cos^2(k \sin\theta\, y). \tag{9.8}$$

Here $k = 2\pi/\lambda$ is the magnitude of the wave vectors.

In the above derivations, the phases of the waves have not been taken into consideration. That is, eqs 9.5 and 9.8 represent the intensity distribution when the two waves have no phase difference. If they have different phases denoted as φ_1 and φ_2, the respective amplitude vectors are given by $\mathbf{E}_{01} = k E_0 e^{i\varphi 1}$ and $\mathbf{E}_{02} = k E_0 e^{i\varphi 2}$. Their complex conjugates are $\mathbf{E}^*_{01} = k E_0 e^{-i\varphi 1}$ and $\mathbf{E}^*_{02} = k E_0 e^{-i\varphi 2}$. Note that $\mathbf{k}$ represents the unit vector along the z-axis, not any wave vector. By inserting these amplitudes and $(\mathbf{k}_1 - \mathbf{k}_2)\cdot \mathbf{r} = (y\,4\pi \sin\theta)/\lambda$ into eq 9.3, we have

$$I(\mathbf{r}) = 2I_0\{1 + \cos(y4\pi \sin\theta/\lambda + \varphi_{12})\} = 4I_0 \cos^2(y2\pi \sin\theta/\lambda + \varphi_{12}/2). \tag{9.9}$$

Here, $\varphi_{12} = \varphi_1 - \varphi_2$ represents the phase difference between the two waves. Equation 9.9 shows that the phase difference between the waves does not alter the shape and period of the 1D interference pattern. The phase difference φ_{12} simply translates the pattern in space without changing its shape. This is graphically illustrated in Figure 9.3. When one of the waves undergoes a relative phase change, its wavefronts will shift accordingly. This will translate the interference fringes. The relation of eq 9.9 can also be derived by adding phase terms into eq 9.6. The derivation remains a problem (Problem 9.1).

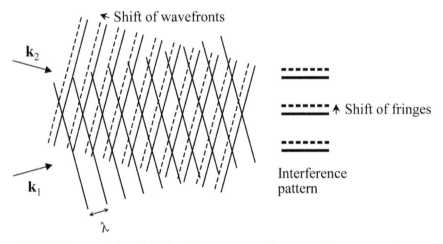

FIGURE 9.3 Translation of the interference pattern when one of the waves undergoes a relative phase change.

In a 1D periodic intensity pattern created by the superposition of two plane waves, the contrast (i.e., the difference between intensity maxima and minima) of the pattern is maximized when the waves have the same polarization direction. Waves that are orthogonally polarized do not interfere with each other and will have zero contrast. It is also to be noted that even if the interference pattern is formed inside a material, its period is still determined by eq 9.4. In other words, when a 1D pattern is recorded inside a material, its period can be calculated from the external intersecting angle and wavelength of the two waves. This is because the tangential components of the wave vectors are conserved at the air–material interface. Suppose that two plane-wave beams with the same wavelength λ are symmetrically incident into a material of refractive index n at an external angle of θ_i and generates an interference pattern inside the material (Fig. 9.4a). They are refracted at the air-material interface. Therefore, the incident and refracted waves will have different propagation directions. When the incident wave vector $\mathbf{k}_1$ has a magnitude of $2\pi/\lambda$, the magnitude of the refracted wave vector $\mathbf{k}_{1r}$ becomes $2\pi n/\lambda$ because the wavelength is decreased to λ/n inside the material (Fig. 9.4b). Snell's law states that the tangential components of the wave vectors are conserved. This means that the magnitude of the grating vector given by $\mathbf{g} = \mathbf{k}_1 - \mathbf{k}_2$ remains unchanged upon refraction, that is, $|\mathbf{k}_{1r} - \mathbf{k}_{2r}| = |\mathbf{k}_1 - \mathbf{k}_2|$. As a result, the interference pattern formed inside the material has the same period as in air.

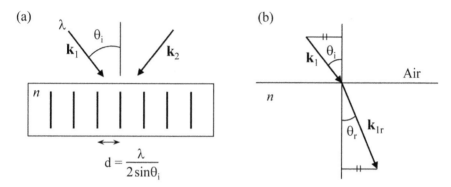

FIGURE 9.4 (a) 1D interference pattern formed inside a material. (b) Wave vectors of the incident and refracted waves.

9.2.2 THREE-BEAM INTERFERENCE

Three-beam interference can be described in a similar way. When three non-coplanar plane waves are superimposed, the resultant wave is given by

$$\mathbf{E}_t(\mathbf{r},t) = \mathbf{E}_1(\mathbf{r},t) + \mathbf{E}_2(\mathbf{r},t) + \mathbf{E}_3(\mathbf{r},t)$$
$$= \mathrm{Re}\{(\mathbf{E}_{01}e^{i\mathbf{k}_1\cdot\mathbf{r}} + \mathbf{E}_{02}e^{i\mathbf{k}_2\cdot\mathbf{r}} + \mathbf{E}_{03}e^{i\mathbf{k}_3\cdot\mathbf{r}})e^{-i\omega t}\} \qquad (9.10)$$

The intensity distribution $I(\mathbf{r})$ of the total wave is then

$$I(\mathbf{r}) = |\mathbf{A}(\mathbf{r})|^2 = (\mathbf{E}_{01}e^{i\mathbf{k}_1\cdot\mathbf{r}} + \mathbf{E}_{02}e^{i\mathbf{k}_2\cdot\mathbf{r}} + \mathbf{E}_{03}e^{i\mathbf{k}_3\cdot\mathbf{r}})$$
$$\cdot(\mathbf{E}_{01}^*e^{-i\mathbf{k}_1\cdot\mathbf{r}} + \mathbf{E}_{02}^*e^{-i\mathbf{k}_2\cdot\mathbf{r}} + \mathbf{E}_{03}^*e^{-i\mathbf{k}_3\cdot\mathbf{r}}) \qquad (9.11)$$

Equation 9.11 can be rewritten as

$$I(\mathbf{r}) = I_1 + I_2 + I_3 + \sum_{i,j(i\neq j)}^{3} (\mathbf{E}_{0i} \cdot \mathbf{E}_{0j}^*)e^{i(\mathbf{k}_i - \mathbf{k}_j)\cdot\mathbf{r}}. \qquad (9.12)$$

The intensity pattern generally consists of three constant terms and six interference terms. 2D patterns can be fabricated directly by interfering three beams. As an example, Figure 9.5a shows an umbrella-like geometry that can be used to make 2D hexagonal structures. In the geometry, three plane-wave beams are symmetrically placed (azimuthal angle $\phi = 0$, $120°$, $240°$) about the central normal z-axis with the same polar angle θ. The wave vector of the jth beam can be expressed as $\mathbf{k}_j = -k\,(\cos\phi_j \sin\theta,\ \sin\phi_j \sin\theta,\ \cos\theta)$, where $k = 2\pi/\lambda$ is the magnitude of the wave vectors. The differences between the wave vectors lie in the x-y plane, producing a 2D interference pattern with basis vectors of $\mathbf{k}_{12}$, $\mathbf{k}_{23}$, and $\mathbf{k}_{31}$ (Fig. 9.5b). These three difference vectors form a regular triangle within the x-y plane and have the same magnitude of $\sqrt{3}k\sin\theta$. The strict requirement to obtain intensity variations only in two dimensions is that the difference vectors must be coplanar (all lying within the same plane). From elementary inspection of the 2D intensity function $I(x, y)$, it is evident that the maximum intensity points form a lattice. The interference pattern is a real-space lattice generated by the reciprocal-space vectors $\mathbf{k}_{12}$, $\mathbf{k}_{23}$, and $\mathbf{k}_{31}$. Any two of the difference vectors (e.g., $\mathbf{k}_{12}$ and $\mathbf{k}_{23}$) may be defined as the primitive vectors of the reciprocal lattice. The third vector is simply obtained by the summation of the two chosen vectors. There are five different types of 2D lattices: *oblique, rectangular, rhombic, square,* and *hexagonal.*[1] Figure 9.5c shows the most general 2D lattice where vectors $\mathbf{b}_1$ and $\mathbf{b}_2$ define the primitive unit cell of the lattice. The two vectors generally have different magnitudes, and the angle γ between them is arbitrary. This type of lattice is known as *oblique lattice*. When $\mathbf{b}_1$ and $\mathbf{b}_2$ represent the primitive vectors of a reciprocal lattice, the primitive vectors $\mathbf{a}_1$ and $\mathbf{a}_2$ of its real lattice can be derived from the relations of $\mathbf{a}_1 \cdot \mathbf{b}_1 = \mathbf{a}_2 \cdot \mathbf{b}_2 = 2\pi$ and $\mathbf{a}_1 \cdot \mathbf{b}_2 = \mathbf{a}_2 \cdot \mathbf{b}_1 = 0$ (Fig. 9.5d). The vectors $\mathbf{a}_1$ and $\mathbf{a}_2$ are perpendicular to $\mathbf{b}_2$ and $\mathbf{b}_1$, respectively. If $\mathbf{b}_1$ and $\mathbf{b}_2$ have the same magnitude (i.e., $b_1 = b_2$) and the angle γ is either $120°$ or $60°$, both the reciprocal and real lattices have hexagonal symmetry. Of course, the real lattice of a square reciprocal lattice ($b_1 = b_2$, $\gamma = 90°$) is also a square lattice with $a_1 = a_2$ and $\gamma = 90°$. The real-space primitive vectors $\mathbf{a}_1$ and $\mathbf{a}_2$ have units of meters, while the reciprocal lattice vectors $\mathbf{b}_1$ and $\mathbf{b}_2$ have units of inverse meters because they represent wave vector differences. In the geometry of Figure 9.5a, we have $b_1 = b_2 = \sqrt{3}k\sin\theta$ and $\gamma = 120°$. The real lattice is hexagonal with a lattice constant of $a_1 = a_2 = 2\lambda/(3\sin\theta)$. Thus, the interference pattern has a period of $d = 2\lambda/(3\sin\theta)$. The period can also be derived by interpreting the reciprocal lattice vectors as grating vectors. Since the wave vectors of the incident beams are symmetric about the normal z-axis, the difference vectors $\mathbf{k}_{12}$, $\mathbf{k}_{23}$, and $\mathbf{k}_{31}$ form a regular triangle in the x–y plane (Fig. 9.6a). Just as in two-beam interference, each of the difference vectors induces a 1D grating. As the difference vectors have the same magnitude of $\sqrt{3}k \sin \theta$, the 1D gratings also have the same period of $2\pi/(\sqrt{3}k \sin \theta)$. If we draw the lines of the grating perpendicular to the corresponding difference vector, a 2D interference pattern of hexagonal symmetry is obtained, as illustrated in Figure 9.6b. The obtained pattern has a period of $d = 2\lambda/(3 \sin \theta)$. It is to be noted that the period of the pattern depends only on the incident angle θ and wavelength λ of the beams.

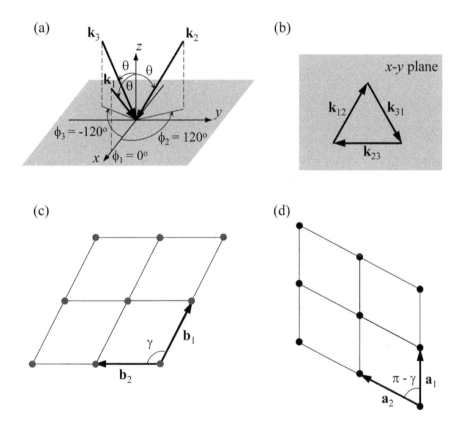

FIGURE 9.5 (a) A three-beam configuration (umbrella-like geometry). (b) Wave vector differences. (c) Two primitive vectors $\mathbf{b}_1$ and $\mathbf{b}_2$ in a reciprocal lattice. (d) Primitive vectors $\mathbf{a}_1$ and $\mathbf{a}_2$ of the real-space lattice of the reciprocal lattice shown in (c).

The polarization of the beams do not influence the period of the interference pattern but strongly affect the pattern morphology. For simplicity, we assume that the three beams are of equal amplitude E_0. The amplitude vectors can be described as $\mathbf{E}_{01} = \mathbf{e}_1 E_0 e^{i\varphi_1}$, $\mathbf{E}_{02} = \mathbf{e}_2 E_0 e^{i\varphi_3}$, and $\mathbf{E}_{03} = \mathbf{e}_3 E_0 e^{i\varphi_3}$, where $\mathbf{e}_1$, $\mathbf{e}_2$, and $\mathbf{e}_3$ are the unit polarization vectors. Equation 9.12 is then simplified to

$$I(\mathbf{r}) = 3I_0 + 2I_0[e_{12}\cos(\mathbf{k}_{12}\cdot\mathbf{r}+\varphi_{12}) + e_{23}\cos(\mathbf{k}_{23}\cdot\mathbf{r}+\varphi_{23})$$
$$+ e_{31}\cos(\mathbf{k}_{31}\cdot\mathbf{r}+\varphi_{31})]. \qquad (9.13)$$

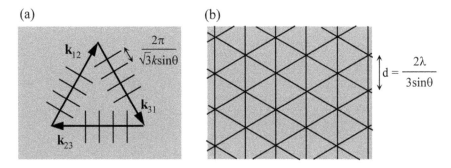

FIGURE 9.6 (a) A regular triangle formed by three difference vectors $\mathbf{k}_{12}$, $\mathbf{k}_{23}$, and $\mathbf{k}_{31}$. (b) 2D interference pattern of hexagonal symmetry obtained with the fringe lines perpendicular to these difference vectors.

Here, $e_{ij} = \mathbf{e}_i\cdot\mathbf{e}_j$ is the dot product of two linear polarization vectors. The differences between the wave vectors, $\mathbf{k}_{ij}$, determine the spatial period of the interference pattern, while the morphology and contrast of the pattern are affected by the polarization and phase of each beam. The configuration where the three beams are in phase ($\varphi_1 = \varphi_2 = \varphi_3 = 0$) is widely used to fabricate 2D hexagonal arrays. In the geometry of Figure 9.5a, we have $\phi_1 = 0$, $\phi_2 = 120°$, and $\phi_3 = 240°$. Assuming that the initial phase of each beam is zero, the intensity distribution is

$$I(x,y) = 3I_0 + 2I_0\left\{ e_{12}\cos k\left[\left(\frac{3}{2}x - \frac{\sqrt{3}}{2}y\right)\sin\theta\right]\right.$$
$$\left. + e_{23}\cos k\left[\sqrt{3}y\sin\theta\right] + e_{31}\cos k\left[\left(\frac{3}{2}x + \frac{\sqrt{3}}{2}y\right)\sin\theta\right]\right\} . \qquad (9.14)$$

Equation 9.14 shows that the interference pattern is significantly affected by the product of the polarization vectors e_{ij}. In two-beam interference, the polarizations of two beams can always be chosen to be perpendicular to a common plane. However, there is no common plane for three non-coplanar beams. This means that the product $e_{ij} = \mathbf{e}_i \cdot \mathbf{e}_j$ has a magnitude less than unity. To achieve a uniform 2D interference pattern, the polarization state of each beam must be determined such that the three products e_{12}, e_{23}, and e_{31} have the same or at least similar magnitudes. Let us examine how the beam polarization influences the interference pattern. As an example, Figure 9.7a illustrates an interference setup in which the polarization of incident beams have the same direction when projected onto the x–y plane. The components (i.e., direction cosines) of the wave vectors depend on the polar angle θ. For instance, when it is $\theta = 30°$, the wave vectors are given by

$$\mathbf{k}_1 = -\frac{2\pi}{\lambda}\left(\frac{1}{2},0,\frac{\sqrt{3}}{2}\right), \mathbf{k}_2 = -\frac{2\pi}{\lambda}\left(-\frac{1}{4},\frac{\sqrt{3}}{4},\frac{\sqrt{3}}{2}\right), \mathbf{k}_3 = -\frac{2\pi}{\lambda}\left(-\frac{1}{4},-\frac{\sqrt{3}}{4},\frac{\sqrt{3}}{2}\right).$$

Each polarization vector should satisfy two inherent conditions of $|\mathbf{e}_j| = 1$ and $\mathbf{e}_j \cdot \mathbf{k}_j = 0$. In the given setup, it has no y-component. Then, the polarization vectors can be expressed as

$$\mathbf{e}_1 = 0.500(\sqrt{3},0,-1), \mathbf{e}_2 = 0.277(2\sqrt{3},0,1), \mathbf{e}_3 = 0.277(2\sqrt{3},0,1). \quad (9.15)$$

From eq 9.15, we have $e_{23} = 1$ and $e_{12} = e_{31} = 0.693$. The intensity profile calculated using eq 9.14 is shown in Figure 9.7b. A 2D hexagonal array of pillars is obtained. It is to be noted that the cross-section of the pillars is not perfectly circular because the dot products e_{ij} have different magnitudes. In the case where the polarization of one beam is perpendicular to the other two (Fig. 9.7c), the polarization vector of the first beam in eq 9.15 is replaced by $\mathbf{e}_1 = (0, 1, 0)$, while the other two remain unchanged. Then we have $e_{23} = 1$ and $e_{12} = e_{31} = 0$. This results in a 1D profile (Fig. 9.7d). Figure 9.8 shows how the relative magnitude of the dot product e_{ij} changes the pattern morphology. If all the products are of equal value ($e_{12} = e_{23} = e_{31}$), a hexagonal array of circular pillars is produced. When these products have different values, the cross-section of the pillars is not circular but rather elongated along one direction. As the difference increases, the pillars are more elongated. When the polarization of one beam is perpendicular to the other two, two of the three products becomes zero. This makes the produced intensity pattern essentially one-dimensional. Equation 9.14 has been derived assuming that the incident beams have an identical intensity. It contains three interference terms, where the contribution of each term is determined by the amplitude factor e_{ij}. For the parallel polarization configuration shown in Figure 9.7a, not all the amplitude factors are equal in magnitude. Therefore, the shape of intensity pillars will be slightly elongated along one direction. Nevertheless, each repeat unit is a regular hexagon. Completely circular pillars may be obtained by adjusting the relative intensity of each beam.

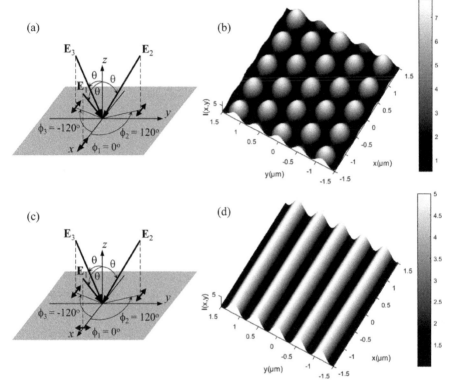

FIGURE 9.7 (a) A three-beam setup in which the beams are polarized in the same direction. (b) Calculated intensity distribution in the case of "(a)." (c) An interference setup in which the polarization of one beam is perpendicular to the other two. (d) Calculated intensity profile in the case of "(c)."

As another example, suppose that the projected polarization of each beam is parallel to its projected wave vector (Fig. 9.9a). In this radial configuration, the polarization vectors are given by

$$\mathbf{e}_1 = 0.500(\sqrt{3},0,-1), \mathbf{e}_2 = 0.433(-1,\sqrt{3},-2/\sqrt{3}), \mathbf{e}_3 = 0.433(-1,-\sqrt{3},-2/\sqrt{3}). \quad (9.16)$$

The dot products of these vectors are all of equal value ($e_{12} = e_{23} = e_{31} = -0.125$). Equation 9.14 then becomes

$$I(x,y) = 3I_0 - 0.25I_0 \left\{ \cos k\left(\frac{3}{4}x - \frac{\sqrt{3}}{4}y\right) + \cos k\left(\frac{\sqrt{3}}{2}y\right) + \cos k\left(\frac{3}{4}x + \frac{\sqrt{3}}{4}y\right) \right\}. \quad (9.17)$$

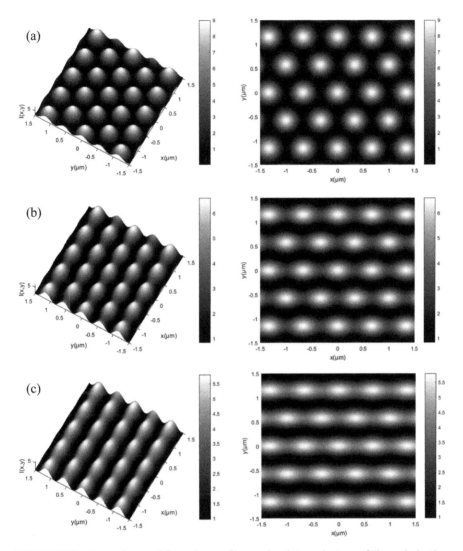

FIGURE 9.8 Dependence of intensity profile on the dot product e_{ij} of the polarization vectors. (a) $e_{12} = e_{23} = e_{31}$. (b) $e_{12} = e_{31} = 0.4e_{23}$. (c) $e_{12} = e_{31} = 0.2e_{23}$.

The resulting intensity profile is shown in Figure 9.9b. Compared to the profile of Figure 9.7b, it is an inverse structure with circular holes. In this radial polarization configuration, destructive interference occurs at the center of the x–y plane when the incident beams are all in phase. Therefore, circular holes rather than pillars are periodically created. It will be interesting to see what happens if a relative phase change is introduced into one of the beams. Suppose that the phase of the first beam in Figure 9.9a is shifted by π. Then we have the relations of $\varphi_{12} = \pi$, $\varphi_{31} = -\pi$, $\varphi_{23} = 0$. Inserting the phase differences into eq 9.13 gives

$$I(x,y) = 3I_0 - 0.25I_0\left\{-\cos k\left(\frac{3}{4}x - \frac{\sqrt{3}}{4}y\right) + \cos k\left(\frac{\sqrt{3}}{2}y\right) - \cos k\left(\frac{3}{4}x + \frac{\sqrt{3}}{4}y\right)\right\}. \quad (9.18)$$

The intensity pattern is depicted in Figure 9.9c. As illustrated, this phase shift does not change the pattern morphology but leads only to a translation of $I(x, y)$ within the x–y plane. On the contrary, when the phase shift is $\pi/2$, Equation 9.13 reduces to

$$I(x,y) = 3I_0 - 0.25I_0\left\{-\sin k\left(\frac{3}{4}x - \frac{\sqrt{3}}{4}y\right) + \cos k\left(\frac{\sqrt{3}}{2}y\right) + \sin k\left(\frac{3}{4}x + \frac{\sqrt{3}}{4}y\right)\right\}. \quad (9.19)$$

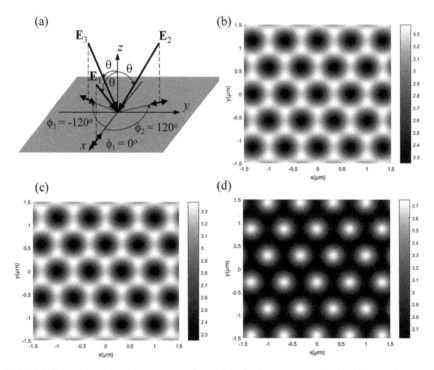

FIGURE 9.9   (a) A three-beam setup in which the beams are polarized along the radial directions. (b) Intensity profile obtained with the setup of "(a)." (c) Intensity profile obtained when the phase of the beam $\mathbf{E}_1$ is shifted by π. (d) Intensity profile obtained when the phase of the beam $\mathbf{E}_1$ is shifted by $\pi/2$.

The resulting intensity profile is plotted in Figure 9.9d. It is an inverse structure to that obtained when the beams are all in phase. As illustrated above, the pattern morphology can be controlled by properly adjusting the polarization state and phase of each beam.

Three-beam interference can be employed to fabricate other 2D patterns such as square and rectangular patterns. In order to produce a 2D square lattice, two basis vectors should have an equal magnitude and be perpendicular to one another. This requirement can be fulfilled with the configuration shown in Figure 9.10a. In the configuration where $\phi_1 = \pi/4$, $\phi_2 = 3\pi/4$, and $\phi_3 = -3\pi/4$, the differences between the wave vectors are

$$\mathbf{k}_{12} = -\frac{2\pi}{\lambda}(\sqrt{2}\sin\theta, 0, 0), \quad \mathbf{k}_{23} = -\frac{2\pi}{\lambda}(0, \sqrt{2}\sin\theta, 0),$$

$$\mathbf{k}_{31} = -\frac{2\pi}{\lambda}(-\sqrt{2}\sin\theta, -\sqrt{2}\sin\theta, 0)$$

The difference vectors $\mathbf{k}_{12}$ and $\mathbf{k}_{23}$ have the same magnitude of $(2\sqrt{2}\,\pi\sin\theta)/\lambda$ and are mutually orthogonal. They can be taken as two primitive vectors $\mathbf{b}_1$ and $\mathbf{b}_2$ in reciprocal space. The primitive vectors $\mathbf{a}_1$ and $\mathbf{a}_2$ in real space are also orthogonal and have the same magnitude of $\lambda/(\sqrt{2}\sin\theta)$, thereby forming a square lattice (Fig. 9.10b). If the beam arrangement has $\phi_1 = \pi/3$, $\phi_2 = 2\pi/3$, and $\phi_3 = -2\pi/3$, $\mathbf{k}_{12}$ and $\mathbf{k}_{23}$ are taken as the primitive reciprocal lattice vectors $\mathbf{b}_1$ and $\mathbf{b}_2$. The magnitudes of these vectors are $b_1 = k\sin\theta$ and $b_2 = \sqrt{3}k\sin\theta$, respectively. The unit cell of the real lattice, defined by $\mathbf{a}_1$ and $\mathbf{a}_2$, have different sides of $a_1 = \lambda/\sin\theta$ and $a_2 = \lambda/(\sqrt{3}\sin\theta)$, thereby forming a rectangular lattice. Once the azimuthal angles have the relations of $\phi_1 = (\pi/2 - \alpha)$, $\phi_2 = (\pi/2 + \alpha)$, and $\phi_3 = (3\pi/2 - \alpha)$, a 2D rectangular intensity pattern is produced. The lattice constant ratio a_2/a_1 can be controlled by varying the angle. When $\alpha = \pi/4$, a 2D square pattern is obtained. In three-beam interference lithography, the difference between the wave vectors, $\mathbf{k}_{ij}$, determines the translational periodicity (and hence the point lattice) of the resultant 2D interference pattern. All five 2D lattices can be created by adjusting the geometrical configuration of three coherent beams.[2]

9.2.3 FOUR-BEAM INTERFERENCE

The intensity distribution produced by multi-beam interference is generally written as

$$I(\mathbf{r}) = I_1 + I_2 + .. + I_N + \sum_{i,j(i\neq j)}^{N} \left(\mathbf{E}_{0i} \cdot \mathbf{E}_{0j}^*\right) e^{i(\mathbf{k}_i - \mathbf{k}_j)\cdot\mathbf{r}}. \tag{9.20}$$

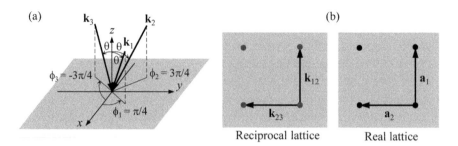

Reciprocal lattice   Real lattice

FIGURE 9.10   (a) A beam configuration to produce a 2D square intensity profile. (b) Reciprocal lattice (wave vector differences) and real lattice (intensity pattern) associated with this configuration.

Here, N is the number of interfering beams. To fabricate a 3D periodic structure by interference lithography, we need at least four beams. The larger the number of interfering beams, the larger the number of different structures that can be fabricated. However, increasing the number of beams makes it much more difficult to assemble and position the required optical components properly. We here consider the superposition of four beams only. In fact, all fourteen 3D Bravais lattices can be formed by interference of four non-coplanar beams.[3] For an arbitrary Bravais lattice defined by the primitive vectors $\mathbf{a}_1$, $\mathbf{a}_2$, and $\mathbf{a}_3$, the primitive vectors of its reciprocal lattice are given by

$$\mathbf{b}_1 = 2\pi \frac{\mathbf{a}_2 \times \mathbf{a}_3}{\mathbf{a}_1 \cdot (\mathbf{a}_2 \times \mathbf{a}_3)}, \mathbf{b}_2 = 2\pi \frac{\mathbf{a}_3 \times \mathbf{a}_1}{\mathbf{a}_2 \cdot (\mathbf{a}_3 \times \mathbf{a}_1)}, \mathbf{b}_3 = 2\pi \frac{\mathbf{a}_1 \times \mathbf{a}_2}{\mathbf{a}_3 \cdot (\mathbf{a}_1 \times \mathbf{a}_2)}. \tag{9.21}$$

The denominator is a scalar quantity that is identical for $\mathbf{b}_1$, $\mathbf{b}_2$, and $\mathbf{b}_3$. The above definition is called the physics definition, as the factor "2π" comes from the study of periodic structures. In the crystallographer's definition, the factor of 2π is dropped. This can simplify some mathematical manipulations, and expresses reciprocal lattice dimensions in units of spatial frequency. It is a matter of choice which definition to use, as long as the two are not mixed. The physics definition is used in this book. There are various ways to choose a unit cell in space lattice. The smallest repeat unit is referred to as the *primitive cell* and it contains only one lattice point. However, it is more convenient to consider a larger repeat unit in many cases because the periodicity and symmetry of the lattice can be more easily visualized with this *non-primitive, conventional cell*. Although the primitive cell of a face-centered cubic (FCC) lattice is rhombohedral, a cubic cell is conventionally taken as its unit cell. The primitive cell of a body-centered cubic (BCC) lattice is also rhombohedral. The primitive cell vectors of the reciprocal lattice of a real lattice can be derived using eq 9.21. Figure 9.11 shows three cubic lattices (simple cubic, BCC, and FCC) and their reciprocal lattices. The reciprocal lattice of a simple cubic lattice of side "*a*" is a simple cubic lattice of side "$2\pi/a$." The reciprocal lattice of an FCC lattice of side "*a*" is a BCC lattice of side "$4\pi/a$." The reciprocal lattice of a BCC lattice of side "*a*" is an FCC lattice of side "$4\pi/a$."

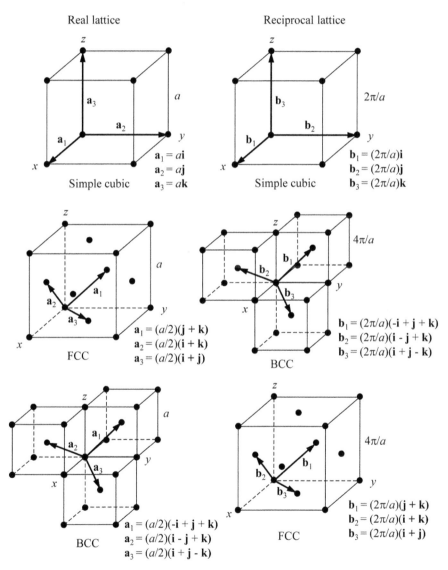

FIGURE 9.11   Three cubic lattices (simple cubic, BCC, and FCC) and their reciprocal lattices.

For four-beam interference, eq 9.20 has 12 interference terms determined by k_{ij}. Consider the four-beam geometry illustrated in Figure 9.12. Each pair of wave vectors induces one sinusoidal interference pattern characterized by a grating vector. The number of gratings is given by $(N^2 - N)/2$, where N is the number of beams. The interference of four beams thus produces six gratings characterized by k_{12}, k_{13}, k_{14}, k_{23}, k_{24}, and k_{34}. Note that k_{ij} induces the same grating as k_{ji}. The grating vectors can be interpreted as reciprocal lattice vectors. Any of the fourteen Bravais lattices can be uniquely described by three lattice vectors. Therefore we need to choose any three of the six reciprocal lattice vectors. These three determine the symmetry of the induced lattice. The most common choice is k_{12}, k_{13}, and k_{14}. The choice of k_{12}, k_{13}, and k_{23} is invalid because it contains no information on the wave vector k_4. The strict requirement for intensity variations in three dimensions is that the chosen difference vectors must be non-coplanar. When these difference vectors (i.e., k_{12}, k_{13}, and k_{14}) are the primitive vectors of a BCC lattice, the produced intensity pattern forms an FCC lattice in real space. If they are the primitive vectors of an FCC lattice, the resulting intensity profile forms a BCC lattice.

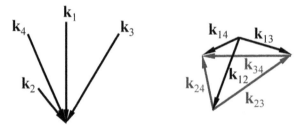

FIGURE 9.12   A four-beam geometry with six wave vector differences.

Suppose that we intend to generate a 3D intensity profile of FCC symmetry. In this case, the wave vector differences k_{12}, k_{13}, and k_{14} should be the primitive vectors of a BCC reciprocal lattice, as shown in Figure 9.13a. When this BCC lattice has a lattice constant of $4\pi/a$, the difference vectors are written as

$$k_{12} = \frac{2\pi}{a}(-1,1,1), \quad k_{13} = \frac{2\pi}{a}(1,-1,1), \quad k_{14} = \frac{2\pi}{a}(1,1,-1). \tag{9.22}$$

All four-wave vectors have the same magnitude. If one wave vector is known, the rest can be immediately calculated. Since the difference vectors are symmetric about the [111] direction of the lattice, the first wave vector is parallel to this direction. The four-wave vectors are then found to be

$$\mathbf{k}_1 = \frac{\pi}{a}(3,3,3), \mathbf{k}_2 = \frac{\pi}{a}(5,1,1), \mathbf{k}_3 = \frac{\pi}{a}(1,5,1), \mathbf{k}_4 = \frac{\pi}{a}(1,1,5). \tag{9.23}$$

Figure 9.13b shows the required arrangement of the four beams. In the well-known umbrella configuration (Fig. 9.13c), the three side beams ($\mathbf{k}_2$, $\mathbf{k}_3$, and $\mathbf{k}_4$) should be symmetrically placed about the center beam ($\mathbf{k}_1$) with the same polar angle of $\theta = 39°$. This configuration results in a 3D intensity profile of FCC structure. The lattice constant of the structure is $a = 3\sqrt{3}\lambda/2$. If we intend to generate a 3D intensity profile of BCC symmetry, the wave vector differences $\mathbf{k}_{12}$, $\mathbf{k}_{13}$, and $\mathbf{k}_{14}$ should be the primitive vectors of an FCC lattice (Fig. 9.14a). Then the difference vectors are given by

$$\mathbf{k}_{12} = \frac{2\pi}{a}(0,1,1), \ \mathbf{k}_{13} = \frac{2\pi}{a}(1,0,1), \ \mathbf{k}_{14} = \frac{2\pi}{a}(1,1,0). \tag{9.24}$$

This leads to the following four-wave vectors.

$$\mathbf{k}_1 = \frac{\pi}{a}(1,1,1), \mathbf{k}_2 = \frac{\pi}{a}(1,-1,-1), \mathbf{k}_3 = \frac{\pi}{a}(-1,1,-1), \mathbf{k}_4 = \frac{\pi}{a}(-1,-1,1) \tag{9.25}$$

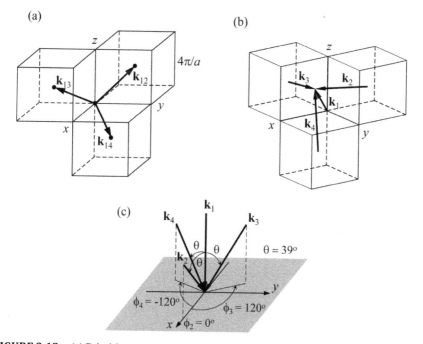

FIGURE 9.13 (a) Primitive vectors of a BCC reciprocal lattice. (b) Wave vectors of the four beams in a BCC reciprocal lattice to produce a real-space FCC lattice. (c) Umbrella geometry for an FCC intensity pattern.

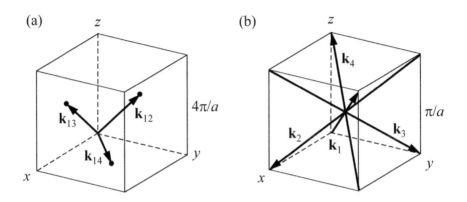

FIGURE 9.14   (a) Primitive vectors of an FCC reciprocal lattice. (b) Beam geometry for a BCC intensity pattern.

The required beam geometry is shown in Figure 9.14b. The wave vectors of the four interfering beams have the same direction and magnitude as the four body diagonals of a cube with side "π/a." This produces a BCC intensity pattern whose lattice constant is given by $a = \sqrt{3}\lambda/2$. The beam arrangement required to generate a simple-cubic intensity pattern can also be easily derived. Consider four beams with the following wave vectors.

$$\mathbf{k}_1 = \frac{\pi}{a}(1,1,1), \mathbf{k}_2 = \frac{\pi}{a}(-1,1,1), \mathbf{k}_3 = \frac{\pi}{a}(1,-1,1), \mathbf{k}_4 = \frac{\pi}{a}(1,1,-1) \qquad (9.26)$$

The wave vector differences are then given by

$$\mathbf{k}_{12} = \frac{2\pi}{a}(1,0,0), \mathbf{k}_{13} = \frac{2\pi}{a}(0,1,0), \mathbf{k}_{14} = \frac{2\pi}{a}(0,0,1). \qquad (9.27)$$

These difference vectors are the lattice vectors of a simple cubic lattice whose lattice constant is $2\pi/a$. The reciprocal lattice of a simple cubic lattice with lattice constant "a" is simple cubic with a lattice constant of $2\pi/a$. Therefore, the four beams given in eq 9.26 produce a simple-cubic intensity pattern whose lattice constant is $a = \sqrt{3}\lambda/2$.

As shown in three-beam interference, the polarization of the beams can have a dramatic impact on the morphology and contrast of the interference pattern. Since 3D patterns are more complex, the intensity, polarization, and relative phase of each beam have profound effects on the final patterns.[4-8] To maximize the contrast of the three chosen difference vectors ($\mathbf{k}_{12}$, $\mathbf{k}_{13}$, and $\mathbf{k}_{14}$) and minimize the contrast of the others ($\mathbf{k}_{23}$, $\mathbf{k}_{24}$, and $\mathbf{k}_{34}$), the center beam ($\mathbf{k}_1$) is often circularly or elliptically polarized while the three side beams are linearly polarized. In this section, we described the beam arrangements required to construct interference patterns of three highest symmetries; simple cubic, FCC, and BCC. It is to be noted that there exists more than one arrangement to obtain a particular Bravais lattice. Periodic interference patterns with any desired Bravais lattice can be produced by interference of four beams with appropriate wave vectors.[3]

9.3 FABRICATION OF PERIODIC STRUCTURES

9.3.1 MULTI-BEAM INTERFERENCE CONFIGURATIONS

When we want to fabricate a periodic structure by interference lithography, multiple light beams should be properly arranged so that they can generate a spatial distribution of intensity that resembles the structure. The desired beam configuration (i.e., wave vectors and polarization directions) can be obtained by a number of different ways. In dealing with the basic concepts of multi-beam interference in the previous section, it has been implicitly assumed that the interfering beams are monochromatic and coherent. The light source for interference lithography should be coherent over both the spatial extent of the sample and the duration of time for which the intensity pattern is recorded into the chosen material. Otherwise, the recorded pattern will be smoothened due to the random fluctuations in the amplitude and phase of the interfering beams. Continuous-wave lasers, pulsed lasers, and lamps (with filters to narrow the frequency spectrum) are the typical light sources employed for interference lithography. Once a light source is chosen, the next task is to achieve the desired beam configuration. The experimental parameters that must be considered for each beam are its direction given by the wave vector k_j, and the amplitude, polarization, and phase of its electric field. The beam configurations may be classified into amplitude-splitting and wavefront-splitting methods.

Amplitude-splitting configurations divide a single source beam into two or more beams using beam splitters, which are eventually combined through the use of mirrors, lenses, and other optical elements to produce an interference pattern. A typical two-beam configuration to record a 1D grating is depicted in Figure 9.15a. In any two-beam configuration, multiple exposures can be employed to generate complex 2D and 3D patterns in a photosensitive material. This is often accomplished by incorporating a rotating sample stage. After recording a 1D grating, another grating is superimposed onto the existing grating by rotating the sample. Typical three-beam and four-beam configurations are shown in Figure 9.15b,c, respectively. In these multi-beam configurations, the output laser beam is split into the required number of beams by beam splitters. Lasers usually emit a linearly polarized beam. Therefore, the output beam is in general linearly polarized with its electric field oscillating along a fixed direction. The intensity of each beam can be controlled by using a polarizing beam splitter. Such a beam splitter divides an incident linearly polarized beam into two linearly polarized beams whose polarization directions are orthogonal. By adjusting the direction of oscillation of the electric field of the incoming beam, we can control the relative amplitudes (i.e., the relative intensities) of the two linearly polarized beams that come out of the splitter. The oscillation direction of the incident beam can be controlled by placing a half-wave plate in front of the beam splitter. Before the interfering beams are incident on the material to be fabricated, the polarization of each beam must also be adjusted. Light beams with specific linear polarizations are obtained by rotating the oscillation direction of the electric field of each beam using a half-wave plate. Light beams with circular or elliptical polarizations are also achievable if a half-wave plate and a quarter-wave plate are used in sequence.

One drawback of the amplitude-splitting configurations is the potential for interferometric instability. Any perturbations to the optical path lengths or relative phases of the interfering beams may cause a translation of the recorded pattern or change in the symmetry of the unit cell. Wavefront-splitting configurations can mitigate these effects. The most common and widely used wavefront-splitting system incorporates a Lloyd's mirror, as shown in Figure 9.16a. In this configuration, half of a diverging laser beam is made to strike the surface of a mirror, so that the beam reflecting off the mirror interferes with the beam directly traveling to the sample. A lens and a pinhole are placed in the beam path to create a diverging beam. The flat mirror is typically aligned perpendicular to the substrate. The angle between two interference beams is modulated by rotating the sample stage on which the mirror and substrate holder are placed. Compared with other two-beam configurations, Lloyd's mirror interferometer provides a simple way to fabricate linear gratings, where the period of the grating can be easily controlled by rotating the sample stage. The interference of two beams results in the formation of standing waves in the horizontal and vertical directions, generally registered on a photosensitive material such as photoresist. The horizontal wave is regarded as the major wave as it determines the period of the interference fringes along the substrate. When the incoming beams interfere with the light reflected from the film/substrate interface, a vertical standing wave can be generated and recorded within the film along its thickness direction. A single exposure produces 1D grating patterns, while multiple exposures with an interim rotation of the sample are capable of producing patterns of varying lattice.[9]

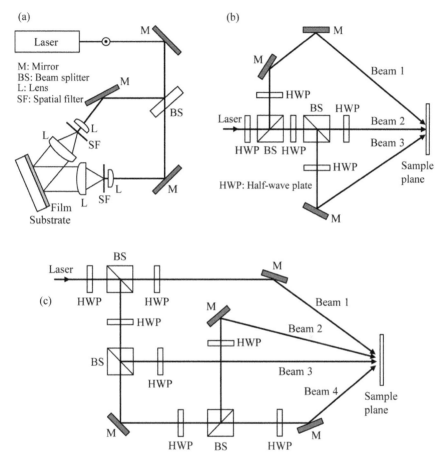

FIGURE 9.15 Typical amplitude-splitting configurations for (a) two-beam interference, (b) three-beam interference, and (c) four beam-interference.

Another common wavefront-splitting configuration incorporates a specially designed prism, which simultaneously splits and recombines the incident beam (Fig. 9.16b). A trigonal pyramid-shaped prism is employed for three-beam interference, while a truncated pyramid prism is utilized for four-beam interference. The prism-based method is particularly useful for implementing the umbrella configuration described in the previous section. Diffraction by gratings may also be used as a wavefront-splitting scheme (Fig. 9.16c). In this configuration, a single expanded beam is diffracted by two or more gratings incorporated into one package mask. The first-order diffracted beams produce an interference pattern in the sample plane. An opaque layer may be deposited onto the center of the mask to block the transmission of the zero-order diffracted beam. Since the interference pattern depends on the period of the gratings, different grating masks are required to change the pattern. An alternative way of utilizing diffraction incorporates a periodic dielectric structure known as a *phase mask*. When exposed to a single beam of light, this phase mask produces diffracted beams that are traveling in the specific directions (Fig. 9.16d). The relative intensity and phase of the beam that travels along each particular direction are determined by many factors including the geometry and refractive index of the dielectric structure. The diffracted beams may be recombined in the far field by focusing them using a lens. A self-interference pattern between the 0, +1, and −1 diffraction orders can be recorded by placing a photosensitive material in close proximity to the mask. A single 2D diffractive optical element can be used to produce multiple beams with a single exposure. Multiple exposures with a 1D phase mask may also be used to fabricate complex 3D structures. Like the other wavefront-splitting configurations, this phase mask approach has limited control over the amplitude and polarization of each individual beam.

Most laser systems have Gaussian beam profiles, which leads to a significant variation in the light intensity between the center and the periphery of the beam. This intensity variation can result in a spatial variation of the resultant structure. This issue can be alleviated by two approaches. The first approach is to expand the output beam sufficiently and then use only a central portion of the expanded beam; this inevitably weakens the intensity. Another approach is to employ a beam homogenizer to convert the Gaussian profile into a top-hat profile.

9.3.2 FABRICATION OF PERIODIC STRUCTURES

A wide range of periodic structures has been fabricated via interference lithography using near-infrared, visible, and ultraviolet light sources.[10–22] Once the desired beam configuration is obtained, the next step is to transfer the interference pattern into a photosensitive material. There are many material platforms that facilitate such a transfer, including photoresists, hybrid organic–inorganic materials, and holographic polymer-dispersed liquid crystals. In fact, an interference pattern may be recorded on or in any materials (e.g., metals and glasses), as long as the optical intensity is sufficiently high. The most common photosensitive materials are photoresists, which were developed primarily for use in the semiconductor industry. There are two different types of photoresists: positive and negative. In a positive photoresist, the regions that are exposed to light above a certain intensity threshold are chemically changed and become soluble to the developing solvent. This means that the irradiated regions are removed by the solvent. In a negative photoresist, the regions exposed to light remain insoluble to the developer, forming the resultant structure. SU-8 is a common negative photoresist, which is an epoxy-based monomer that undergoes cationic photopolymerization. Highly cross-linked polymers, which are usually obtained when negative photoresists are exposed, are difficult to dissolve.

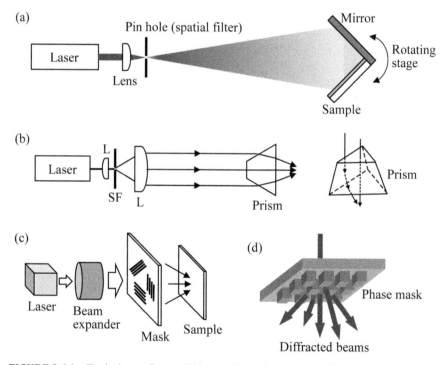

FIGURE 9.16 Typical wavefront-splitting configurations. (a) Lloyd's mirror configuration. (b) Prism-based umbrella configuration. (c) Diffractive mask-based configuration. (d) Phase mask lithography.

One of the advantages of interference lithography is that it can provide precise control over the size and shape of the resultant 3D structure in a single laser shot. Campbell et al.[23] have demonstrated that nearly perfectly ordered 3D structures with sub-micrometer periodicity can be produced using an interference pattern formed by four laser beams where one beam is directed normal to the SU-8 photoresist surface. Polymeric structures made by this technique have also been used as templates to create complementary structures with higher refractive-index contrast. The approach is a viable technique of fabricating 3D photonic crystal structures. Figure 9.17a shows a calculated surface of constant intensity corresponding to a laser interference pattern with FCC symmetry. The wave vectors of the four beams that interfere to produce this pattern are the same as in Figure 9.13b and eq 9.23. The difference wave vectors, which are shown in Figure 9.13a and eq 9.22, generate a BCC reciprocal lattice corresponding to a real-space FCC interference pattern with a lattice constant $a = 3\sqrt{3}\lambda/2 = 922$ nm for a laser wavelength $\lambda = 355$ nm. The repeated "abc" pattern of closed packed (111) planes that make up an FCC lattice is indicated on the side of the cube in Figure 9.17a. Figure 9.17b is an equivalent surface of constant intensity, generated by one of many alternative four-beam configurations that can use the same laser wavelength to create distinct FCC lattices. Figure 9.18a shows a polymeric photonic crystal produced by the interference pattern of Figure 9.17a. Refraction at the film surface changes the incident wave vectors, stretching the interference pattern in the [111] direction. Figure 9.18b is a magnified image of the (111) surface. An inverse replica in titania (TiO_2) is shown in Figure 9.18d, which was produced by filling the pores with titanium (iv) ethoxide, and then heating to burn off the polymer template and sinter the ceramic. Figure 9.18e shows an image of the structure obtained when a polymeric film is exposed to a BCC interference pattern.

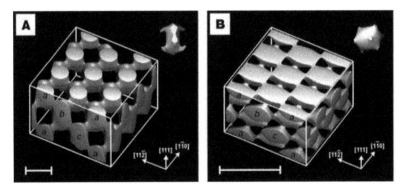

FIGURE 9.17 Calculated constant-intensity surfaces in four-beam laser interference patterns designed to produce photonic crystals. (a) FCC pattern with lattice constant 922 nm. The close-packed layers of the FCC lattice are indicated on one side of the cube. (b) FCC pattern with lattice constant 397 nm. Scale bars are 500 nm. Reprinted with permission from Ref. [23]. Copyright© 2000 Springer Nature.

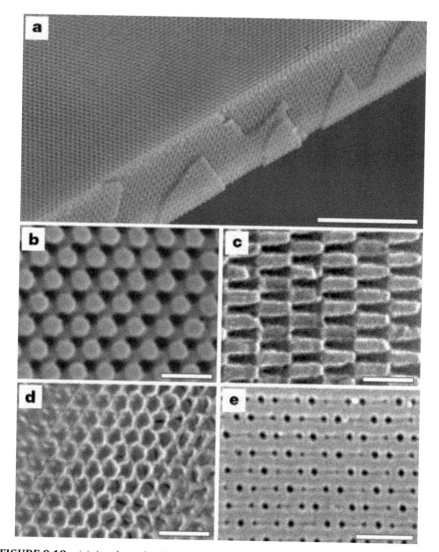

FIGURE 9.18 (a) A polymeric photonic crystal generated by exposure of a 10 μm-film of photoresist to the interference pattern shown in Figure 9.17a. Scale bar is 10 μm. (b) Close-up of a (111) surface. Scale bar is 1 μm. (c) Close-up of a (11$\bar{1}$) surface. Scale bar is 1 μm. (d) Inverse replica in titania made by using the polymeric structure as a template. Scale bar is 1 μm. (e) ($\bar{1}$02) surface of a BCC polymeric photonic crystal. Scale bar is 1 μm. Reprinted with permission from Ref. [23]. Copyright© 2000 Springer Nature.

The structure of a crystal can be described by combining its lattice with the motif associated with each lattice point. While the translational symmetry defined by the lattice is determined by the wave vector differences, the motif can be varied by the polarization and relative phase of each beam. In photo-lithography, a desired pattern is printed in a photoresist layer by exposing the layer to an intensity distribution of light, that is, an image. There are different ways to form such an image. In its simplest form, the photoresist-coated substrate is put in contact with or in close proximity to a photomask that has the desired pattern etched in an opaque (e.g., Cr) film. The mask is then illuminated with a parallel beam of UV light, with its pattern printed in the photoresist. While this technique is relatively simple, diffraction effects limit its resolution to about 1 μm. Projection methods using steppers enable high-resolution printing but they require complex and expensive imaging optics. The simplicity and cost advantage of contact/proximity printing make it still attractive for certain applications, such as for the fabrication of periodic micro or nanostructures used in photonics or optoelectronics. The use of phase shifting masks, simply called phase masks, is a well-known method of improving resolution in proximity lithography. This method takes advantage of interference effects to form a desired intensity distribution. A phase mask relies on the fact that light passing through a transparent medium will undergo a phase shift as a function of its optical thickness. In a phase mask, certain regions are made thicker or thinner. This induces a phase difference in the light transmitting through different regions of the mask. For instance, two transparent regions of a mask that induces a phase difference of 180° in the transmitted light lead to the formation of a deep intensity minimum, improving the resolution of an image pattern formed on the photoresist. The intensity dip obtained in this way has a strong dependence on the distance from the mask. When exposed to a single beam of light, the phase mask produces diffracted beams that are traveling in the specific directions. Phase masks may be designed to produce only two diffraction orders ($m = 1$ and -1) so that the resultant fringe pattern becomes practically independent of the distance from the mask. When mutually coherent multiple beams interfere, the resultant intensity distribution depends strongly on the relative phases of the beams. The advantage of using a phase mask is that it simplifies the experimental setup significantly. In the amplitude-splitting configurations shown in Figure 9.15, it is very difficult to adjust the phases of individual beams. Phase masks make it possible to create beams with precisely controlled relative phases. In other words, the relative phases of the diffracted beams are already encoded in the mask. Figure 9.19 show some patterns printed in 300-nm-thick photo-resists with different mask designs. The sharp and smooth features in the photoresist indicate good image quality, that is, high contrast. The narrow features measuring as small as 0.25 μm in width imply the formation of deep intensity minima at these positions.

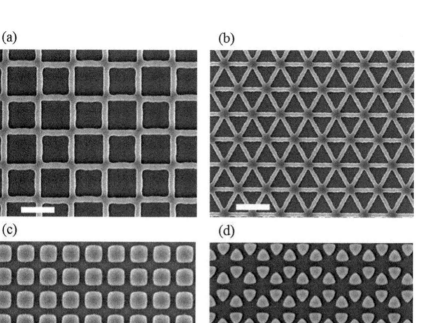

FIGURE 9.19 (a) and (b) SEM images of patterns obtained in a 300-nm-thick positive photoresist. Scale bars are 3 μm. (c) and (d) SEM images of patterns obtained in a 300-nm-thick negative photoresist. Scale bars are 1 μm. Reprinted with permission from Ref. [24]. Copyright© 2015 Elsevier.

9.4 APPLICATIONS OF INTERFERENCE LITHOGRAPHY

The most prominent applications of periodic structures are photonic crystals. Photonic crystals can be described as dielectric composites with periodically varying refractive indices. The periodic structure usually consists of two media with different dielectric constants; one medium has a higher dielectric constant than the other. One of the fundamental properties exhibited by photonic crystals is that electromagnetic waves in a certain range of frequencies are not allowed to propagate within the periodic structure. This range of forbidden frequencies is called a *phonic band gap*, analogous to the band gaps of electrons in solids. The periodic dielectric structure affects the propagation of electromagnetic waves in the same way that the periodic potential in a semiconductor affects the motion of electrons by defining allowed and forbidden energy bands. Intuitively, the photonic band gap can be understood to arise from the interference of light reflected from the interfaces between the two different media. If the multiply reflected light interferes constructively, the incident electromagnetic wave is not allowed to propagate within the structure, since all the electromagnetic energy is reflected backward. On the contrary, when the light interferes destructively, the electromagnetic wave can propagate through the structure. In 1887, the English physicist L. Rayleigh experimented with periodic multilayer dielectric stacks, showing that they had a band gap in one dimension. However, the term "photonic crystal" had not been used until E. Yablonovitch published two milestone papers on 3D photonic crystals in the late 1980s.[25,26] Photonic crystals are attractive optical materials for controlling and manipulating light flow. The use of such crystals holds the promise of numerous applications in optics and optoelectronics. One-dimensional photonic crystals are already in widespread use in the form of multilayer thin-film structures, as discussed in Chapter 8. Such structures are utilized in a diverse range of applications, from coatings to enhance the efficiency of photovoltaic and light-emitting devices to highly reflective mirrors in certain laser cavities. 2D and 3D photonic crystals are of great interest for both fundamental and applied research. However, these higher-dimensional photonic crystals (especially 3D crystals) are still far from commercialization because of the difficulty in fabricating uniform and defect-free 3D periodic structures over large areas. 3D photonic crystals can be fabricated by various techniques. For instance, layer-by-layer photolithography and etching can generate the desired 3D structures.[27] However, this approach is expensive and slow because it relies on optical projection lithography and each layer requires a different photomask. Self-assembly techniques involving the use of colloidal particles[28,29] or copolymer phase separation[30] are relatively simple and cost-effective. However, the overall synthetic processes are not rapid and random defects are inevitably introduced in large-area samples. Since the pioneering work of M. Campbell et al.,[23] multi-beam interference has been extensively studied as a simple technique for fabricating 3D photonic crystals, with ongoing efforts to combine it with other techniques such as laser direct writing,[20,31] mask lithography,[32] pattern-integrated interference lithography,[33] and two-photon polymerization.[34]

The intensity pattern produced by multi-beam interference is not always recorded in a photosensitive material. Interference lithography is a quick and cost-effective method for producing periodic/aperiodic surface structures. Such surface patterns can be effectively utilized for a range of applications including tribology (wear and friction reduction), photovoltaics, and biotechnology. Interfering high-power pulsed laser beams can induce periodic surface structures on any materials. Using this approach, the material can be directly structured just in a few seconds. As described in Section 8.2.2, surface structures at the subwavelength scales significantly reduce light reflection over a broad range of wavelengths and angles, mimicking the so-called moth's eye structure. Laser interference metallurgy has been receiving increasing attention from the tribology community, especially in the macro-mechanical applications.[35–37] The tribological behavior of metallic substrates can be strongly improved by introducing laser-induced surface structures. Laser surface texturing is also a powerful and fast method to control the wetting properties of surfaces,[38–40] which finds applications in corrosion resistance, microfluidics, and self-cleaning.

PROBLEMS

9.1 If phase terms φ_1 and φ_2 are added, eq 9.6 reduces to the following expression.

$$E_t = E_0 e^{i(\mathbf{k}_1 \cdot \mathbf{r} - \omega t + \varphi_1)} + E_0 e^{i(\mathbf{k}_2 \cdot \mathbf{r} - \omega t + \varphi_2)}$$

Derive the intensity distribution of eq 9.9 using this relation.

9.2 Show that eq 9.3 can be alternatively expressed as

$$I(\mathbf{r}) = I_1 + I_2 + 2\mathrm{Re}(\mathbf{E}_{01} \cdot \mathbf{E}_{02}^*) \times \cos(\mathbf{k}_{12} \cdot \mathbf{r}) - 2\mathrm{Im}(\mathbf{E}_{02} \cdot \mathbf{E}_{01}^*) \times \sin(\mathbf{k}_{12} \cdot \mathbf{r})$$

Here Re and Im mean the real and imaginary parts of the amplitude product $\mathbf{E}_{01} \cdot \mathbf{E}_{02}^*$, respectively. The amplitude product is generally complex.

9.3 Show that a rhombic 2D lattice ($a_1 = a_2$ and γ is arbitrary) is equivalent to a based-centered rectangular lattice.

9.4 Figure 9.5a shows an umbrella configuration for three-beam interference. Suppose that the three incident beams have the following wave vectors.

$$\mathbf{k}_1 = \frac{2\pi}{\lambda}\left(\frac{1}{2}\mathbf{i} + \frac{\sqrt{3}}{6}\mathbf{j} - \frac{2}{\sqrt{6}}\mathbf{k}\right)$$

$$\mathbf{k}_2 = \frac{2\pi}{\lambda}\left(-\frac{1}{2}\mathbf{i} + \frac{\sqrt{3}}{6}\mathbf{j} - \frac{2}{\sqrt{6}}\mathbf{k}\right)$$

$$\mathbf{k}_3 = \frac{2\pi}{\lambda}\left(-\frac{2\sqrt{3}}{6}\mathbf{j} - \frac{2}{\sqrt{6}}\mathbf{k}\right)$$

Here, **i**, **j**, and **k** are the unit vectors along the x, y, and z directions, respectively.

(a) What are the azimuthal angles of the three beams?
(b) Do they have the same polar angle with respect to the central normal z-axis?
(c) The interfering three beams will produce an interference pattern. For $\lambda = 500$ nm, what are the shape and period of the pattern?

REFERENCES

1. Lee, M. *X-Ray Diffraction for Materials Research*; Apple Academic Press/CRC Press: Oakville, Canada, 2016.
2. Escuti, M.; Crawford, G. Holographic Photonic Crystals. *Opt. Eng.* **2004**, *43*, 1973–1987.
3. Cai, L.; Yang, X.; Wang, Y. All Fourteen Bravais Lattices Can Be Formed by Interference of Four Noncoplanar Beams. *Opt. Lett.* **2002**, *27*, 900–902.
4. Cai, L.; Yang, X.; Wang, Y. Formation of Three-dimensional Periodic Microstructures by Interference of Four Noncoplanar Beams. *J. Opt. Soc. Am. A* **2002**, *19*, 2238–2244.
5. Ullal, C.; Maldovan, M.; Thomas, E.; Chen, G.; Han, Y.; Yang, S. Photonic Crystals Through Holographic Lithography: Simple Cubic, Diamond-like, and Gyroid-like Strictures. *Appl. Phys. Lett.* **2004**, *84*, 5434–5436.
6. Maldovan, M.; Urbas, A.; Yufa, N.; Carter, W.; Thomas, E. Photonic Properties of Bicontinous Cubic Microphases. *Phys. Rev. B* **2002**, *65*, 165123.
7. Wang, D., Wang, Z.; Zhang, Z.; Yue, Y.; Li, D.; Maple, C. Effects of Polarization on Four-beam Laser Interference Lithography. *Appl. Phys. Lett.* **2013**, *102*, 081903.
8. Behera, S.; Joseph, J. Design and Realization of Functional Metamaterial Basis Structures Through Optical Phase Manipulation Based Interference Lithography. *J. Opt.* **2017**, *19*, 105103.
9. Hung, Y.; Chang, P.; Lin, Y.; Lin, J. Compact Mirror-tunable Laser Interference System for Wafer-scale Patterning of Grating Structures with Flexible Periodicity. *J. Vac. Sci. Tecnhnol. B* **2016**, *34*, 040601.
10. George, M.; Nelson, E.; Rogers, J.; Braun, P. Direct Fabrication of 3D Periodic Inorganic Microstructure Using Conformal Phase Masks. *Angew. Chem. Int. Ed.* **2009**, *48*, 144–148.
11. Shin, H.; Kim, H.; Lim, K.; Lee, M. Step-wise Ag Thin Film Patterns Fabricated by Holographic Lithography. *Thin Solid Films* **2009**, *517*, 3273–3275.
12. Lee, J.; Yoon, J.; Jin, M.; Lee, M. Holographic Modification of TiO_2 Nanostructure for Enhanced Charge Transport in Dye-sensitized Solar Cell. *J. Appl. Phys.* **2012**, *112*, 043110.
13. Jang, J. et al., 3D Micro- and Nanostructures via Interference Lithography. *Adv. Funct. Mater.* **2007**, *17*, 3027–3041.
14. Yang, Y.; Wang, G. Realization of Periodic and Quasiperiodic Microstructures with Sub-diffraction-limit Feature Sizes by Far-field Holographic Lithography. *Appl. Phys. Lett.* **2006**, *89*, 111104.
15. Zhang, Y.; Zhou, J.; Wong, K. Two-photon Fabrication of Photonic Crystals by Single-beam laser Holographic Lithography. *J. Appl. Phys.* **2010**, *107*, 074311.

16. Stay, J.; Burrow, G.; Gaylord, T. Three-beam Interference Lithography Methodology. *Rev. Sci. Instrum.* **2011**, *82*, 023115.

17. Lemme, M.; Moormann, C.; Lerch, H.; Moller, M.; Vratzov, B.; Kurz, H. Triple-gated Metal-oxide-Semiconductor Field Effect Transistors Fabricated by Interference Lithography. *Nanotechnology* **2004**, *15*, 208–210.

18. Capeluto, M. et al., Nanopatterning with Interferometric Lithography Using a Compact Lambda = 46.9 nm Laser. *IEEE Nanotechnol.* **2006**, *5*, 3–7.

19. Lu, C.; Lipson, R. Interference Lithography: A Powerful Tool for Fabricating Periodic Structures. *Laser Photonics Rev.* **2010**, *4*, 568–580.

20. Ghosh, S.; Ananthasuresh, G. Single-photon-multi-layer-interference Lithography for High-aspect-ratio and Three-dimensional SU-8 Micro-/nanostructures, *Sci. Reports* **2016**, *6*, 18428.

21. Kim, H.; Jung, H.; Lee, D.; Lee, K.; Jeon, H. Period-chirped Gratings Fabricated by Laser Interference Lithography with a Concave Lloyd's Mirror. *App. Optics* **2016**, *55*, 354–359.

22. Chen, X.; Yang, F.; Zhang, C.; Zhou, J.; Guo, L. Large-area High Aspect Ratio Plasmonic Interference Lithography Utilizing a High-*k* mode. *ACS Nano* **2016**, *10*, 4039–4045.

23. Campbell, M.; Sharp, D.; Harrison, M.; Denning, R.; Turberfield, A. Fabrication of Photonic Crystals for the Visible Spectrum by Holographic Lithography. *Nature* **2000**, *404*, 53–56.

24. Solak, H.; Dais, C.; Clube, R.; Wang, L. Phase Shifting Masks in Displacement Talbot Lithography for Printing Nano-grids and Periodic Motifs. *Microelectronic. Eng.* **2015**, *143*, 74–80.

25. Yablonovitch, E. Inhibited Spontaneous Emission in Solid State Physics and Electronics. *Phys. Rev. Lett.* **1987**, *58*, 2059–2062.

26. Yablonovitch, E.; Gmitter, T. Photonic Band Structure: The Face-centered Cubic Case. *Phys. Rev. Lett.* **1987**, *63*, 1950–1953.

27. Maldovan, M. Layer-by-layer Photonic Crystal with a Repeating Two-layer Sequence. *Appl. Phys. Lett.* **2004**, *85*, 911–913.

28. Blanco, A. et al., Large-scale Synthesis of a Silicon Photonic Crystal with a Complete 3-Dimensional Bandgap Near 1.5 Micrometres. *Nature* **2000**, *405*, 437–440.

29. Checoury, X.; Enoch, S.; Lopez, C.; Blanco, A. Stacking Patterns in Self-assembly Opal Photonic Crystals. *Appl. Phys. Lett.* **2007**, *90*, 161131.

30. Kang, Y.; Walish, J.; Gorishnyy, T.; Thomas, E. Broad-wavelength-range Chemically Tunable Block-Copolymer Photonic Gels. *Nat. Mater.* **2007**, *6*, 957–960.

31. Yuan, L.; Herman, P. Laser Scanning Holographic Lithography for Flexible 3D Fabrication of Multi-scale Integrated Nano-structures and Optical Biosensors. *Sci. Reports* **2016**, *6*, 22294.

32. Lin, Y.; Harb, A.; Lozano, K.; Xu, D.; Chen, K. Five Beam Holographic Lithography for Simultaneous Fabrication of Three Dimensional Photonic Crystal Templates and Line Defects Using Phase Tunable Diffractive Optical Element. *Opt. Express* **2009**, *17*, 16625–16631.

33. Burrow, G.; Leibovici, M.; Gaylord, T. Pattern-integrated Interference Lithography: Single-exposure Fabrication of Photonic-crystal Structures. *Appl. Optics* **2012**, *51*, 4028–4041.

34. Ramanan, V.; Nelson, E.; Brzezinski, A.; Braun, P.; Wiltzius, P. Three Dimensional Silicon-air Photonic Crystals with Controlled Defects Using Interference Lithography. *Appl. Phys. Lett.* **2008**, *92*, 173304.

35. Duarte, M.; Lasagni, A.; Giovanelli, R.; Narcisco, J.; Louis, E.; Mucklich, F. Increasing Lubricant Film Lifetime by Grooving Periodic Patterns Using Laser Interference Metallurgy. *Adv. Eng. Mater.* **2008**, *10*, 554–558.

36. Li, J.; Xiong, D.; Wu, H.; Huang, Z.; Dai, J.; Tyagi, R. Tribological Properties of Laser Surface Texturing and Molybdenizing Duplex-treated Ni-base Alloy, *Tribol. Trans.* **2010**, *53*, 195–202.

37. Hu, T.; Hu, L. The Study of Tribological Properties of Laser-textured Surface of 2024 Aluminium Alloy Under Boundary Lubrication *Lubrication Sci.* **2012**, *24*, 84–93.

38. Yang, Y.; Hsu, C.; Chang, T.; Kuo, L.; Chen, P. Study on Wetting Properties of Periodical Nanopatterns by a Combinative Technique of Photolithography and Laser Interference Lithography. *Appl. Surf. Sci.* **2010**, *256*, 3683–3687.

39. Trdan, U.; Hocevar, M.; Gregorcic, P. Transition from Superhydrophilic to Superhydrophobic State of Laser Textured Stainless Steel Surface and its Effect on Corrosion Resistance. *Corros. Sci.* **2017**, *123*, 21–36.

40. Boinvich, L.; Emelyanenko, A.; Modestov, A.; Domantovsky, A.; Emelyanenko, K. Synergistic Effect of Superhydrophobicity and Oxidized Layers on Corrosion Resistance of Aluminum Alloy Surface Textured by Nanosecond Laser Treatment *ACS Appl. Mater. Interfaces* **2015**, *7*, 19500–19508.

CHAPTER 10

Metal Plasmonics

The interaction of metals with electromagnetic radiation is largely dictated by the behavior of free electrons in the metal. Plasmons play an important role in the optical properties of metals. A plasmon, which is a quantum of plasma oscillation, can be described as an oscillation of free electron gas density with respect to the fixed positive ions in a metal. Just as light (an optical oscillation) consists of photons, the plasma oscillation consists of plasmons. The free electron gas of the metal can sustain surface and volume charge density oscillations, with distinct resonance frequencies. The resonance frequency of the volume charge density oscillation is known as the plasma frequency ω_p. Most of the physics of the light-metal interaction is hidden in the frequency dependence of the metal's complex dielectric function. According to the approximation given in eq 1.123, the dielectric constant (i.e., the squared refractive index) is negative when the frequency (ω) of the incident electromagnetic wave is lower than ω_p. This means that the free electrons oscillate 180° out of phase with respect to the driving electric field. In most metals, the plasma frequency is in the UV region. Therefore, visible light cannot propagate through bulk metals since their refractive index is pure imaginary. In the opposite case of $\omega > \omega_p$. the dielectric constant is positive and the refractive index is real. As a consequence, the incident wave (deep UV and X-rays) can penetrate into the metal with a small attenuation. When the frequency of the incident radiation coincides with the plasma frequency, a strong absorption, known as *plasma resonance absorption*, is observed. Note that this is a result of the interaction that occurs inside the volume of metals

Plasmons are collective oscillations of the free electron gas density. Surface plasmons (SPs) are those plasmons that are confined to the surfaces of metals. They exist at the interface of a metal whose real part of dielectric constant is negative and a dielectric medium (e.g., air, glass, and other dielectrics) that has a positive real dielectric constant. In order for such plasmons to exist, the real part of the dielectric constant of the metal must be negative, with its magnitude greater than that of the dielectric. SPs have smaller energies than bulk (or volume) plasmons. Therefore, they can be resonant with visible and/or infrared light. Surface plasmon resonance (SPR) refers to the resonant oscillation of conduction electrons at the interface between negative- and positive-permittivity media excited by incident light. The surface charge density oscillations associated with SPR can lead to strongly enhanced optical fields that are spatially confined near the metal surface. SPs can also be found on nonplanar interfaces. Localized surface plasmons (LSPs) exist at the surface of metallic nanostructures such as nanoparticles, nano-holes, and grooves, where the structures have dimensions comparable to or smaller than the wavelength of light. Localized surface plasmon resonances (LSPRs) are collective electron charge oscillations in metallic nanostructures that are excited by incident light. The strong interactions of light with LSPs give rise to high extinction coefficients and significant amplification of the electric field near the metallic structure. The enhanced field is highly localized at the metal/dielectric interface and decays rapidly into the dielectric surrounding. The resonant properties of plasmonic nanostructures are very sensitive to the geometry (shape and size) of these structures and the local dielectric environment. The study of plasmonic phenomena in metallic structures is termed *metal plasmonics*. This rapidly growing active field finds a wide range of application areas including electronic/photonic devices, solar cells, biosensors, and catalysis. SPR and LSPR can also be used to control the color of materials. Many innovative concepts and applications of metal plasmonics have been proposed over the past decade. We begin this chapter with a description of the fundamental optical properties of metals.

10.1 OPTICAL PROPERTIES OF METALS

The optical properties of a metal can be described by its complex dielectric function that depends on the frequency of light. The properties are mainly determined by the response of free electrons in the conduction band. The contribution of these free carriers to the dielectric function can be understood in terms of a simple classical model, called the *Drude model*. This model is based on the equation of motion of a free electron in an oscillating electric field given by

$$m\frac{d^2y}{dt^2} + m\gamma\frac{dy}{dt} = -eE_0 e^{-i\omega t}. \tag{10.1}$$

Here γ is a damping factor that reflects the effect of many collisions between free electrons, and m is the effective mass of the free electrons. We can anticipate that the electron oscillates with the same frequency as the electric field. Solving eq 10.1 using the displacement $y(t) = y_0 e^{-i\omega t}$ yields

$$y(t) = \frac{eE_0}{m(\omega^2 + i\gamma\omega)} e^{-i\omega t}. \tag{10.2}$$

If there are N free electrons per unit volume, the electric polarization P becomes

$$P(t) = -eyN = \frac{-Ne^2}{m(\omega^2 + i\gamma\omega)} E_0 e^{-i\omega t} = \frac{-Ne^2}{m(\omega^2 + i\gamma\omega)} E(t). \tag{10.3}$$

Since the proportionality constant between $P(t)$ and $E(t)$ is complex, the resulting polarization may be *out of phase* with the applied electric field. From eqs 1.58 and 1.59, we have the following dielectric function:

$$\varepsilon(\omega) = 1 - \frac{Ne^2}{\varepsilon_0 m(\omega^2 + i\gamma\omega)} = 1 - \frac{\omega_p^2}{\omega^2 + i\gamma\omega}. \tag{10.4}$$

Here ω_p, which is defined in eq 1.121, is known as the volume plasma frequency. The dielectric function $\varepsilon(\omega)$ of eq 10.4 represents the relative permittivity (i.e., dielectric constant), not the absolute permittivity. The frequency dependence of the dielectric function describes the dispersive optical properties of metals. For large frequencies close to the plasma frequency ($\omega \gg \gamma$), the damping can be neglected. Then eq 10.4 reduces to eq 1.123. The dielectric function is generally written as $\varepsilon(\omega) = \varepsilon_1(\omega) + i\varepsilon_2(\omega)$, where $\varepsilon_1(\omega)$ is the real part and $\varepsilon_2(\omega)$, the imaginary part:

$$\varepsilon(\omega) = 1 - \frac{\omega_p^2}{\omega^2 + i\gamma\omega} = 1 - \frac{\omega_p^2}{\omega^2 + \gamma^2} + i\frac{\gamma\omega_p^2}{\omega(\omega^2 + \gamma^2)} = \varepsilon_1(\omega) + i\varepsilon_2(\omega). \tag{10.5}$$

In typical metals, the damping factor γ is in the infrared frequency range. This indicates that the real part of the dielectric constant, $\varepsilon_1(\omega)$, is negative over the extended visible range ($\omega < \omega_p$), where metals maintain their metallic features. The imaginary part, $\varepsilon_2(\omega)$, is always positive. The refractive index of metals is generally a complex number consisting of real and imaginary parts: $n = n_R + in_I$. Since $\varepsilon = n^2$, $\varepsilon_1 = n_R^2 - n_I^2$ and $\varepsilon_2 = 2n_R n_I$. An obvious consequence of the negative real dielectric constant is that light cannot penetrate deeply into metal because it leads to a strong imaginary part of the refractive index. The imaginary part of the dielectric function accounts for the dissipation of energy associated with the motion of electrons in the metal. At optical frequencies, $\varepsilon(\omega)$ can be experimentally determined via reflectivity measurements, which provide information on the optical constants n_R and n_I. The dielectric function can be derived from the measured optical constants.

Figure 10.1 shows the real and imaginary parts of the dielectric function experimentally measured for silver.[1] Note that the abscissae have different scales. The experimental data for gold are also shown in Figure 10.2. Drude's free-electron model can fit the data only at low frequencies. Although the free-electron model accurately describes the optical properties of metals in the infrared range, it should be supplemented in the UV–visible range by the response of bound electrons. For instance, gold has much larger imaginary dielectric constants than those predicted by the Drude model at wavelengths shorter than ~550 nm (i.e., at photon energies higher than ~2.25 eV). This is because high-energy photons can excite electrons occupying lower bands into the conduction band. In a classical picture, such interband transitions can be described by the resonant oscillations of bound electrons. The equation of motion for a bound electron is already given in eq 1.91. This equation describes the electron as a damped harmonic oscillator driven at frequency ω. The resulting dielectric function given in eq 1.96 holds for metals, although the resonance frequency ω_o and the damping term γ vary from material to material. As discussed in Section 1.4, clear resonance behavior is observed near the resonance frequency ω_o. The interband excitations of bound electrons can make a significant contribution to the dielectric function of a metal, particularly to its imaginary part. Therefore, the free-electron expression of $\varepsilon(\omega)$ is useful only for photon energies below the threshold energy for the onset of such excitations. Above this threshold energy, the form of the $\varepsilon_2(\omega)$ curve depends on the specific band structure of the metal. The complex dielectric function of metals can be well explained by adding the interband absorption contribution to the free-electron model. While most bulk metals are gray or silvery white, some metals exhibit particular colors. This color is related to the interband transitions. Gold is slightly reddish-yellow because it absorbs green/blue wavelengths and reflects yellow ones more strongly. Copper absorbs green wavelengths and reflects orange/red ones.

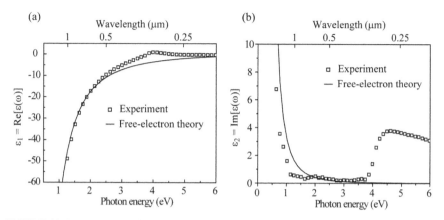

FIGURE 10.1 (a) Real and (b) imaginary parts of the dielectric functions of silver. Experimental data adapted from Ref. [1].

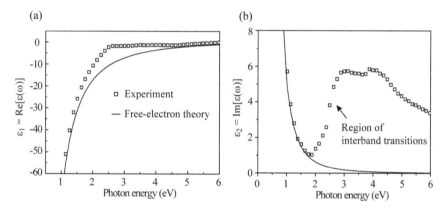

FIGURE 10.2 (a) Real and (b) imaginary parts of the dielectric functions of gold. Experimental data adapted from Ref. [1].

10.2 SURFACE PLASMONS

10.2.1 SURFACE PLASMON POLARITONS AT PLANAR INTERFACES

The study of propagating surface plasmons (SPs) is important for understanding the interaction between light and metallic structures. One of the key aspects related to SPs is the dispersion relation, that is, the energy versus wave vector relation. This relation provides the basis for understanding the coupling of light to SPs, which employs special approaches to match the wave vector. The behavior of SPs can be derived from Maxwell's equations. In plasmonics, it is common to deal with structures without any external current sources and/or charges. When there is no free charge or current, Maxwell's equations can be expressed as

$$\nabla \cdot \mathbf{D} = 0$$
$$\nabla \cdot \mathbf{B} = 0$$
$$\nabla \times \mathbf{E} = -\partial \mathbf{B} / \partial t \cdot \tag{10.6}$$
$$\nabla \times \mathbf{H} = \partial \mathbf{D} / \partial t$$

Maxwell's equations hold for any arbitrary time-dependence of the electric field. The time-dependent electric and magnetic fields can be written as

$$\mathbf{E}(\mathbf{r},t) = \mathbf{E}(\mathbf{r})e^{-i\omega t}, \quad \mathbf{H}(\mathbf{r},t) = \mathbf{H}(\mathbf{r})e^{-i\omega t}. \tag{10.7}$$

The phasors $\mathbf{E}(\mathbf{r})$ and $\mathbf{H}(\mathbf{r})$ contain information on the magnitude and direction of the corresponding field. Using the vector identity of eq 1.69, we can get equations for the time-independent fields.

$$\nabla^2 \mathbf{E}(\mathbf{r}) + k^2 \mathbf{E}(\mathbf{r}) = 0$$
$$\nabla^2 \mathbf{H}(\mathbf{r}) + k^2 \mathbf{H}(\mathbf{r}) = 0 \tag{10.8}$$

Here $k = nk_o$ is the propagation constant in a medium with refractive index n. $k_o = \omega/c$ is the free-space propagation constant. The above two equations are called the *Helmholtz equations*. They describe the spatial distributions of electric and magnetic fields when an electromagnetic wave with frequency ω propagates. Suppose that a plane wave is incident into a planar interface between two media with dielectric constants ε_1 and ε_2 (Fig. 10.3). The medium 1 characterized by $\varepsilon_1 = n_1^2$ is a dielectric and the medium 2 of $\varepsilon_2 = n_2^2$ is a metal. Here the x–z plane is defined as the plane of incidence. The interface is located at the point of $z = 0$, across which the dielectric function changes in a stepwise manner from ε_1 to ε_2. For a plane wave propagating in the x–z plane, the magnitudes of the phasor fields $\mathbf{E}(\mathbf{r})$ and $\mathbf{H}(\mathbf{r})$ are dependent on the coordinates x and z, but independent of the y coordinate. We are looking for SPs, which are waves bound to the interface. Then we can write the magnitudes of the phase fields as $\mathbf{E}(\mathbf{r}) = \mathbf{E}(x, y, z) = \mathbf{E}(z)\exp(ik_x x)$ and $\mathbf{H}(\mathbf{r}) = \mathbf{H}(x, y, z) = \mathbf{H}(z)\exp(ik_x x)$. This leads to

$$\frac{\partial^2 \mathbf{E}(z)}{\partial z^2} + (\varepsilon k_o^2 - k_x^2)\mathbf{E}(z) = 0. \tag{10.9}$$

$$\frac{\partial^2 \mathbf{H}(z)}{\partial z^2} + (\varepsilon k_0^2 - k_x^2)\mathbf{H}(z) = 0. \tag{10.10}$$

When the incident wave is TE-polarized (Fig. 10.3a), only the field components $E_y(z)$, $H_x(z)$, and $H_z(z)$ are nonzero. For this TE polarization mode, $E_x(z)$ and $E_z(z)$ components do not exist. Then eq 10.9 can be simply written as

$$\frac{\partial^2 E_y}{\partial z^2} + k_z^2 E_y = 0. \tag{10.11}$$

Here $k_z^2 = \varepsilon k_0^2 - k_x^2$ and k_z is defined as the wave vector component along the z-direction. k_x is the same for both media (i.e., $k_x = k_{x1} = k_{x2}$), since the wave vector component parallel to the interface should always be conserved. The E_y fields in the media 1 and 2 are $E_1(z) = E_1\exp(ik_{z1}z)$ and $E_2(z) = E_2\exp(ik_{z2}z)$, respectively. The boundary condition for the E_y field yields $E_1 = E_2$. From the Faraday's law, we have $H_x = (-1/i\mu\omega)(\partial E_y/\partial z)$. The boundary condition for the H_x field sets another requirement of $\mu_1 k_{z2} = \mu_2 k_{z1}$. We are looking for a wave bound to the interface, which rapidly vanishes away from it. For such a SP to exist, k_{z1} ($z > 0$, medium 1) must have a positive imaginary value, whereas k_{z2} ($z < 0$, medium 2) should be negative imaginary. The materials found in nature have permeability values close to unity (i.e., $\mu_1 = \mu_2 \approx 1$) at optical frequencies. This means that SPs cannot exist in TE polarization. However, magnetic SPs can be achieved with metamaterials,[2-6] for which the magnetic permeability can be engineered to be negative in the frequency range from visible to microwave.

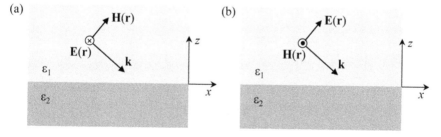

FIGURE 10.3 (a) TE and (b) TM polarized plane wave incident into a planar interface between two media with dielectric constants ε_1 and ε_2. The x–z plane is the plane of incidence and the x–y plane is the interface.

For the case of a TM-polarized wave (Fig. 10.3b), only the components $H_y(z)$, $E_x(z)$, and $E_z(z)$ are nonzero. Then eq 10.10 is simplified to

$$\frac{\partial^2 H_y}{\partial z^2} + k_z^2 H_y = 0. \tag{10.12}$$

The $H_y(z)$ components in the two media are $H_1(z) = H_1\exp(ik_{z1}z)$ and $H_2(z) = H_2\exp(ik_{z2}z)$. The following relations hold for the wave vector components.

$$k_{z1}^2 = \varepsilon_1 k_o^2 - k_x^2, \quad k_{z2}^2 = \varepsilon_2 k_o^2 - k_x^2 \tag{10.13}$$

Since $H_1(0) = H_2(0)$, we also have $H_1 = H_2$. From the last one of Maxwell's equations, we have $E_x = (1/i\varepsilon\omega)(\partial H_y/\partial z)$. The boundary condition for the E_x field yields

$$\frac{k_{z2}}{k_{z1}} = \frac{\varepsilon_2}{\varepsilon_1}. \tag{10.14}$$

This requirement can be satisfied if the dielectric constants of the two media have opposite signs, that is, one of these media is a dielectric and another one is a metal. Equation 10.14, combined with eq 10.13, leads to a dispersion relation.

$$k_x^2 = \frac{\varepsilon_1\varepsilon_2}{\varepsilon_1+\varepsilon_2}k_o^2 = \frac{\varepsilon_1\varepsilon_2}{\varepsilon_1+\varepsilon_2}\frac{\omega^2}{c^2} \tag{10.15}$$

We also obtain an expression for the normal component of the wave vector.

$$k_{z1}^2 = \frac{\varepsilon_1^2}{\varepsilon_1+\varepsilon_2}k_o^2$$
$$k_{z2}^2 = \frac{\varepsilon_2^2}{\varepsilon_1+\varepsilon_2}k_o^2 \tag{10.16}$$

We now turn to the condition that should be fulfilled for a SP to exist. For simplicity, we assume that the imaginary parts of the complex dielectric functions are negligibly small compared with the real parts. Since we are looking for waves that propagate along the interface, k_x should be real. Based on eq 10.15, this requirement can be met if the sum $(\varepsilon_1+\varepsilon_2)$ and the product $(\varepsilon_1\varepsilon_2)$ of the dielectric functions are either both positive or both negative. To obtain a wave that rapidly vanishes away from the interface, the z-components of the wave vector should be purely imaginary in both media. This can be achieved only when the sum of the dielectric functions is negative. Therefore, the conditions for a SP mode to exist are

$$\varepsilon_1\varepsilon_2 < 0 \text{ and } \varepsilon_1 + \varepsilon_2 < 0. \qquad (10.17)$$

Equation 10.17 means that one of the dielectric functions must be negative with an absolute magnitude exceeding that of the other. Metals (especially noble metals such as silver and gold) have a large negative real part of the dielectric function along with a small imaginary part in the visible range. Therefore, SPs can exist at the interface between a metal and a dielectric such as air, glass, and so on. Such plasmon modes are possible in TM polarization only. SPs are collective electron oscillations on the surface of a metal. They are longitudinal charge-density waves propagating along the interface between a metal and a dielectric. An oscillation of the electron charge density produces an electromagnetic wave in the dielectric medium (also in the metal), as illustrated in Figure 10.4. They are altogether referred to as *surface plasmon polariton* (SPP). The electromagnetic field intensity exponentially decays away from the interface.

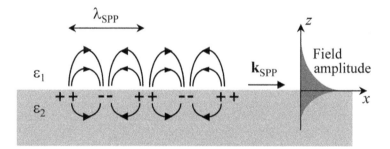

FIGURE 10.4 An oscillation of the electron charge density produces an electromagnetic wave in both media. The electromagnetic field intensity exponentially decays away from the interface.

We now discuss some properties of SPPs. Metals play a more crucial role in plasmonics than dielectric media. The dispersion relation of eq 10.15 is also valid even if the metal has a complex dielectric function. As an SPP propagates along the surface, it loses energy to the metal due to absorption. It may also lose energy due to scattering into other directions. The dielectric function of metal has both real and imaginary parts. This means that the propagation constant k_x in eq 10.15 also has both parts. The real part of k_x determines the wavelength of a propagating SPP, while the imaginary part accounts for the damping of the SPP and thus determines its propagation length. The $1/e$ decay length of the SPP electric field is given by the inverse of the imaginary part of k_x. The dielectric function of silver is $\varepsilon_{Ag} = -18 +$ 0.5i at a wavelength of 633 nm (photon energy = 1.96 eV). This leads to a 1/e propagation length of ~120 μm when the surrounding medium is air (ε_1 = 1). For gold ($\varepsilon_{Au} = -11.5 + 1.2i$), the propagation length is about 20 μm at the same wavelength. The decay of the SPP field away from the interface can be deduced from eq 10.16 by $1/k_{z1}$ and $1/k_{z2}$. Neglecting the very small imaginary part in ε_2, the decay lengths into silver and air are 23 nm and 415 nm, respectively. The decay lengths into gold and air are 28 nm and 325 nm, respectively. We note that the decay into the metal is much shorter than into the dielectric and that the SPP electric field can penetrate through a thin enough metal film. If silver is in contact with a lossless dielectric with $\varepsilon_1 = 4$ (i.e., $n_1 = 2$), we have decay lengths of 21 nm (silver) and 95 nm (dielectric). As the refractive index of the surrounding dielectric increases, the decay length into the dielectric is significantly reduced.

In the infrared and visible frequency range, most metals have a very small imaginary part of the dielectric function compared with the real part. Neglecting the damping of SPP along the interface, the SPP dispersion relation can be expressed as

$$k_{SPP} \approx \frac{\omega}{c} \sqrt{\frac{\varepsilon_d \varepsilon_m(\omega)}{\varepsilon_d + \varepsilon_m(\omega)}}. \tag{10.18}$$

Here $\varepsilon_m(\omega)$ is the real part of the metal's dielectric function and ε_d is the dielectric constant of the dielectric medium ($\varepsilon_d = 1$ for air). The SPP wavelength is given by $\lambda_{SPP} = 2\pi/k_{SPP}$. Applying the Drude's free electron model without damping ($\gamma = 0$), we have $\varepsilon_m(\omega) = 1 - \omega_p^2/\omega^2$. Equation 10.18 then becomes

$$k_{SPP} \approx \frac{\omega}{c} \sqrt{\varepsilon_d} \sqrt{\frac{\omega^2 - \omega_p^2}{(1+\varepsilon_d)\omega^2 - \omega_p^2}}. \tag{10.19}$$

For very small frequencies, we have $k_{SPP} \approx n_d\omega/c$, which is the dispersion relation of light in the dielectric medium, that is, the light line. The condition for the propagation of SPP is that k_{SPP} is real. As $\varepsilon_m(\omega) = 1 - \omega_p^2/\omega^2$ is negative, this condition is fulfilled for frequencies $\omega < \omega_p/(1 + \varepsilon_d)^{1/2} = \omega_{SP}$. The frequency ω_{SP} is called the *surface plasmon frequency*. The higher the refractive index of the dielectric medium, the lower the surface plasmon frequency is. The SPP dispersion relation for the metal-air system is plotted in Figure 10.5. The SPP dispersion curve approaches the surface plasmon frequency. For frequencies higher than the bulk plasma frequency ($\omega > \omega_p$),

$\varepsilon_m(\omega)$ is positive and the index of refraction of a metal is real. Therefore, the propagation of transverse electromagnetic waves is allowed with the dispersion law given by $\omega^2 = c^2k^2 + \omega_p^2$. In this frequency region, the metal is transparent and the wave propagates freely. In the range $\omega_p/\sqrt{2} < \omega < \omega_p$, however, no propagation mode exists. The straight line in Figure 10.5 represents the dispersion law of electromagnetic waves in vacuum, that is, the light line in vacuum ($\omega = ck$). For $\omega > \omega_p$, transverse electromagnetic waves can propagate inside a metal. When $\omega < \omega_{SP}$, longitudinal SPP waves are allowed to propagate along its surface. An important feature of SPs is that for a given energy $\hbar\omega$, the propagation constant k_{SPP} is always larger than that of light in free space. The SPP dispersion curve emerges from the light line and then deviates from it, asymptotically approaching the surface plasmon frequency. Within the Drude's ideal free electron model, the frequency range between ω_p and ω_{SP} represents a plasmonic gap where no propagating electromagnetic mode exists, as shown in Figure 10.5. In real metals, however, the contribution from interband transitions modifies the SPP dispersion relation. The value of k_{SPP} does not diverge infinitely even if ω approaches ω_{SP}. It bends backward filling the plasmonic gap.[7,8] The back-bending effect imposes a limit to the maximum k_{SPP} value that can be achieved in an experiment.

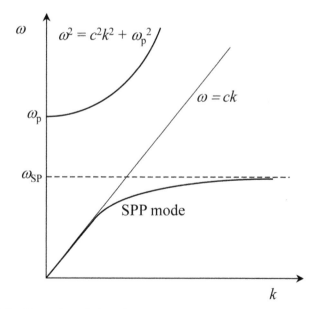

FIGURE 10.5 Dispersion relation of SPPs at a metal/air interface. The straight line represents the light line in free space.

10.2.2   EXCITATION OF SPP

As mentioned above, the propagation constant k_{SPP} is always larger than that of light in free space at the same energy $\hbar\omega$. In order to excite SPs, both energy and momentum (i.e., wave vector) should be matched. Therefore, SPs cannot be excited by direct illumination of light because the energy and momentum matching ($\omega_{\text{light}} = \omega_{\text{SPP}}$ and $k_{\text{light}} = k_{\text{SPP}}$) cannot be fulfilled at the same time. Suppose that a light wave of wave vector $k_o = \omega/c$ is incident onto a planar metal surface at an angle θ with the surface normal, as shown in Figure 10.6a. Here the x–z plane is the plane of incidence. The momentum component of the incident light parallel to the surface is $k_o \sin\theta$. When this component equals the momentum (k_{SPP}) of SPs, the conservation of energy and momentum can be simultaneously satisfied. However, k_{SPP} is always larger than $k_o \sin\theta$ regardless of the incident angle θ, since k_{SPP} is larger than k_o. This is graphically depicted in Figure 10.6b, where ω/c is drawn as a function of k_x for both light ($k_x = k_o \sin\theta$) and SPs ($k_x = k_{\text{SPP}}$). The in-plane momentum of the incident light (i.e., the parallel component of its wave vector) should be increased to excite SPs at a planar air-metal interface. There are several techniques to realize this momentum matching process, which include the Otto and Kretschmann configurations[9-12] and the application of a periodic grating structure.

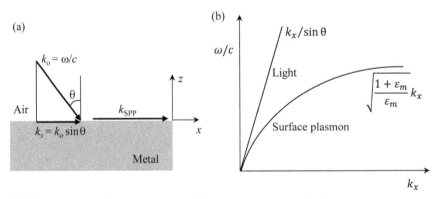

FIGURE 10.6   (a) The momentum of SPP, k_{SPP}, compared with the in-plane momentum, $k_o \sin\theta$, of incident light. (b) In-plane momentum k_x versus frequency diagram. At a given frequency (i.e., energy), the in-plane momentum of free-space light is always lower than the momentum of SPP.

(i)   Otto and Kretschmann configurations

The Otto and Kretschmann configurations utilize evanescent waves that arise from frustrated total internal reflection. Both configurations are fundamentally based on the fact that the magnitude of the wave vector increases by a factor of n within a material of refractive index n. Figure 10.7a illustrates the typical Otto configuration, where a glass prism is placed very close to the surface of a thick metal film (or the surface of a bulk metal). In this configuration, a light beam (typically a laser beam) is incident onto the prism and refracted towards its bottom surface. If the incident angle θ is sufficiently large, the light in the prism will undergo total internal reflection and an evanescent field will exist within the air gap below the prism. The width of the air gap is adjusted to be very small, usually several hundred nanometers, so that the tail of the evanescent field can reach the metal surface. The wave vector of the incident light inside the prism is $k = n_p k_o$, where n_p is the refractive index of the prism material. The component of this wave vector parallel to the bottom surface of the prism is $k_x = k \sin \theta = n_p k_o \sin \theta$, which is also the wave vector of the evanescent wave (see Section 2.5.3). When this component equals the wave vector of SPs, an SPP can be excited. The resonance condition is then

$$k_{\text{SPP}} = \frac{\omega}{c} n_p \sin \theta. \tag{10.20}$$

At a certain angle of incidence, the in-plane wave vector of the incident light coincides with the SPP wave vector. Under this resonant condition, some energy of the evanescent field is tunneled into the metal. Therefore, a sharp minimum is observed in the reflectivity spectrum, as illustrated in Figure 10.7b. The momentum matching is obvious from the dispersion relations shown in Figure 10.7c. The dashed line ($\omega/c = k_x/\sin \theta$) represents the energy versus in-plane momentum relation when the light is directly incident on a planar metal surface at an angle θ. As a prism of refractive index n_p is employed, the in-plane momentum k_x increases by a factor of n_p. As a consequence, the light line now has the relation of $\omega/c = k_x/(n_p \sin \theta)$. For a certain incident angle, the energy and momentum of the incident light are matched to those of the SPP. The slope of the light line depends on the incident angle θ, as depicted in Figure 10.8. One might think that there are multiple incident angles satisfying the resonant condition. In the Otto configuration, the incident light is usually monochromatic. This means that the incident light has a fixed frequency of ω_{inc}. Then, the plasmonic resonance occurs only at a specific incident angle θ_{res}. For other angles such as θ_1 and θ_3, the energy and momentum conservation cannot be simultaneously fulfilled. In Figure 10.7a, the SPP propagates in the x-direction, which is the direction of surface charge oscillation. The incident light should have an electric field component in the x-direction in order to couple to this surface charge oscillation. Therefore, it should be TM-polarized. TE-polarized light, whose electric field is orthogonal to the surface charge oscillation, cannot excite an SPP. In the Otto configuration, the reflectivity is measured as a function of the incident angle for different air gap widths. Equation 10.20 does not have any dependence on the gap width. However, the gap width significantly influences the efficiency of energy transfer from the incident light into the SPP. For a large air gap, the SPP cannot be efficiently excited and the resonance becomes hardly observable. Because of the difficulty in controlling the tiny air gap accurately, the Otto configuration is inconvenient to use experimentally. The strong dependence of the SPP coupling efficiency on the air gap width is the primary reason that this configuration has not been popular. A low-index dielectric layer may be used instead of the air gap once it has a lower refractive index than the prism. In this case, the SPP resonance is shifted to a much larger angle of incidence, which may be located outside of the detection range.

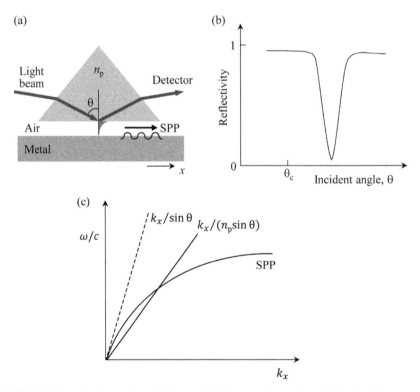

FIGURE 10.7   (a) Illustration of the typical Otto configuration. (b) Reflectivity versus incident angle. (c) Dispersion relations in the Otto configuration. For a certain incident angle, the energy and momentum of the incident light are matched to those of the SPP. The dashed line ($\omega/c = k_x/\sin\theta$) represents the energy versus in-plane momentum relation when the light is directly incident on a planar metal surface.

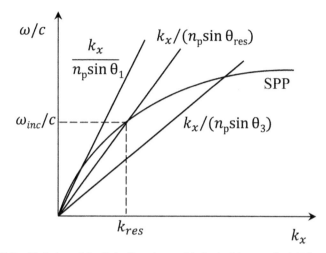

FIGURE 10.8 Variation of the light-line slope with the incident angle θ. When the incident light has a fixed frequency of ω_{inc}, the plasmonic resonance occurs only at a specific incident angle θ_{res}.

Immediately after Otto developed his optical configuration for exciting SPPS, Kretschmann and Raether proposed another prism-based configuration.[10] This configuration, typically called the *Kretschmann configuration*, has become the most popular configuration for SPP excitation. The geometry is sketched in Figure 10.9. The light is again incident through a prism. But a metal film is deposited on the bottom surface of the prism. To excite an SPP at the metal/air interface, an evanescent wave created at the prism/metal interface has to penetrate through the metal layer. If the metal film is too thick, the SPP cannot be efficiently excited due to absorption in the metal. That is, the evanescent wave does not reach the bottom surface of the metal. On the contrary, if the metal is too thin, the SPP is strongly damped because of radiation damping into the prism. The reflectivity of the system is measured as a function of the angle of incidence for different metal thicknesses. The condition for resonance is identical to that for the Otto configuration and is also given by eq 10.20. As before, the resonant excitation is characterized by a dip in the reflectivity curve. In contrast to the Otto configuration, the excited SPP propagates along the bottom surface of the metal film. This exposed surface is easily accessible for interactions and detections. It is to be noted that the prism/metal interface cannot support any SPP mode because the SPP wave vector at this interface is greater than the photon wave vector in the prism at all angles of incidence. To excite an SPP on the internal metal interface, an additional dielectric layer whose refractive index is smaller than that of the prism should be deposited between the prism and the metal film. In such a two-layer configuration, SPP modes can be excited on both interfaces of the metal film at different angles of incidence.

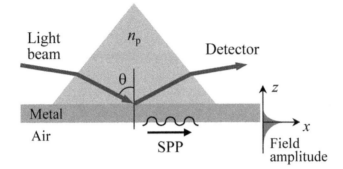

FIGURE 10.9   Kretschmann configuration.

(ii) Diffraction gratings

Another way to accomplish the coupling of light into SPPs is to use diffraction effects.[12–18] A periodic texture formed on the metal surface acts as a diffraction grating. Diffraction gratings provide the incident light with an additional momentum, satisfying the resonance condition. Figure 10.10a shows a 1D grating of period d, where the grating grooves run along the y-direction and the grating vector $\mathbf{g}$ whose magnitude is given by $g = 2\pi/d$ is parallel to the x-direction. Suppose that a light wave of wave vector $\mathbf{k}_i$ is incident onto the grating at an angle θ_i with the surface normal in the x–z plane. The incident light will be diffracted into multiple beams according to the diffraction relation of eq 5.34.

$$k_i(\sin\theta_m - \sin\theta_i) = mg \qquad (10.21)$$

Here m represents the diffraction order ($m = 0, \pm1, \pm2,..$). The $m = \pm1$ terms can cause a strong diffraction effect. The diffracted beams have a wave vector component parallel to the surface that is given by $k_i \sin\theta_m$. If this component equals the wave vector (i.e., momentum) of SPs, the conservation of energy and momentum can be simultaneously satisfied. This condition is mathematically given by

$$\pm k_{SPP} = k_i \sin\theta_m = k_i \sin\theta_i + mg. \qquad (10.22)$$

This is alternatively expressed as

$$\pm k_{SPP} - mg = k_i \sin \theta_i. \tag{10.23}$$

The $\pm$ sign was added to k_{SPP} because SPs can propagate either in the positive or negative x-direction. As illustrated in Figure 10.10b, the vector form of eq 10.23 is $\pm \mathbf{k}_{SPP} - m\mathbf{g} = \mathbf{k}_{||}$, where $k_{||} = k_i \sin \theta_i$. In the absence of a grating (i.e., $g = 0$), k_{SPP} is always larger than $k_i \sin \theta_i = (\omega/c) \sin \theta_i$ at the same energy (i.e., at the same ω). Therefore, SPs cannot be excited by light directly impinging on a planar metal surface. This is graphically depicted in Figure 10.10c, where ω/c is drawn as a function of k_x for both light ($k_x = k_i \sin \theta_i$) and SPs ($k_x = k_{SPP}$). The in-plane momentum (i.e., the parallel component of the wave vector) of light is modified by a diffraction grating. This is equivalently viewed as a shift of the dispersion relation of SPs by mg, as described in eq 10.23 and also illustrated in Figure 10.10d. As shown, surface plasmon resonance (SPR) occurs at two different frequencies ω_1 and ω_2. This means that for a given incident angle, two different wavelengths $\lambda_1 (= 2\pi c/\omega_1)$ and $\lambda_2 (= 2\pi c/\omega_2)$ of light can be coupled into SPs and thus be strongly absorbed. As the incident angle changes, the slope of the line "$k_x/\sin \theta_i$" also changes. This consequently shifts the wavelengths at which SPR occurs. The SPR wavelengths also depend on the dielectric constant ε_m of the metal and the period d of the grating.

Figure 10.10b,d depicts the momentum-matching condition when the light is incident in the x–z plane. Figure 10.10e shows a more general case in which the plane of incidence makes an azimuthal angle ϕ with the grating grooves. Then, the x-component of the momentum is

$$k_x = k_i \sin \theta_i \sin \phi + mg. \tag{10.24}$$

The second term on the right-hand side of eq 10.24 arises from the diffraction grating. The y-component is

$$k_y = k_i \sin \theta_i \cos \phi. \tag{10.25}$$

The conservation of energy and momentum can be simultaneously satisfied if the momentum of SPs equals the vector summation of these two components, that is, $k_{SPP} = (k_x^2 + k_y^2)^{1/2}$. This leads to

$$\sin \theta_i = -(\frac{m\lambda}{d}) \sin \phi \pm \sqrt{\frac{\varepsilon_m}{1+\varepsilon_m} - (\frac{m\lambda}{d})^2 \cos^2 \phi}. \tag{10.26}$$

Here λ is the vacuum wavelength of the incident light. The resonant incident angle θ_i depends on the azimuthal angle ϕ of the plane of incidence.

When $\phi = 0°$, both k_x and k_y have a non-zero value: $k_x = mg$ and $k_y = k_i \sin\theta_i$. Equation 10.26 then reduces to

$$\sin\theta_i = \sqrt{\frac{\varepsilon_m}{1+\varepsilon_m} - (\frac{m\lambda}{d})^2}. \qquad (10.27)$$

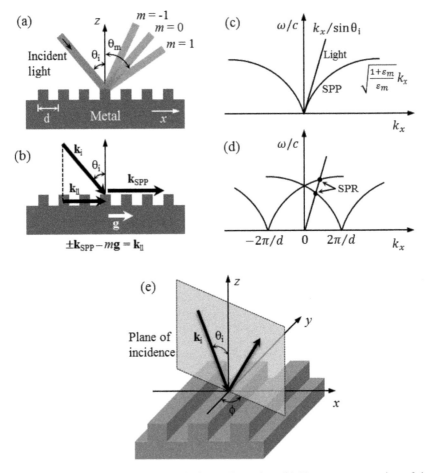

FIGURE 10.10 (a) Diffraction of light by a 1D grating. (b) Vector representation of the momentum-matching condition. (c) Dispersion relations of light and surface plasmons on a planar metal surface. (d) Dispersion relations of light and surface plasmons when a diffraction grating of period "d" is present. (e) General incidence of light into a 1D grating.

In this case, the grating vector $\mathbf{g}$ is perpendicular to the plane of incidence and we have the vector relation of $\mathbf{k}_{SPP} = mg\mathbf{i} + k_i \sin\theta_i \mathbf{j}$. Therefore, the SPs propagates at some angle with respect to the grating grooves; its propagation direction is neither the x-direction nor the y-direction. Since the summation magnitude of k_x and k_y, $(k_x^2 + k_y^2)^{1/2}$, is independent of whether m is 1 or -1, only a single resonance mode can be supported. It can be seen from eq 10.27 that when θ_i is fixed, the resonance wavelength λ is also fixed. When the plane of incidence is perpendicular to the grating grooves, that is, $\phi = 90°$, eq 10.26 then becomes

$$\sin \theta_i = -(\frac{m\lambda}{d}) \pm \sqrt{\frac{\varepsilon_m}{1+\varepsilon_m}}. \qquad (10.28)$$

Note that eq 10.28 is equivalent to eq 10.23. At $\phi = 90°$, k_y does not exist. Therefore, SPs can propagate only in the x-direction, as described in Figure 10.10b. In this case, the grating vector $\mathbf{g}$ that lies in the plane of incidence can be added to or subtracted from the parallel component of $\mathbf{k}_i$, depending on whether m is 1 or -1. For each case, there exists a wavelength satisfying the momentum-matching condition of eqs 10.23 and 10.28. As a consequence, SPR can be supported at two different wavelengths for a given incident angle. It is to be noted that the above equations mention nothing on the effect of the polarization of the incident light. To support SPR, the incident light should have an electric field component that is parallel to the grating vector.[12,17,18] A 2D texture of square lattice can be regarded as a superposition of two orthogonal 1D gratings. Consider a 2D texture illustrated in Figure 10.11, where two orthogonal grating vectors are oriented in the x- and y-directions. The two grating vectors have the same magnitude of $g = 2\pi/d$. We here consider the case when the y–z plane is the plane of incidence. Compared with Figure 10.10e, this is the case of $\phi = 0°$. However, the result will be the same for $\phi = 90°$ due to a rotational symmetry. With respect to the geometry of Figure 10.11, the x- and y-components of the momentum are

$$k_x = m_1 g$$
$$k_y = k_i \sin \theta_i + m_2 g.$$

Here m_1 and m_2 are integers representing the diffraction order for each grating. As in the 1D grating, the conservation of energy and momentum can be simultaneously satisfied if the momentum of SPs equals the vector summation of these two components, that is, $k_{SPP} = (k_x^2 + k_y^2)^{1/2}$. This leads to

$$\sin \theta_i = -(\frac{m_2\lambda}{d}) \pm \sqrt{\frac{\varepsilon_m}{1+\varepsilon_m} - (\frac{m_1\lambda}{d})^2}. \qquad (10.29)$$

Equation 10.29 can be solved unless m_1 and m_2 are both zero. This means that when m_1 and m_2 are simultaneously zero, SPR does not occur.

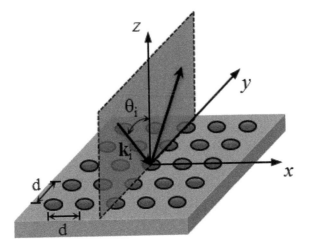

FIGURE 10.11   Light incidence into a 2D texture of square lattice.

Theoretical and experimental works on metal plasmonics have mostly been carried out with gold, silver, and aluminum, which are all good plasmonic materials. Grating-coupled SPR in stainless steel was also reported recently.[18] Figure 10.12 shows the experimental dielectric function of bulk stainless steel (316L-type), together with a 1D grating structure of $d = 500$ nm fabricated on its surface. Figures 10.13 and 10.14 shows reflectance spectra measured as a function of the incident angle θ_i. At $\phi = 90°$, SPR does not occur regardless of the θ_i value when the incident light is TE-polarized (Fig. 10.13a). For TM polarization, SPR absorption is observed at two different wavelengths. The spectral separation between two absorption peaks increases with increasing θ_i (Fig. 10.13b). For the azimuthal angle of $\phi = 0°$, a single resonance peak is observed only when the incident light is TE-polarized. No SPR occurs for TM polarization, as illustrated in Figure 10.14. These results indicate that in order to support SPR, the incident light should have an electric-field component perpendicular to the grating grooves. Figure 10.15 compares the experimental peak positions with those predicted by eqs 10.27 and 10.28. The theoretical branches given in Figure 10.15 correspond to $m = \pm 1$. They start from ~530 nm at $\theta_i = 0°$. According to the theory, there is another set of theoretical branches that correspond to $m = \pm 2$. SPR peaks for these higher order branches are inherently too weak to be experimentally detected.

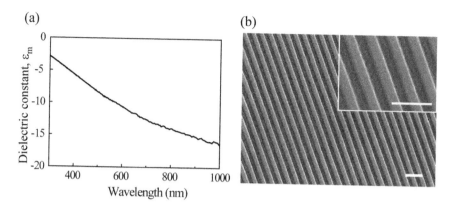

FIGURE 10.12 (a) Experimental dielectric function (real part) of bulk stainless steel (data adapted from Ref. [18]). (b) A fabricated 1D grating. Scale bars are 1 μm.

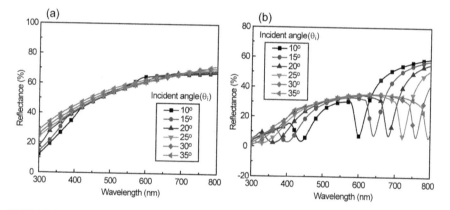

FIGURE 10.13 Specular reflectance at $\phi = 90°$: (a) TE polarization, (b) TM polarization. Reprinted with permission from Ref. [18]. Copyright© 2017 Optical Society of America.

In the Otto and Kretschmann configurations, the thickness of the metal film is a critical factor to influence the efficiency of light coupling. Likewise, there are some critical parameters for efficiently launching the grating-coupled SPR. The shape and depth of the groove should be adjusted to obtain the most efficient coupling. An optimum groove profile can cause the reflectivity to drop to nearly zero at the resonance angle.[12] It is important to note that the theoretical derivations given in this section are purely based on the mechanism of grating-assisted light coupling. They may not apply for gratings with periods smaller than the wavelength of incident light. As the period of the grating decreases, the scale of its features also diminishes. Nanoscale features themselves can absorb light strongly due to localized surface plasmon resonance (LSPR). For gratings of a small period, LSPR may thus be dominant over the grating-coupled SPR.[19-21] Even if the grating has a period comparable to visible wavelengths, LSPR may be significantly depending on the geometric details. A work on silver gratings of 400 nm period[16] showed that many factors including film properties (e.g., grain size and roughness) have an influence on SPR generation. The grating width and height also had a substantial effect on the SPR wavelength and coupling strength.

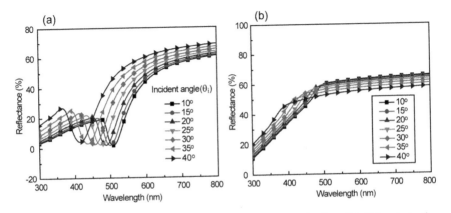

FIGURE 10.14 Specular reflectance at $\phi = 0°$: (a) TE polarization, (b) TM polarization. Reprinted with permission from Ref. [18]. Copyright© 2017 Optical Society of America.

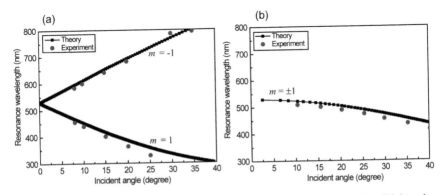

FIGURE 10.15 Dependence of resonance wavelength on the incident angle of light, when (a) $\phi = 90°$ and (b) $\phi = 0°$. Reprinted with permission from Ref. [18]. Copyright© 2017 Optical Society of America.

10.3 LOCALIZED SURFACE PLASMONS

10.3.1 INTRODUCTION

SPs can be divided into propagating SPs and localized surface plasmons (LSPs). While the former propagate relatively large distances from tens to hundreds of micrometers, the latter are confined to nanostructures with dimensions smaller than the wavelength of light. LSPs exist at the surface of metallic nanostructures such as nanoparticles, nano-holes, and nano-grooves. A LSP is a collective oscillation of the free electrons in a metallic nanostructure. LSPR refers to the light-excited, resonant oscillation of these free electrons. As discussed in the previous section, the propagating SPs can be excited by coupling the incident light through a prism or a grating on the metal surface. Unlike this SPR, the LSPR does not require any momentum-matching procedure, making its application more convenient. The LSPs can be directly excited by the oscillating electric field of the incoming light. When a SP is confined to a particle whose size is comparable to or smaller than the wavelength of light, the free electrons of the particle participate in the collective oscillation, as illustrated in Figure 10.16. It is termed a LSP. The strong interaction of light with this LSP has two distinct effects. First, the electric field is greatly enhanced near the particle, with the enhanced field highly localized at the surface and rapidly decaying into the surrounding dielectric medium. Second, the particle's optical extinction is maximized at the plasmon resonant frequency, which occurs at visible wavelengths for noble metal nanoparticles. As in the SPR, the LSP resonant frequency (i.e., the extinction peak) depends on the refractive index of the surrounding medium. This is the basis for sensing applications. The resonant properties of plasmonic nanostructures are also very sensitive to the material, shape, and size of the structures.

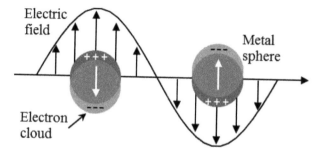

FIGURE 10.16 Illustration of the excitation of a localized surface plasmon.

LSPR is a long-studied nanoscale phenomenon of considerable recent interest. These resonances, associated with noble metal nanostructures, create sharp spectral absorption and scattering peaks as well as strong electromagnetic near-field enhancements. People in the Roman Empire and the Middle Ages employed gold and silver nanoparticles to manufacture colored glasses, although they might have not known about the LSPR phenomenon. These noble metal particles preferentially absorb certain wavelengths of visible light, producing specific colors. Over the past two decades, there have been significant improvements in the fabrication of metal nanostructures, which has led to advances in the science and technology of LSPR. This ever-growing field finds a wide range of application areas including surface-enhanced Raman spectroscopy (SERS), optoelectronics, photovoltaics, chemical and biological sensors, biomedical diagnostics, plasmonic structural colors, and novel optical devices. Some of these applications are described in Section 10.4. While the LSPR properties of spherical particles can be analytically calculated using Mie theory,[22,23] there is no analytic solution for other structures. Therefore, a variety of numerical methods have been developed to predict the plasmonic behaviors of arbitrary structures, including finite-difference time-domain (FDTD), discrete dipole approximation (DDA), finite element method (FEM), and finite integration technique. The numerical techniques are also useful for studying the effect of substrates. In FDTD simulations, the time-dependent Maxwell's equations are discretized using central-difference approximations to space and time partial derivatives. The resulting finite-difference equations are solved using the software. The electric field vector components in a volume of space are solved at a given instant in time and the magnetic field components in the same spatial volume are solved at the next instant in time. The process is repeated over and over again until the desired transient or steady-state electromagnetic field behavior is fully evolved. The nature of Maxwell's differential equations is that the time derivative of the H-field is dependent on the curl of the E-field, and the time derivative of the E-field is dependent on the curl of the H-field. This results in the basic FDTD time-stepping relation that, at any point in space, the updated value of the E-field in time is dependent on the stored value of the E-field and the numerical curl of the local distribution of the H-field in space.[24] The FDTD method is very versatile and widely used for plasmonics since the interaction of an electromagnetic wave with matter can be mapped into the space lattice by assigning appropriate values of permittivity to each E-field component, and permeability to each H-field component. It can cover a wide frequency range with a single simulation run.

10.3.2 PHYSICS OF LSPR

When a plane electromagnetic wave is incident on a metallic particle, it is partially scattered and partially absorbed by the particle. The scattering cross section C_{sca} is the power of the scattered light normalized by the power per unit area (i.e., intensity) of the incident wave, which has dimensions of the area. The scattering coefficient Q_{sca}, which is a dimensionless quantity, is the ratio of the scattering cross-section to the geometric cross section of the particle. Thus, for a spherical particle of radius R, $Q_{sca} = C_{sca}/\pi R^2$. Similarly, the absorption cross section C_{abs} is the ratio of the power absorbed by a particle to the intensity of the incident wave. The absorption coefficient Q_{abs} of a spherical particle has also the relation of $Q_{abs} = C_{abs}/\pi R^2$. The extinction coefficient Q_{ext} is the sum of Q_{sca} and Q_{abs}. It means the power fraction of an incident light wave that is not transmitted due to scattering and/or absorption. In the early 20th century, Gustav Mie developed an analytic solution to Maxwell's equations that describe the scattering and absorption of light by spherical particles.[22] The Mie theory gives the following scattering and extinction cross sections.

$$C_{sca} = \frac{2\pi}{k^2} \sum_{l=1}^{\infty} (2l+1)(|a_l|^2 + |b_l|^2) \tag{10.30}$$

$$C_{ext} = \frac{2\pi}{k^2} \sum_{l=1}^{\infty} (2l+1) \, \mathrm{Re}\{a_l + b_l\} \tag{10.31}$$

Here k is the incoming wave vector and l is an integer representing the dipole, quadruple, and higher multipoles of the scattering. In the above expressions, a_l and b_l are the parameters represented by the Reccati-Bessel functions. These parameters depend on the radius of the metal particle and also the refractive indices of the metal and the surrounding medium. For large particles, the electric field of the incident wave is not constant across the entire particle. Retardation effects across the particle enable higher-order multipoles to be excited. Although the Mie theory provides an exact solution to the scattering problem, it is also rather complex, making it difficult to obtain physical insight into the theoretical results. The complex Reccati-Bessel functions can be approximated by power series if the particle size is much smaller than the wavelength of the incoming light ($R \ll \lambda$). This approximation gives the following scattering and absorption coefficients.[23]

$$Q_{sca} = \frac{8}{3} k^4 R^4 \frac{(\varepsilon_1 - \varepsilon_d)^2 + (\varepsilon_2)^2}{(\varepsilon_1 + 2\varepsilon_d)^2 + (\varepsilon_2)^2} \tag{10.32}$$

$$Q_{abs} = 12kR\varepsilon_d \frac{\varepsilon_2}{(\varepsilon_1 + 2\varepsilon_d)^2 + (\varepsilon_2)^2} \qquad (10.33)$$

Here k is the wave vector of light in the dielectric medium surrounding the particle, given by $k = 2\pi n_d/\lambda$. The complex dielectric function of the metal is $\varepsilon(\lambda) = \varepsilon_1(\lambda) + i\varepsilon_2(\lambda)$, and the medium's dielectric constant is $\varepsilon_d = n_d^2$. While these small particle approximations strictly apply only to very small particles (<10 nm diameter), their predictions of dielectric sensitivity are still accurate for larger particles.[25] The extinction coefficient Q_{ext} is the sum of Q_{sca} and Q_{abs}. If extinction is dominated by absorption, the extinction spectrum will vary as $1/\lambda$. When it is dominated by scattering, the spectrum will vary as $1/\lambda^4$. The extinction coefficient will be maximized when the denominator in eqs 10.32 and 10.33 is minimized. Then, the resonance condition is $\varepsilon_1 = -2\varepsilon_d$. This explains the dependence of the LSPR peak on the surrounding dielectric environment. For instance, the expected LSPR wavelength for gold particles in water is ~515 nm, according to the real dielectric function for gold. Indeed, the experimental absorption spectrum of gold colloid exhibits a peak near this wavelength. The frequency-dependent form for ε_1 from the Drude model is

$$\varepsilon_1(\omega) = 1 - \frac{\omega_p^2}{\omega^2 + \gamma^2}. \qquad (10.34)$$

Here ω_p is the bulk plasma frequency and γ is the damping factor. The Drude model is a purely classical model of electron transport in conductors and describes the collision between freely moving electrons and heavy, stationary ion cores. It provides a good approximation of the conductivity of noble metals. For visible and near-infrared frequencies ($\omega \gg \gamma$), eq 10.34 can be simplified to

$$\varepsilon_1(\omega) \approx 1 - \frac{\omega_p^2}{\omega^2}. \qquad (10.35)$$

Applying this expression for the resonance condition ($\varepsilon_1 = -2\varepsilon_d$), we have the following relation.

$$\omega_{res} = \frac{\omega_p}{\sqrt{2\varepsilon_d + 1}} \qquad (10.36)$$

Here ω_{res} is the LSPR peak frequency. Using the relations of $\lambda = 2\pi c/\omega$ and $\varepsilon_d = n_d^2$, eq 10.36 can be alternatively written as

$$\lambda_{\text{res}} = \lambda_p \sqrt{2n_d^2 + 1}. \tag{10.37}$$

Here λ_{res} is the LSPR peak wavelength and n_d is the refractive index of the surrounding medium. λ_p is the wavelength corresponding to the bulk plasma frequency. This expression describes the dependence of LSPR peak wavelength on the dielectric environment. As the refractive index of the surrounding medium increases, the LSPR peak shifts to longer wavelengths. The resonance condition ($\varepsilon_1 = -2_d$) derived from eqs 10.32 and 10.33 holds when the particle size is much smaller than the vacuum wavelength of the incoming light ($R \ll \lambda$). This is the case when the incident light can couple only to the dipole moment ($l = 1$) of the sphere. For large particles, the electric field of the incident light is not uniform throughout the entire particle. Retardation effects across the particle enable higher-order multipoles to be excited. It is also possible to determine the resonance condition for all higher multipole orders,[12] which is given by $\varepsilon_1 \approx -\varepsilon_d(l + 1)/l$. Then, eq 10.36 is more generalized to

$$\omega_{\text{res}} = \omega_p \sqrt{\frac{l}{l + (l+1)\varepsilon_d}}. \tag{10.38}$$

As the multipole order l increases, the resonance frequency also increases. That is, higher-order multipole resonances occur at shorter wavelengths. For a vacuum dielectric environment, the maximum resonance frequency is $\omega_p/\sqrt{2}$. As illustrated in Figure 10.16, when a very small particle is irradiated by light, the oscillating electric field causes the conduction electrons to oscillate coherently. Since the electric field is uniform across the particle, all the conduction electrons move in-phase. This collective oscillation of the electrons is called the dipole resonance of the particle (Fig. 10.17a). The dipole resonance is characterized by a single, narrow peak in the extinction spectrum. As the particle size increases, the field across the particle becomes nonuniform. The resulting phase retardation broadens the dipole resonance and also excites higher multipole resonances such as the quadrupole, octupole, and so on. (Fig. 10.17b,c), leading to several peaks in the spectrum. As the particle size changes, the relative contributions of Q_{sca} and Q_{abs} also change.

A nanohole is a complementary structure to a nanoparticle and consists of a void surrounded by a metallic medium. LSPs can also be excited by such a cavity. For spherical voids inside a bulk metal, the resonance condition can be found by interchanging the dielectric constants in the resonance condition for spherical particles: $\varepsilon_d \approx -\varepsilon_1(l + 1)/l$. Then, the resonance frequencies for spherical voids are given by

$$\omega_{res} = \omega_p \sqrt{\frac{l+1}{l(\varepsilon_d+1)+1}}. \tag{10.39}$$

By comparison with eq 10.38, we find that for each l value, the resonance frequency (ω_{np}) of the nanosphere is lower than that (ω_{nv}) of the complimentary nanovoid. In vacuum ($\varepsilon_d = 1$), the resonance frequencies of these two structures are related to each other as $\omega_{np}^2 + \omega_{nv}^2 = \omega_p^2$.

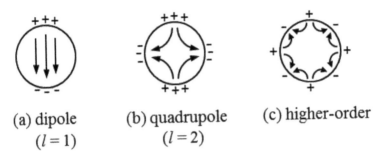

(a) dipole   (b) quadrupole   (c) higher-order

($l = 1$)   ($l = 2$)

FIGURE 10.17   Surface modes in a small sphere.

10.3.3 EFFECTS OF PARTICLE SIZE, SHAPE, AND MATERIAL

LSP resonances are affected by several factors, most importantly by the dielectric functions of the material and the surrounding medium. It can be easily seen from the resonance condition of $\varepsilon_1 \approx -\varepsilon_d(l+1)/l$ that the resonance frequencies depend both on ε_1 and ε_d. Since the absolute value of $\varepsilon_1(\omega)$ for metals increases as ω decreases, the resonance frequency also decreases as the dielectric constant ε_d of the surrounding medium increases. A number of reports show shifts in the SP absorption peak for metallic particles dispersed in various media. Similar shifts are also induced by coating metallic particles with dielectric layers. The peak shift induced by changes in the surrounding material can be used to develop various kinds of chemical and biological sensors. Particle size and shape are also important factors affecting LSPR behaviors.[26] The real (ε_1) and imaginary (ε_2) parts of the dielectric function of a metal describe polarizability and energy dissipation, respectively. The imaginary part of the dielectric function, which accounts for losses, must be small at the LSPR frequency to provide efficient electron oscillations. Several processes can damp the oscillations, including electron scattering by lattice phonon modes, inelastic electron–electron interactions, electron scattering at the particle surface, and excitation of bound electrons into the conduction band (interband transitions). While electron–phonon interactions account for a majority of the imaginary part of the dielectric function, inelastic electron–electron interactions and surface scattering are less significant unless the particle size is extremely small (<5 nm). Interband transitions can cause a substantial decrease in the efficiency of plasmon excitation because the incident electromagnetic energy is used for this electronic transition, not for the plasmonic excitation. In Au and Cu, the interband absorption edge significantly overlaps the plasmon resonance. For Ag, however, the absorption edge is in the UV region (320 nm) and has little impact on the excitation of LSPs, which occurs at wavelengths longer than 370 nm. This accounts for the fact that the excitation of LSPs in Ag nanoparticles is more efficient than for Au and Cu.

For metal nanoparticles of a given material and shape, the LSPR properties depend strongly on the particle size. According to the Mie theory, for spherical particles of radius R much smaller than the wavelength of light, the scattering coefficient is proportional to R^4, while the absorption coefficient is proportional to R. Therefore, for very small particles, LSPR extinction is dominated by absorption. As the particle size increases, scattering becomes more dominant. For gold nanospheres, the transition occurs at around 80 nm in particle diameter.[27] It is also well known that the plasmon resonance wavelength depends on the particle size. For gold nanospheres, the LSPR wavelength can be tuned over 60 nm by varying the particle size from 10 nm to 100 nm.[28] The plasmon resonance line width also varies with the particle size, due to a combination of interband transitions and higher order (non-dipole) plasmon modes. The Mie theory is strictly applicable only to spherical particles. Many nanoparticles exhibiting LSPR are not strictly spherical. Gans generalized the Mie result to spheroid particles of any aspect ratio in the small particle approximation.[29] The absorption cross section for a prolate spheroid was found analogous to that for a sphere, except that the resulting LSPR peak frequencies are different at different directions. Elongated nanoparticles such as nanorods are well described by the spheroid model. The polarizability for an elongated particle depends on the orientation of the particle to the incident field. The extinction spectrum of nanorods has two peaks, one corresponding to the transverse plasmon mode and the other corresponding to the longitudinal mode (Fig. 10.18). Particle shape plays a significant role in determining the LSPR spectrum. The effect of particle shape is well demonstrated in Figure 10.19, in which the spectra of silver nanoparticles of different shapes (spheres, triangles, and pentagons) but similar volume are correlated with their structure. For nanoparticles other

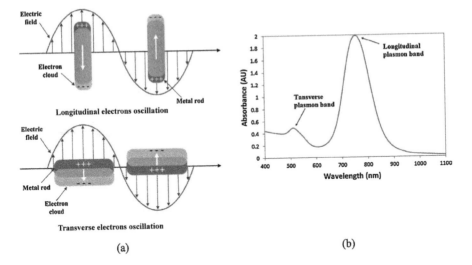

(a)   (b)

FIGURE 10.18   (See color insert.) (a) Schematic illustration of LSPR excitation for gold nanorods. (b) LSPR absorption bands of gold nanorods; longitudinal and transverse plasmon bands correspond to the electron oscillation along the long and short axes of the nanorod, respectively. Reprinted with permission from Ref. [30]. Copyright© 2015 Elsevier.

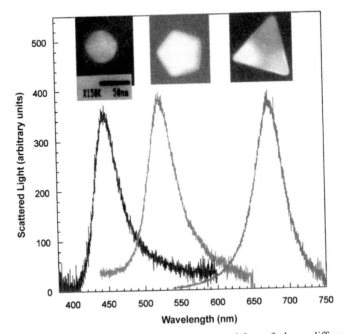

FIGURE 10.19   Scattering spectra of silver nanoparticles of three different shapes. Reprinted with permission from Ref. [31]. Copyright© 2002 American Institute of Physics.

than spheres and spheroids, the shape effect cannot be found analytically and must be studied numerically or experimentally. Numerical methods include FDTD, DDA, and FEM. Figure 10.20 shows simulation results for gold nanorods of different aspect ratios, where the surrounding medium has a fixed dielectric constant of 4. It can be seen that two LSPR peaks are present in the absorption spectra, corresponding to the transverse and longitudinal resonance modes. As the aspect ratio increases, the absorption maximum of the transverse mode shifts to shorter wavelength, while the longitudinal mode red-shifts. The effect of the aspect ratio is much more profound on the longitudinal mode, exhibiting a shift of over 100 nm. Furthermore, the intensity of the longitudinal mode relative to the transverse mode increases with increasing aspect ratio. In Figure 10.21, absorption spectra are plotted for different dielectric constants of the medium with a fixed aspect ratio of 3.3. In this case, both resonance peaks shift to longer wavelengths with an

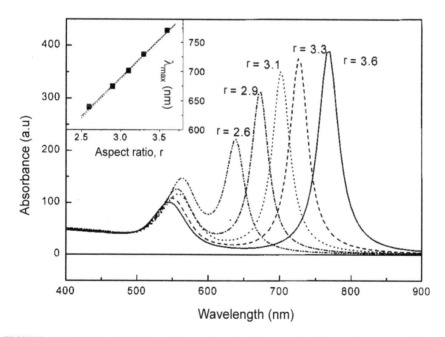

FIGURE 10.20   Simulated absorption spectra of gold nanorods with varying aspect ratio *r*. The surrounding medium was assumed to have a dielectric constant of 4. The inset shows a plot of the maximum of the longitudinal plasmon band determined from the calculated spectra as a function of the aspect ratio. The solid line is a linear fit to the data points. Reprinted with permission from Ref. [32]. Copyright© 1999 American Chemical Society.

increasing dielectric constant of the medium. Again, the longitudinal mode is more sensitive. In the insets of Figures 10.20 and 10.21, the absorption maximum of the longitudinal resonance is plotted against the aspect ratio and the medium dielectric constant, respectively. The solid lines represent linear fits to the points, which are determined from the absorption spectra. Figure 10.22 presents extinction spectra for silver oblate spheroids having a ratio of major to minor axes, r, ranging from $r = 1$ (a sphere) to $r = 10$ (highly oblate), where the total volume of each spheroid is equal to that for a sphere of 80 nm radius and the field polarization is along the major axis of the spheroid. As the particle becomes more oblate, that is, the ratio r increases, the dipole plasmon resonance gradually redshifts. The second effect seen in Figure 10.22 is that the quadrupole resonance peak, which for the case of a sphere is larger than the dipole peak, is much less important as r is increased. This indicates that the quadrupole mode can be "quenched" by particle asymmetry.

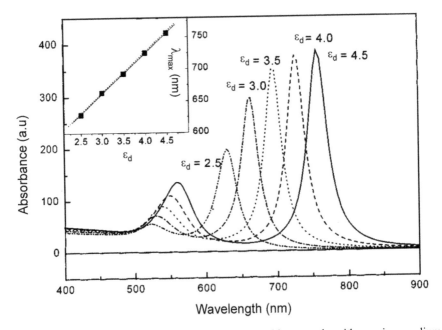

FIGURE 10.21 Simulated absorption spectra of gold nanorods with varying medium dielectric constant ε_d. The aspect ratio was fixed at a value of 3.3. The inset shows a plot of the maximum of the longitudinal plasmon band determined from the calculated spectra as a function of the medium dielectric constant. The solid line is a linear fit to the data points. Reprinted with permission from Ref. [32]. Copyright© 1999 American Chemical Society.

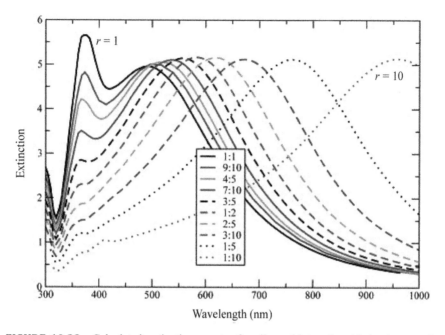

FIGURE 10.22 Calculated extinction spectra for silver oblate spheroids having a ratio of major to minor axes, r, ranging from $r = 1$ (a sphere) to $r = 10$ (highly oblate). In the calculation, the total volume of each spheroid is equal to that for a sphere of 80 nm radius and the field polarization is along the major axis of the spheroid. Reprinted with permission from Ref. [26]. Copyright© 2003 American Chemical Society.

There has been a great deal of interest in nanoparticle shapes with sharp features or tips, which have been developed through both bottom–up (chemical synthesis) and top–down (lithography) methods. Some of these structures include nanocubes, nanostars, nanotriangles, and bipyramids.[33–37] One effect of the sharp tips is to produce a red shift in the plasmon resonance, increasing the refractive index sensitivity. As another example of the numerical calculation, Figure 10.23 shows extinction spectra for Ag trigonal prisms based on a 100 nm edge dimension, in which the prism thickness is 16 nm. The perfect prism exhibits three peaks, a long wavelength peak at 770 nm, a weaker peak at 460 nm, and a small but sharp peak at 335 nm. The redmost peak is very sensitive to snipping, with the 20 nm snipped prism giving a peak that is blue-shifted by 200 nm as compared to the perfect prism. The vast majority of LSPR experiments have been carried out on gold or silver nanoparticles. Gold is often chosen because of its high chemical stability and resistance to oxidation. Instead, silver has stronger and sharper resonances and higher refractive index sensitivity. The real dielectric function of silver varies with wavelength more than that of gold over the visible light region, especially in the 400–600 nm region (see Figs. 10.1 and 10.2). Also, because the imaginary part of the dielectric function of silver is smaller than that of gold across the visible region, less plasmon damping occurs, resulting in higher scattering efficiency and narrower plasmon linewidths. Gold and silver exhibit resonances at different wavelengths; gold nanoparticles have a resonance peak at longer wavelengths than silver particles of similar size and shape. Bimetallic alloy particles may have a wider tuning range than the single-element particles because the change in the composition provides another dimension in tailoring the LSPR properties. Experiments on gold–silver bimetallic alloys showed a linear relationship between the LSPR peak wavelength and the mole fraction of gold in the alloy.[38,39] The effect of material composition was found to be much greater than that of particle size. Core/shell structures, such as dielectric core/metal shell, metal core/dielectric shell, and metal core/metal shell, also enable the LSPR peak wavelength to be tuned over a larger wavelength range than for spherical metal particles. A dielectric core/metal shell structure can support SPs on both sides of the shell, which can interact with each other if the shell is thin enough. By adjusting the core and shell thicknesses, the resonance wavelength can be varied from the UV to the IR range.[40–42]

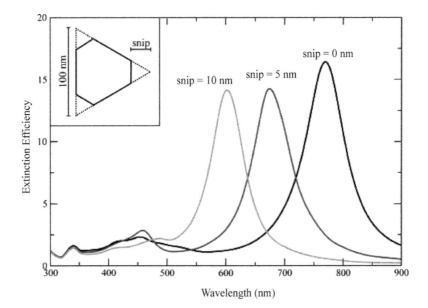

FIGURE 10.23 Extinction spectra for Ag trigonal prisms based on a 100 nm edge dimension. Reprinted with permission from Ref. [26]. Copyright © 2003 American Chemical Society.

10.3.4   *COUPLING OF NANOPARTICLES*

In the previous subsections, the optical responses and LSPR properties of a single particle were described. We found that the LSPR frequency can be shifted by altering the size and shape of the particle and the local dielectric environment. However, it is not easy to handle single particles in actual experiments and applications. It is more common to work with various kinds of ensembles of particles. In such a situation, interparticle interactions come into play. The coupling of particles makes the LSPR behaviors polarization-dependent and also shifts the resonance frequency. Coupled nanoparticles can be considered as a "plasmonic molecule," in contrast to isolated particles viewed as "plasmonic atoms." In real samples consisting of small particles, the particles often exist in the form of aggregates. The aggregation of particles is also an important factor in determining the optical properties of the sample. The simplest aggregate form is a dimer, that is, a pair of particles. When the particles are small enough, the aggregation effect can be considered in terms of the interaction between the dipole modes of the particles. As illustrated in Figure 10.24, the individual dipole modes in a dimer are coupled and split into longitudinal and transverse modes, which oscillate parallel and perpendicular to the dimer axis, respectively. The two dipoles oscillate either in phase (bonding state) or out of phase (antibonding state) in the coupled modes and the in-phase mode is optically active. The dipole–dipole interaction has no analog in a single particle and makes the resonance polarization-dependent. The longitudinal mode is excited by light polarized along the dimer axis and the transverse mode, by light polarized perpendicular to the dimer axis. The resonance frequency of the longitudinal mode is located at lower energy than for an isolated particle and shifts to lower energies as the distance between the particles decreases, while the resonance frequency of the transverse mode is located at higher energy and shifts to higher energies with decreasing interparticle distance. This can be understood by the charge distribution of the dimer particles. When the incident light is polarized along the dimer axis and excites the longitudinal mode, the charges induced on the inside surfaces of the particles are attractive. It means that energy is required to separate the particles and reduce their interaction. Therefore, as the interparticle distance decreases, the energy of the entire system is reduced and the resonance shifts to lower energies (i.e., longer wavelengths). On the other hand, the surface charges induced in the transverse mode make the particles repulsive. Consequently, the energy of the system is increased as the inter-particle distance decreases. This causes the resonance to shift to shorter wavelengths (i.e., higher energies). A remarkable effect of bringing two particles in close proximity to each other is the generation of a localized region of the extremely large electric field, often referred to as a hot spot. As illustrated in Figure 10.25, the electric field is strongly enhanced in the gap between the particles when the longitudinal mode is excited. When the transverse mode is excited, the field enhancement is not very large. The enhancement is often several orders of magnitude larger than that for a single particle, making the system useful for many applications including sensors and SERS. Hot spots can also be formed at the junction regions of randomly arrayed metal nanowires.[44] They can act as very efficient plasmonic heating sources.

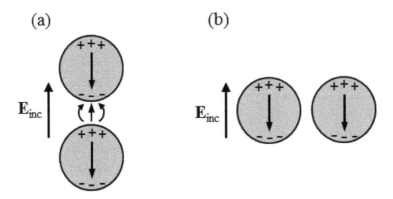

FIGURE 10.24   (a) Longitudinal and (b) transverse modes for a dimer of particles. When the longitudinal mode is excited, the gap between the particles becomes a hot spot.

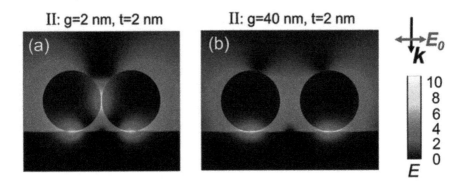

FIGURE 10.25   **(See color insert.)** Local electric field distributions in a dimer of particles, when the gap width is (a) $g = 2$ nm and (b) $g = 40$ nm. Reprinted with permission from Ref. [43]. Copyright© 2016 Macmillan Publishers Limited.

Rechberger et al.[45] studied the resonance properties of gold nanodisks on an indium tin oxide (ITO)-coated glass substrate. The disk arrays were fabricated by e-beam lithography and a lift-off process. With the disk diameter and height fixed to 150 nm and 17 nm, respectively, only the interdisk distance (center-to-center) was varied from 450 nm to 150 nm. The experimental extinction spectra are shown in Figure 10.26. When the polarization direction of the incident light is parallel to the disk-pair axis, the extinction peak remarkably redshifts with decreasing inter-disk distance (Fig. 10.26a). A smaller but distinct blue-shift of the resonance peak is also observed for the orthogonal polarization. Arrays of gold nanodisk pairs were also investigated by Jain et al. in the process of determining a universal scaling law for the resonance shift.[46] The gold nanodisk pairs were fabricated on quartz slides using e-beam lithography. Each nanodisk had a diameter of 88 nm and a thickness of 25 nm. The edge-to-edge gap separation (s) was varied from 212 nm down to 2 nm. An image of the array is shown in Figure 10.27, where the edge-to-edge gap is $s = 12$ nm. Two different polarization directions of the incident light were chosen, one parallel to the pair axis (Fig. 10.28a) and the other perpendicular to the axis (Fig. 10.28c). Under parallel polarization, the plasmon resonance strongly shifts to longer wavelengths as the separation gap is reduced. Conversely, there is a very weak blue-shift for orthogonal polarization. The resonance shift results from the coupling of the single-particle plasmons, the polarization dependence of which can be explained by the dipole–dipole interaction mechanism depicted in Figure 10.24. The interparticle interaction is clearly stronger for the parallel polarization, as seen from the larger wavelength shift. At extremely small gaps, a new shoulder appears at shorter wavelengths. This new band is attributed to higher order interactions, possibly the quadrupole mode, and cannot be explained by the dipolar interaction model. The results of DDA simulations are shown for the light polarization parallel (Fig. 10.28c) and perpendicular (Fig. 10.28d) to the pair axis. Scaling law of the form of $y = a \exp(-x/\tau)$ was derived, where $y = \Delta\lambda/\lambda_o$ is the fractional resonance shift, $x = s/D$ is the gap separation normalized by the disk diameter (D). The decay constant, τ, was ~0.23 from the simulation results and ~0.18 from the experimental measurements. This indicates that the interparticle coupling and the resulting resonance shift become very significant when the gap separation is less than 0.2 times the disk diameter, that is, $s < 0.2D$.

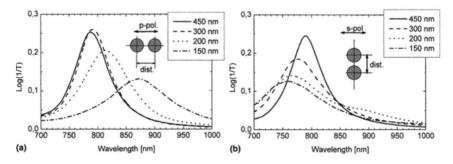

FIGURE 10.26 Extinction spectra of a 2D array of the Au nanoparticle pairs with the interparticle center-to-center distances as the parameter. The polarization direction of the exciting light is (a) parallel to the long particle pair axis and (b) orthogonal to it. Reprinted with permission from Ref. [45]. Copyright© 2003 Elsevier.

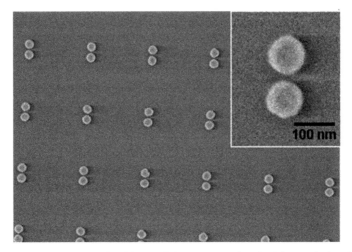

FIGURE 10.27 An image of gold nanodisk array. Reprinted with permission from Ref. [46]. Copyright© 2007 American Chemical Society.

10.4 APPLICATIONS

10.4.1 SERS

Raman scattering refers to the inelastic scattering of a photon by molecules that are excited to higher vibrational or rotational energy levels. Raman scattering is so named after C. Raman who first observed the effect in 1928.[47] When photons are scattered from an atom or molecule, most of them are

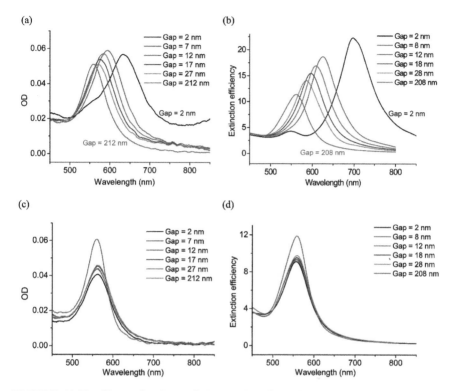

FIGURE 10.28 (See color insert.) (a, c) Microabsorption and (b, d) DDA-simulated extinction efficiency spectra of Au nanodisc pairs for varying interparticle separation gap for incident light polarization direction (a, b) parallel and (c, d) perpendicular to the interparticle axis. Reprinted with permission from Ref. [46]. Copyright© 2007 American Chemical Society.

elastically scattered such that the scattered photons have the same energy (frequency and wavelength) as the incident photons, which is known as Rayleigh scattering. A small fraction of the incident photons (approximately 1 in 10 million) is inelastically scattered by such an excitation, with the scattered photons having a frequency and energy different from (usually lower than) those of the incident photons. This inelastic scattering is the Raman scattering or effect. Molecular vibrations have a frequency range of 10^{12}–10^{14} Hz. These frequencies correspond to the infrared (IR) region of the electromagnetic spectrum. Molecules can be excited to a higher vibrational mode through the direct absorption of a photon with appropriate energy. In Raman scattering, an absorbed photon is reemitted (scattered) with lower energy. The difference in energy between the incident and scattered photons corresponds to the energy required to excite a molecule to a higher vibrational mode. Raman spectroscopy utilizes the Raman effect for substances analysis. In typical Raman spectroscopy, a laser beam whose wavelength lies in the visible or near-IR region is incident onto a sample and a spectrum of the scattered photons, called the Raman spectrum, is measured. The spectrum shows the intensity of the scattered light as a function of its frequency difference with the incident light. The Raman spectrum depends on the molecular constituents present in the sample and their state, allowing it to be employed for material identification and analysis. Raman spectroscopy can be used to analyze a wide range of materials, including highly complex materials such as biological organisms and human tissues.

It is important to note that (conventional) Raman scattering is an extremely weak effect. Surface enhanced Raman spectroscopy (SERS) is a surface-sensitive technique that significantly enhances Raman scattering by rough metal surfaces or nanostructures. SERS was discovered in 1974 by Fleischmann et al.[48] They found that the Raman signal of some molecules could be enormously enhanced on a roughened silver electrode. The effect was originally ascribed to an increase in the amount of adsorbed molecules resulting from a larger surface area. However, it was later shown that the enhancement of the Raman signal was too significant to be simply explained by the increased surface area. Typical enhancement factors observed from roughened metal surfaces as compared to bare glass substrates are on the order of 10^6–10^7. For single-molecules, the enhancement factors can reach values up to 10^{14} and 10^{15}.[49,50] Although a theory satisfactorily explaining the underlying mechanism of SERS is still absent, two different theories are generally employed to account for the signal enhancement. The electromagnetic theory, which is more widely accepted, is based on the excitation of LSPs, while the chemical theory relies on the formation of charge-transfer complexes.[51–54] The latter applies only for species that form a chemical bond with the surface. On the other hand, the former can apply regardless of whether the specimen is either chemisorbed or physisorbed to the surface. SERS enhancement can be observed even when an excited molecule is located rather far away from the surface that hosts metallic nanostructures, supporting the electromagnetic theory. This theory states that the increase of the Raman signal occurs due to an enhancement of the electric field provided by the surface. When the incident light strikes the surface, LSPs are excited. SERS experiments employ roughened surfaces or nanostructures because these surfaces provide an area on which the LSPs can be strongly excited. SERS is typically performed by depositing a sample onto a glass or Si substrate with a nanostructured metal surface. The most common metals used for plasmonic surfaces are silver and gold since their plasmon resonance frequencies fall within the visible and near-IR range necessary for SERS experiments. Recently, much attention is given to aluminum as an alternative plasmonic material. Because its plasmonic band is in the UV region, there is great interest in using aluminum for UV SERS.[55-57] The surface morphology of SERS substrates strongly affects the enhancement of the electric field and the Raman signal. Active SERS substrates can be fabricated by various methods, including electrochemistry, e-beam lithography, photolithography, colloidal self-assembly, and printing. An ideal SERS substrate must exhibit a highly uniform surface structure and high field enhancement. The ability to detect the presence of low-abundance molecules makes SERS beneficial for materials science, art and archeological research, drug and explosives detection, and food quality analysis. SERS combined with plasmonic sensing can also be used for high-sensitivity and quantitative detection of biomolecular interactions. The cost of SERS substrates should be as low as possible, in order to compete with other analytical tools. The fabrication of low-cost, disposable SERS substrates is thus of practical importance.[58-60]

10.4.2 PLASMONIC SENSORS

The distinct resonance condition associated with the excitation of SPs finds various sensing applications. The simplest SPR sensor is based on the Kretschmann configuration, as shown in Figure 10.29. In this sensor, the position of a dip in the reflectivity curve can be used as an indicator of environmental changes. The refractive index of the material touching the metal film surface determines the angle of incidence for which an SPP can be excited. With this method, the adsorption or removal of target materials on the metal surface can be detected with sub-monolayer accuracy. This SPR sensor is particularly useful for liquid environments, that is, for materials with a refractive index close to that of water. If the refractive index at the metal surface changes, the position of the resonance peak shifts. When the angle of incidence is adjusted to the dip in the reflectivity curve, the deposition of a small amount of material increases the reflected signal (i.e., reflectivity) drastically. The sensor can provide kinetic data through continuous optical measurements. The LSPR effect on nanoparticles can also be used for sensing. The optical response of a dielectric matrix containing metal nanoparticles is determined not only by the intrinsic properties of the particles but also by the dielectric permittivity of the matrix. The simplest sensing application of LSPR-active particles is to detect changes in the bulk refractive index of their environment through shifts in the LSPR peak wavelength. As discussed in Section 10.3, the resonant wavelength for metal nanoparticles shifts to longer wavelengths as the refractive index of the surrounding medium increases. LSPR peaks are typically detected by spectral extinction measurements on a dense film. The LSPR peak wavelength shift is approximately linear with changes in the refractive index of the surrounding medium. The refractive index sensitivity S of a particular nanoparticle type is usually given by $S = d\lambda_p/dn$, where λ_p is the LSPR peak wavelength. An example of this type of sensing is depicted in Figure 10.30 for silver nanoparticles embedded in a series of solvents with different refractive indices. The precision that can be achieved with respect to changes in the refractive index depends on the sensitivity, S, and the line width (λ) of the LSPR peak. Although larger nanoparticles tend to have higher sensitivities, their peaks are broadened by multipolar excitations and radiative damping. A figure of merit (FOM) that characterizes the capability of a nanoparticle-based sensor is usually obtained by dividing the sensitivity by the line width: FOM = $S/\Delta\lambda$. LSPR sensors are also useful for molecular sensing. As described previously, LSPR-induced field enhancements decay rapidly with distance from the particle surface. Therefore, shifts in the resonance peak only probe a nanoscale region around the particle. This highly localized volume allows us to sense molecular interactions near the particle surface because they result in changes in the local refractive index.

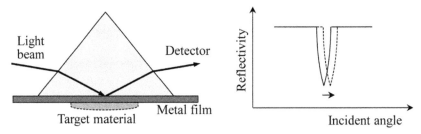

FIGURE 10.29 Schematic of an SPR sensor based on the Kretschmann configuration.

10.4.3 PLASMONICS FOR PHOTOVOLTAICS

Photovoltaics, which refers to the conversion of sunlight to electricity using solar cells, is a promising technology that may allow the production of electric power on a very large scale. Plasmons are free-electron oscillations in a metal that enables light to be manipulated at the nanoscale. The ability of plasmons to guide and confine light opens up new design possibilities for solar cells.[62–71] Photovoltaics can make a considerable contribution to solving the energy problem that our society will face in the next generation. At present, commercial photovoltaic cells are mostly based on crystalline silicon wafers with thicknesses of 150–300 μm, and most of the price of solar cells is due to the costs of silicon materials and processing. To be competitive with fossil-fuel technologies, the cost needs to be reduced by a factor of 2–5. In this respect, there is great interest in thin film solar cells, which are made from a wide variety of semiconductors including amorphous and poly-crystalline Si, GaAs, CdTe, and $CuInSe_2$, as well as organic and perovskite semiconductors. A key advantage of such thin devices is the small amount of material required. However, the reduced thickness should not come at the expense of performance. The most fundamental problem incurred by reduced thickness is poorer light absorption, which is a disadvantage that should be overcome through an approach called "light trapping."

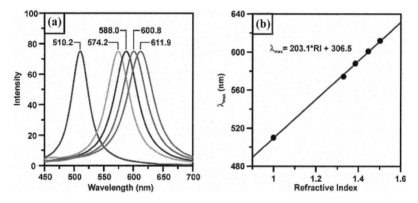

FIGURE 10.30 (a) Single Ag nanoparticle resonant Rayleigh scattering spectrum in various solvent environments (left to right): nitrogen, methanol, 1-propanol, chloroform, and benzene. (b) Plot depicting the linear relationship between the solvent refractive index and the LSPR λ_{max}. Reprinted with permission from Ref. [61]. Copyright© 2003 American Chemical Society.

As sketched in Figure 8.14, a roughened surface reduces reflection by increasing the chance of reflected light bouncing back onto the surface. In conventional thick Si solar cells, light trapping is typically achieved using a micrometer-scale pyramidal surface texture (see Fig. 8.15). The texture causes scattering of light into the solar cell over a large angular range, thereby increasing its effective path length in the cell. However, such large-scale geometries are not suitable for thin-film cells. Figure 10.31a shows the standard AM1.5 solar spectrum together with a graph that illustrates what fraction of the solar spectrum is absorbed on a single pass through a 2-μm-thick crystalline Si film. Obviously, a large fraction of the solar spectrum (in particular, a spectral range of 600–1100 nm) is poorly absorbed. This explains the reason that conventional

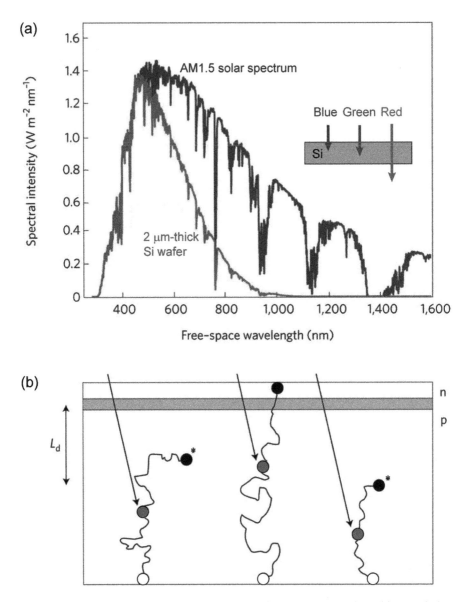

FIGURE 10.31   (See color insert.) (a) AM1.5 solar spectrum, together with a graph that indicates the solar energy absorbed in a 2-μm-thick crystalline Si film (assuming single-pass absorption and no reflection). Clearly, a large fraction of the incident light in the spectral range 600–1100 nm is not absorbed in a thin crystalline Si solar cell. (b) Schematic indicating carrier diffusion from the region where photocarriers are generated to the p–n junction. Charge carriers generated far away (more than the diffusion length L_d) from the p–n junction are not effectively collected, owing to bulk recombination (indicated by the asterisk). Reprinted with permission from Ref. [64]. Copyright© 2010 Springer Nature.

wafer-based crystalline Si solar cells have a large thickness of 150–300 μm. For high-efficiency cells, the absorption layer should be thin enough to ensure that all photocarriers are collected (Fig. 10.31b). This requirement is easily met for thin solar cells. Meanwhile, the absorption layer should also be thick enough to absorb more light. This seemingly contradictory problem may be solved by increasing the path length of light within the absorption layer, that is, its optical thickness. Methods for light trapping are thus important for thin solar devices. A strategy for improving light trapping in thin film solar cells is to use metallic nanostructures that support SPs. LSPs excited in metal nanoparticles and SPPs propagating at the metal/semiconductor interface can both be utilized. These plasmonic structures can offer at least three different ways of increasing the optical thickness of the absorption layer while keeping its physical thickness constant. First, metallic nanoparticles can be used as scattering elements to couple and trap incident sunlight into a thin absorption layer of the cell (Fig. 10.32a). It is to be noted that for very small particles, absorption is more dominant over scattering. Absorption by the particle itself is parasitic absorption, which adversely affects the performance of the cell. It has been shown that the size and shape of metal nanoparticles are both key factors determining the coupling efficiency.[71] Light scattering from a small metal nanoparticle is nearly symmetric in the forward and reverse directions. That is, photons are scattered in all directions with equal probability. Thus, when scattering particles are positioned near the front surface of the cell, some reflection loss is inevitable. One way to overcome this problem is to place particles on the rear surface of the cell. In this case, blue and green light is directly absorbed in the cell, whereas poorly absorbed IR light is scattered and trapped by the nanoparticles.

An alternative use of LSPR in thin film solar cells is to take advantage of the strong field enhancement around the metal nanoparticles (Fig. 10.32b). In this case, the particles act as an effective "antenna" for the incident sunlight and increase absorption in a surrounding semiconductor material. This scheme works for very small particles (5–20 nm diameter). The particles store the incident photon energy in a LSP mode. These plasmonic antennas are particularly useful for materials whose carrier diffusion lengths are short. The excited plasmons rapidly decay. For this scheme to be efficient, the absorption rate of light in the semiconductor layer must be very large. Otherwise, the absorbed energy is dissipated into damping in the metal particle. Such high absorption rates can be achieved in many organic and direct-bandgap inorganic semiconductors. Photocurrent and efficiency enhancements using the plasmonic near-field effect have been reported for organic and inorganic thin-film cells as well as dye-sensitized solar cells.[72–75] In another plasmonic light-trapping scheme (Fig. 10.32c), incident light is converted into SPPs that propagate along the interface between the semiconductor absorption layer and a metal back contact. SPPs excited at the metal/semiconductor interface can efficiently trap and guide light in the semiconductor layer, where the light is absorbed along the lateral direction of the solar cell. As discussed in Section 10.2, SPPs cannot be excited on a planar metal surface due to the momentum mismatch. Therefore, a light-coupling structure must be integrated at the metal/semiconductor interface. A corrugated or textured metal film on the rear surface of a semiconductor absorption layer can couple light into SPP modes supported at the metal/semiconductor interface.[76–78] A common concern with the SPP coupling is whether metal absorption might override the benefit of semiconductor absorption. This scheme is thus efficient when the excited SPPs are more absorbed by the semiconductor than the metal. The evanescent SPP fields decay in both the metal and the semiconductor, but extend further into the semiconductor. A longer decay length in the semiconductor is desirable because it increases absorption there. The light-trapping concepts given in Figure 10.32 are closely related to one another and can be combined together. Indeed, any solar cells with a non-planar metal back reflector will exhibit geometric scattering and high local fields as well as coupling into plasmonic modes. Various other cell designs can benefit from the increased light confinement and scattering from plasmonic nanostructures.

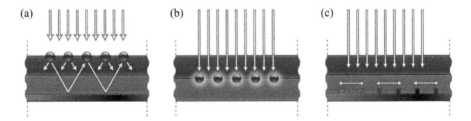

FIGURE 10.32   (See color insert.) Plasmonic light-trapping schemes for thin-film solar cells. (a) Light trapping by scattering from metal nanoparticles at the surface of the solar cell. Light is preferentially scattered and trapped into the semiconductor thin film by multiple and high-angle scattering, causing an increase in the effective optical path length in the cell. (b) Light trapping by the excitation of localized surface plasmons in metal nanoparticles embedded in the semiconductor. The excited particles' near-field causes the creation of electron-hole pairs in the semiconductor. (c) Light trapping by the excitation of surface plasmon polaritons at the metal/semiconductor interface. A corrugated or textured metal film on the rear surface of a semiconductor absorption layer can couple light into SPP modes supported at the metal/semiconductor interface. Reprinted with permission from Ref.[64] Copyright© 2010 Springer Nature.

10.4.4 *GENERATION OF PLASMONIC COLORS*

Plasmonic colors are structural colors that emerge from resonant interactions between light and metallic nanostructures. The engineering of plasmonic colors is a rapidly developing, promising research field that has a large technological impact. Surfaces decorated with structural colors are also receiving tremendous attention due to their widespread use. Plasmonic colors are mainly based on LSPRs, which allow for color printing with sub-wavelength resolution. In relation to the generation of structural colors by LSPRs, it is important to note that metal nanostructures have absorption cross-sections larger than their physical cross-sections. Nevertheless, local-ization of the incident electromagnetic energy is crucial for the production of plasmonic colors with high spatial resolution. Metallic nanostructure-based coloration uses the spectral tunability of LSPRs, so it is highly sensitive to the shape, size, and material of the nanostructure. Common materials are gold, silver, and aluminum. Plasmonic coloration technology has a wide variety of applications, including solar cells, filters for color imaging, solid-state lighting components, flat-panel displays, surface decorations, and anti-counterfeiting. To construct plasmonic color devices, nanostructures are typically grouped into micron-scale arrays known as pixels. The shape of the constituent nanostructures, as well as their patterned arrangement, can both be used to modify and improve pixel color properties.

Plasmonic antennas enable subwavelength localization of the electromag-netic energy and wavelength-selective optical extinction. Arrays of aluminum nanoantennas have been used to demonstrate chromatic plasmonic polar-izers.[79] Their anisotropic cross shape supports two LSPR modes, which can be tuned almost independently from one another. The plasmonic interaction maps the polarization state of incident white light to a visible color. Similarly, the arrays convert unpolarized white light to chromatically polarized light. While conventional color filters transmit a single color that is represented by a single point on the *International Commission on Illumination* (CIE) chroma-ticity diagram, the polarization-dependent color response of chromatic plas-monic polarizer can produce a continuum of colors represented by a line on the chromaticity diagram. This concept was used to demonstrate active color filters, chromatically switchable anti-counterfeiting tags, and polarization imaging. Aluminum-based antennas have also been explored for producing plasmonic pixels.[80] The vivid colors generated by each of the plasmonic pixels are produced by light scattering from an approximately hexagonal array of oriented aluminum nanorods of the same dimension in each case, as shown in Figure 10.33. All samples are prepared using e-beam lithography on an ITO-coated glass substrate. The resulting color is determined by the distance between rows of nanorods and the distance between nanorods, as well as the length and width of each rod. Polarization-independent filters have also been realized with silver and aluminum nanodisk arrays, gold nanoparticle arrays, and gold-based relief metasurfaces.[81–85]

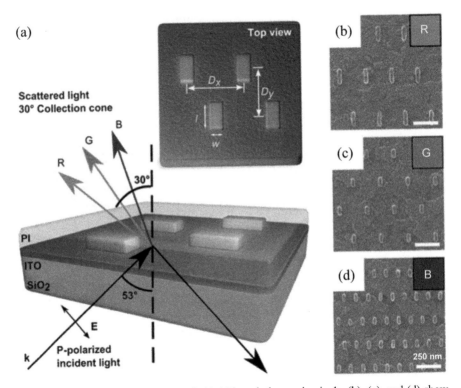

FIGURE 10.33   **(See color insert.)** (a) Al-based plasmonic pixels. (b), (c), and (d) show high-resolution SEM images of portions of a red ($l = 153$ nm, $D_x = 256$ nm, $D_y = 410$ nm), green ($l = 88.5$ nm, $D_x = 272$ nm, $D_y = 272$ nm), and blue ($l = 81$ nm, $D_x = 118$ nm, $D_y = 227$ nm) pixel, respectively. The insets show the color created by a pixel of dimensions shown in each respective SEM image. Reprinted with permission from Ref. [80]. Copyright© 2016 American Chemical Society.

Plasmonic color images with subwavelength resolution can be printed by juxtaposing individual color pixels of dimensions smaller than the wavelength of light. Ink-based traditional color printing is limited in resolution by the size of each ink spot on the substrate as well as the positioning accuracy of multiple color pigments: cyan, magenta, yellow, and black. The resolutions of commercial inkjet and laserjet printers are around 1000 dots per inch (DPI). In contrast, the strong scattering and absorption effects associated with plasmonic nanostructures enable these structures to act as color elements with resolutions up to 100,000 DPI. Subwavelength plasmonic printing typically makes use of gap plasmons and hybridized disk-hole plasmons. An example pixel design, together with a printed color image, is shown in Figure 10.34a. Each pixel consists of four nanodisks that support particle resonances. These disks are raised above equally sized holes on a back-reflector. Crucially, this back-reflector plane functions as a mirror to increase the scattering intensity of the disks. Color information is encoded in the dimensional parameters of metal nanostructures, so that tuning their plasmon resonance determines the colors of the individual pixels. Another example of pixel design and the corresponding color image are shown in Figure 10.34b, where different colors are produced by spatially mixing and adjusting the space between discrete aluminum nanostructures. Metal–insulator–metal structure can be tailored for efficient light absorption, for example, by changing the thickness or the dielectric function of the insulating material separating the two metal films. An example configuration is shown in Figure 10.34c, where gold nanodisks are coupled to a continuous gold film. A fundamental problem with these sub-wavelength plasmonic printings is that color patterns need to be predesigned and printed by e-beam lithography or focused ion beam. While both processes enable the fabrication of nanoscale features with high precision, they are not scalable and also require high cost. From a practical point of view, facile and scalable fabrication methods need to be developed. Recently, a method of laser printing on nanoimprinted plasmonic surfaces has been reported.[89,90] Laser pulses melt and reshape the imprinted nanostructures. Depending on the laser pulse energy density, different surface morphologies can be created. The different morphologies support different plasmonic resonances, ultimately leading to different color appearances. Using this technique, a resolution of $\sim$127,000 DPI is obtained.

PROBLEMS

10.1 The dielectric function of a metal is generally written as $\varepsilon(\omega) = \varepsilon_1(\omega) + i\varepsilon_2(\omega)$. As a consequence, a refractive index is also a complex number given by $n = n_R + in_I$. Then show that

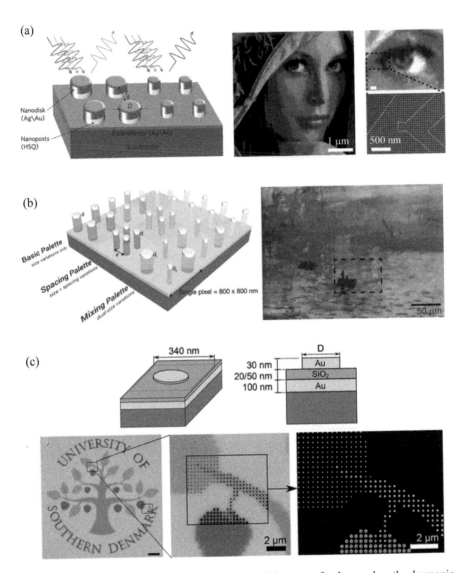

FIGURE 10.34 (See color insert.) Designs and images of sub-wavelength plasmonic printing. (a) Each pixel consists of four nanodisks that support particle resonances. These disks are raised above equally sized holes on a back-reflector by hydrogen silsesquioxane (HSQ) nanoposts. Reprinted with permission from Ref. [86]. Copyright© 2012 Springer Nature. (b) An Al-based pixel design and the corresponding color image, where different colors are produced by spatially mixing and adjusting the space between discrete aluminum nanostructures. Reprinted with permission from Ref. [87]. Copyright© 2014 American Chemical Society. (c) A metal-insulator-metal structure with gold nanodisks coupled to a continuous gold film. Reprinted with permission from Ref. [88]. Copyright© 2014 American Chemical Society.

$$n_R(\omega) = \sqrt{\frac{\sqrt{\varepsilon_1^2 + \varepsilon_2^2}}{2} + \frac{\varepsilon_1}{2}}$$

$$n_I(\omega) = \sqrt{\frac{\sqrt{\varepsilon_1^2 + \varepsilon_2^2}}{2} - \frac{\varepsilon_1}{2}}.$$

10.2 The dielectric function of metal has both real and imaginary parts. This means that the propagation constant k_x in eq 10.15 also has both parts. The real part of k_x determines the wavelength of a propagating SPP, while the imaginary part accounts for the damping of the SPP and thus determines its propagation length. Then express each of the real and imaginary parts of k_x in terms of the real and imaginary parts of the dielectric function.

10.3 To excite an SPP on the internal metal interface, an additional dielectric layer whose refractive index is smaller than that of the prism should be deposited between the prism and the metal film. In such a two-layer configuration, SPP modes can be excited on both interfaces of the metal film at different angles of incidence. Explain the mechanism.

REFERENCES

1. Johnson, P.; Christy, R. Optical Constants of the Noble Metals. *Phys. Rev. B.* **1972**, *6*, 4370–4379.

2. Liu, H.; Genov, D.; Wu, D.; Liu, Y.; Steele, J.; Sun, C.; Zhu, S.; Zhang, X. Magnetic Plasmon Propagation Along a Chain of Connected Subwavelength Resonators at Infrared Frequencies. *Phys. Rev. Lett.* **2006**, *97*, 243902.

3. Pendry, J.; Holden, A.; Robbins, D.; Stewart, W. Magnetism from Conductors and Enhanced Nonlinear Phenomena. *IEEE Trans. Microwave Theory Technol.* **1999**, *47*, 2075–2084.

4. Henzie, J.; Lee, M.; Odom, T. Multiscale Patterning of Plasmonic Metamaterials. *Nat. Nanotechnol.* **2007**, *2*, 549–554.

5. Boltasseva, A.; Atwater, H. Low-loss Plasmonic Metamaterials. *Science* **2011**, *331*, 290–291.

6. Grigorenko, A.; Geim, A; Gleeson, H.; Zhang, Y.; Firsov, A.; Khrushchev, I.; Petrovic, J. Nanofabricated Media with Negative Permeability at Visible Frequencies. *Nature* **2005**, *438*, 335–338.

7. Maier, S. *Plasmonics: Fundamentals and Applications*; Springer Verlag, 2007, New York.

8. Dionee, J.; Sweatlock, L.; Atwater, H.; Polman, A. Planar Metal Plasmon Waveguides: Frequency-dependent Dispersion, Propagation, Localization, and Loss Beyond the Free Electron Model. *Phys. Rev. B* **2005**, *72*, 075405.

9. Otto, A. Excitation of Nonradiative Surface Plasma Waves in Silver by the Method of Frustrated Total Reflection. *Z. Phys.* **1968**, *216*, 398.

10. Kretschmann, E.; Raether, H. Radiative Decay of Nonradiative Surface Plasmons Excited by Light. *Z. Naturforsch.* **1968,** *23a,* 2135.

11. Zayats, A.; Smolyaninov, I. Near-field Photonics: Surface Plasmon Polaritons and Localized Surface Plasmons. *J. Opt. A: Pure Appl. Opt.* **2003,** *5,* 816–850.

12. Sarid, D.; Challenger, W. *Modern Introduction to Surface Plasmons*; Cambridge University Press: New York, 2010,.

13. Dhawan, A.; Canva, M.; Vo-Dinh, T. Narrow Groove Plasmonic Nano-gratings for Surface Plasmon Resonance Sensing. *Opt. Express* **2011,** *19,* 787–813.

14. Ren, F.; Kim, K.; Chong, X.; Wang, A. Effect of Finite Metallic Grating Size on Rayleigh Anomaly-surface Plasmon Polariton Resonances. *Opt. Express* **2015,** *23,* 28868–28873.

15. Guo, J.; Li, Z.; Guo, H. Near Perfect Light Trapping in a 2D Gold Nanotrench Grating at Oblique Angles of Incidence and Its Application for Sensing. *Opt. Express* **2016,** *24,* 17259–17271.

16. Wood, A.; Chen, B.; Pathan, S.; Bok, S.; Mathai, C.; Gangopadhyay, K.; Grant, S.; Gangopadhyay, S. Influence of Silver Grain Size, Roughness, and Profile on the Extraordinary Fluorescence Enhancement Capabilities of Grating Coupled Surface Plasmon Resonance. *RSC Adv.* **2015,** *5,* 78534–78544.

17. Deng, X.; Braun, G.; Liu, S.; Sciortino, P.; Jr., Koefer, B.; Tombler, T.; Moskovits, M. Single-order, Subwavelength Resonant Nanograting as a Uniformly Hot Substrate for Surface-enhanced Raman Spectroscopy. *Nano Lett.* **2010,** *10,* 1780–1786.

18. Seo, M.; Lee, J.; Lee, M. Grating-coupled Surface Plasmon Resonance on Bulk Stainless Steel. *Opt. Express* **2017,** *25,* 26939–26949.

19. Bhatnagar, K.; Pathak, A.; Menke, D.; Cornish, P.; Gangopadhyay, K.; Korampally, V.; Gangopadhyay, S. Fluorescence Enhancement from Nano-gap Embedded Plasmonic Gratings by a Novel Fabrication Technique with HD-DVD. *Nanotechnology* **2012,** *23,* 495201.

20. Ng, R.; Goh, X.; Yang, J. K. All-metal Nanostructured Substrates as Subtractive Color Reflectors with Near-perfect Absorptance. *Opt. Express* **2015,** *23,* 32597–32605.

21. Clausen, J.; Højlund-Nielsen, E.; Christiansen, A.; Yazdi, S.; Grajower, M.; Taha, H.; Levy, U.; Kristensen, A.; Mortensen, N. Plasmonic Metasurfaces for Coloration of Plastic Consumer Products. *Nano Lett.* **2014,** *14,* 4499–4504.

22. Mie, G. *Ann. Phys.* **1908,** *25,* 377.

23. Bohren, C.; Huffman, D. *Absorption and Scattering of Light by Small Particles*; Wiley-Interscience: New York, 2010.

24. Yee, K. Numerical Solution of Initial Boundary Value Problems Involving Maxwell's Equations in Isotropic Media. *IEEE Trans. Antennas Propag.* **1966,** *14,* 302.

25. Anderson, L.; Mayer, K.; Fraleigh, R.; Yang, Y.; Lee, S.; Hafner, J. Quantitative Measurements of Individual Gold Nanoparticle Scattering Cross Sections. *J. Phys. Chem. C* **2010,** *114,* 11127.

26. Kelly, K.; Coronado, E.; Zhao, L.; Schatz, G. The Optical Properties of Metal Nanoparticles: The Influence of Size, Shape, and Dielectric Environment. *J. Phys. Chem. B* **2003,** *107,* 668.

27. Jain, P.; Lee, K.; El-Sayed, I.; El-Sayed, M. Calculated Absorption and Scattering Properties of Gold Nanoparticles of Different Size, Shape, and Composition: Applications in Biological Imaging and Biomedicine. *J. Phys. Chem. B* **2006,** *110,* 7238.

28. Link, S.; El-Sayed, M. Size and Temperature Dependence of the Plasmon Absorption of Colloidal Gold Nanoparticles. *J. Phys. Chem. B* **1999,** *103,* 4212.

29. Gans, R. *Ann. Phys.* **1912**, *37*, 881.
30. Cao, J.; Sun, T.; Grattan, K. Gold Nanorod-based Localized Surface Plasmon Resonance Biosensors: A Review. *Sens. Actuators B* **2014**, *195*, 332.
31. Mock, J.; Barbic, M.; Smith, D.; Schultz, D.; Schultz, S. Shape Effects in Plasmon Resonance of Individual Colloidal Silver Nanoparticles. *J. Chem. Phys.* **2002**, *116*, 6755.
32. Link, S.; Mohamed, M.; El-Sayed, M. Simulation of the Optical Absorption Spectra of Gold Nanorods as a Function of Their Aspect Ratio and the Effect of the Medium Dielectric Constant. *J. Phys. Chem. B* **1999**, *103*, 3073.
33. Sherry, L.; Chang, S.; Schatz, G.; van Duyne, R. Localized Surface Plasmon Resonance Spectroscopy of Single Silver Nanocubes. *Nano Lett.* **2005**, *5*, 2034.
34. Chen, H.; Kou, X.; Yang, Z.; Ni, W.; Wang, J. Shape- and size-Dependent Refractive Index Sensitivity of Gold Nanoparticles. *Langmuir* **2008**, *24*, 5233.
35. Nehl, C.; Liao, H.; Hafner, J. Optical Properties of Star-shaped Gold Nanoparticles. *Nano Lett.* **2006**, *6*, 683.
36. Liu, M. Guyot-Sionnest, P. Mechanism of Silver(I)-assisted Growth of Gold Nanorods and Bipyramids. *Phys. Chem. B* **2005**, *109*, 22192.
37. Lu, X.; Rycenga, M.; Skrabalak, S.; Wiley, B.; Xia, Y. Chemical Synthesis of Novel Plasmonic Nanoparticles. *Annu. Rev. Phys. Chem.* **2009**, *60*, 167.
38. Link, S.; Wang, Z.; El-Sayed, M. Alloy Formation of Gold-Silver Nanoparticles and the Dependence of the Plasmon Absorption on Their Composition. *J. Phys. Chem. B* **1999**, *103*, 3529.
39. Oh, Y.; Lee, J.; Lee, M. Fabrication of Ag-Au Bimetallic Nanoparticles by Laser-induced Dewetting of Bilayer Films. *Appl. Surf. Sci.* **2018**, *434*, 1293.
40. Jackson, J.; Halas, N. Silver Nanoshells: Variation in Morphologies and Optical Properties. *J. Phys. Chem. B* **2001**, *105*, 2743.
41. Prodan, E.; Radloff, C.; Halas, N.; Nordlander, P. A Hybridization Model for the Plasmon Resonance of Complex Nanostructures. *Science* **2003**, *302*, 419.
42. Radloff, C. Halas, N. Plasmonic Properties of Concentric Nanoshells. *Nano Lett.* **2004**, *4*, 1323.
43. Huang, Y.; Ma, L.; Hou, M.; Li, J.; Xie, Z.; Zhang, Z. Hybridized Plasmon Modes and Near-field Enhancement of Metallic Nanoparticle-dimer on a Mirror, *Sci. Reports* **2016**, *6*, 30011.
44. Garnett, E., et al. Self-limited Plasmonic Welding of Silver Nanowire Junctions. *Nat. Mater.* **2012**, *11*, 241.
45. Rechberger, W.; Hohenau, A.; Leitner, A.; Krenn, J.; Lamprecht, B.; Aussenegg, F. Optical Properties of Two Interacting Gold Nanoparticles. *Opt. Commun.* **2003**, *220*, 137.
46. Jain, P.; Huang, W.; El-Sayed, M. On the Universal Scaling Behavior of the Distance Decay of Plasmon Coupling in Metal Nanoparticle Pairs: A Plasmon Ruler Equation. *Nano Lett.* **2007**, *7*, 2080.
47. Raman, C.; Krishnan, K. A New Type of Secondary Radiation. *Nature* **1928**, *121*, 501.
48. Fleischmann, M.; Hendra, P.; McQuillan, A. Raman Spectra of Pyridine Adsorbed at a Silver Electrode. *Chem. Phys. Lett.* **1974**, *26*, 163.
49. Nie, S.; Emory, S. Probing Single Molecules and Single Nanoparticles by SERS. *Science* **1997**, *275*, 1102.
50. Kneipp, K., et al. Single Molecule Detection Using SERS. *Phys. Rev. Lett.* **1997**, *78*, 1667.
51. Gersten, J.; Nitzan, A. Electromagnetic Theory of SERS by Molecules Adsorbed on Rough Surfaces. *J. Chem. Phys.* **1980**, *73*, 3023.

52. Garcia-Vidal, F.; Pendry, J. Collective Theory for SERS. *Phys. Rev. Lett.* **1996**, *77*, 1163.
53. Lombardi, J.; Birke, R.; Lu, T.; Xu, J. Charge-transfer Theory of Surface Enhanced Raman Spectroscopy: Herzberg–Teller Contributions. *J. Chem. Phys.* **1986**, *84*, 4174.
54. Lombardi, J.; Birke, R. A Unified Approach to Surface-enhanced Raman Spectroscopy. *J. Phys. Chem. C* **2008**, *112*, 5605.
55. Ding, T.; Sigle, D.; Herrmann, L.; Wolverson, D.; Baumberg, J. Nanoimprint Lithography of Al Nanovoids for Deep-UV SERS. *ACS Appl. Mater. Interfaces* **2014**, *6*, 17358.
56. Yang, Z.; Li, Q.; Ren, B.; Tian, Z. Tunable SERS from Aluminium Nanohole Arrays in the Ultraviolet Region. *Chem. Commun.* **2011**, *47*, 3909.
57. Knight, M.; King, N.; Liu, L.; Everitt, H.; Nordlander, P.; Halas, N. Aluminum for Plasmonics. *ACS Nano* **2014**, *8*, 834.
58. Liu, L. et al. A High-performance and Low Cost SERS Substrate of Plasmonic Nanopillars on Plastic Film Fabricated by Nanoimprint Lithography with AAO Template. *AIP Adv.* **2017**, *7*, 065205.
59. Suresh, V.; Ding, L.; Chew, A.; Yap, F. Fabrication of Large-area Flexible SERS Substrates by Nanoimprint Lithography. *ACS Appl. Nano Mater.* **2018**, *1*, 886.
60. Park, S.; Mun, C.; Xiao, X.; Braun, A.; Kim, S.; Giannini, V.; Maier, S.; Kim, D. Surface Energy-controlled SERS Substrates for Molecular Concentration at Plasmonic Nanogaps. *Adv. Funct. Mater.* **2017**, *27*, 1703376.
61. McFarland, A.; van Duyne, R. Single Silver Nanoparticles as Real-time Optical Sensors with Zeptomole Sensitivity. *Nano Lett.* **2003**, *3*, 1057.
62. Pillai, S.; Green, M. Plasmonics for Photovoltaic Applications. *Sol. Energy Mater. Sol. Cells* **2010**, *94*, 1481.
63. Green, M.; Pillai, S. Harnessing Plasmonics for Solar Cells. *Nat. Photon.* **2012**, *6*, 131.
64. Atwater, H.; Polman, A. Plasmonics for Improved Photovoltaic Devices. *Nat. Mater.* **2010**, *9*, 205.
65. Schuller, J., et al. Plasmonics for Extreme Light Concentration and Manipulation. *Nat. Mater.* **2010**, *9*, 193.
66. Yue, Z.; Cai, B.; Wang, L.; Wang, X.; Gu, M. Intrinsically Core-shell Plasmonic Dielectric Nanostructures with Ultrahigh Refractive Index. *Sci. Adv.* **2016**, *2*, e1501536.
67. Sha, W., et al. A General Design Rule to Manipulate Photocarrier Transport Path in Solar Cells and Its Realization by the Plasmonic-electrical Effect. *Sci. Report* **2015**, *5*, 8525.
68. Ding, I., et al. Plasmonic Dye-sensitized Solar Cells. *Adv. Energy Mater.* **2011**, *1*, 52.
69. Carretero-Palacios, S.; Jimenez-Solano, A.; Míguez, H. Plasmonic Nanoparticles as Light-harvesting Enhancers in Perovskite Solar Cells: A User's Guide. *ACS Energy Lett.* **2016**, *1*, 323.
70. Lee, K., et al. Highly Efficient Colored Perovskite Solar Cells Integrated With Ultrathin Subwavelength Plasmonic Nanoresonators. *Sci. Report* **2017**, *7*, 10640.
71. Catchpole, K.; Polman, A. Design Principles for Particle Plasmon Enhanced Solar Cells. *Appl. Phys. Lett.* **2008**, *93*, 191113.
72. Kim, S.; Na, S.; Jo, J.; Kim, D.; Nah, Y. Plasmon Enhanced Performance of Organic Solar Cells Using Electrodeposited Ag Nanoparticles. *Appl. Phys. Lett.* **2008**, *93*, 073307.
73. Lindquist, N.; Luhman, W.; Oh, S.; Holmes, R. Plasmonic Nanocavity Arrays for Enhanced Efficiency in Organic Photovoltaic Cells. *Appl. Phys. Lett.* **2008**, *93*, 123308.
74. Konda, R., et al. Surface Plasmon Excitation Via Au Nanoparticles in n-CdSe/p-Si Heterojunction Diodes. *Appl. Phys. Lett.* **2007**, *91*, 191111.
75. Hagglund, C.; Zach, M.; Kasemo, B. Enhanced Charge Carrier Generation in Dye Sensitized Solar Cells by Nanoparticle Plasmons. *Appl. Phys. Lett.* **2008**, *92*, 013113.

76. Dionne, J.; Sweatlock, L.; Atwater, H.; Polman, A. Planar Plasmon Metal Waveguides: Frequency-dependent Dispersion, Propagation, Localization, and Loss Beyond the Free . Electron Model. *Phys. Rev. B* **2005**, *72*, 075405.

77. Ferry, V.; Sweatlock, L.; Pacifici, D.; Atwater, H. Plasmonic Nanostructure Design for Efficient Light Coupling into Solar Cells. *Nano Lett.* **2008**, *8*, 4391.

78. Ferry, V., et al. Improved Red-response in Thin Film a-Si:H Solar Cells with Nanostructured Plasmonic Back Reflectors. *Appl. Phys. Lett.* **2009**, *95*, 183503.

79. Ellenbogen, T.; Seo, K.; Crozier, K. Chromatic Plasmonic Polarizers for Active Visible Color Filtering and Polarimetry. *Nano Lett.* **2012**, *12*, 1026.

80. Olson, J., et al. High Chromaticity Aluminum Plasmonic Pixels for Active Liquid Crystal Displays. *ACS Nano* **2016**, *10*, 1108.

81. Ye, M., et al. Angle-insensitive Plasmonic Color Filters with Randomly Distributed Silver Nanodisks. *Opt. Lett.* **2015**, *40*, 4979.

82. Yue, W.; Gao, S.; Lee, S.; Kim, E.; Choi, D. Subtractive Color Filters Based on a Silicon–Aluminum Hybrid-nanodisk Metasurface Enabling Enhanced Color Purity. *Sci. Reports* **2016**, *6*, 29756.

83. Zhang, J.; Ou, J.; MacDonald, K.; Zheludev, N. Optical Response of Plasmonic Relief Meta-surfaces, *J. Opt.* **2012**, *14*, 114002.

84. Lee, S., et al. Plasmonenhanced Structural Coloration of Metal Films with Isotropic Pinwheel Nanoparticle Arrays. *Opt. Express* **2011**, *19*, 23818.

85. Shrestha, V.; Lee, S.; Kim, E.; Choi, D. Aluminum Plasmonics Based Highly Transmissive Polarization-independent Subtractive Color Filters Exploiting a Nanopatch Array. *Nano Lett.* **2014**, *14*, 6672.

86. Kumar, K., et al. Printing Color at the Optical Diffraction Limit. *Nat. Nanotechnol.* **2012**, *7*, 557.

87. Tan, S., et al. Plasmonic Color Palettes for Photorealistic Printing with Aluminum Nanostructures. *Nano Lett.* **2014**, *14*, 4023.

88. Roberts, A.; Pors, A.; Albrektsen, O.; Bozhevolnyi, S. Subwavelength Plasmonic Color Printing Protected for Ambient Use. *Nano Lett.* **2014**, *14*, 783.

89. Zhu, X.; Vannahme, C.; Højlund-Nielsen, E.; Asger Mortensen, N.; Kristensen, A. Plasmonic Colour Laser Printing. *Nat. Nanotechnol.* **2016**, *11*, 325.

90. Zhu, X.; Yan, W.; Levy, U.; Asger Mortensen, N.; Kristensen, A. Resonant Laser Printing of Structural Colors on High-index Dielectric Metasurfaces. *Sci. Adv.* **2017**, *3*, e1602487.

Index